SELF-ASSEMBLY AND NANOTECHNOLOGY

SELF-ASSEMBLY AND NANOTECHNOLOGY

A Force Balance Approach

Yoon S. Lee
Scientific Information Analyst
Chemical Abstracts Service
A Division of the American Chemical Society
Columbus, Ohio

A JOHN WILEY & SONS, INC., PUBLICATION

Published by John Wiley & Sons, Inc., Hoboken, New Jersey.
Published simultaneously in Canada.

Library of Congress Cataloging-in-Publication Data:

Lee, Yoon Seob.
Self-assembly and nanotechnology : a force balance approach / Yoon Seob Lee.
p. cm.
Includes index.
ISBN 978-0-470-24883-6 (cloth)
1. Nanostructured materials–Design. 2. Nanotechnology. 3. Self-assembly (Chemistry) I. Title.
TA418. 9. N35L44 2008
620′.5—dc22

2007052383

Printed in the United States of America.

10 9 8 7 6 5 4 3 2 1

To my mother

CONTENTS

PREFACE AND ACKNOWLEDGMENTS

The area of nanotechnology has grown tremendously over the past decade and is expected to keep growing rapidly in the future. In following this new megatrend, there is a strong sense of need for education in nanotechnology among the academic community. However, nanotechnology is a huge topic that cannot be covered by a single book. This book covers the topic of self-assembly and its implications for nanotechnology. Self-assembly is now widely identified as one of the major themes in the development of nanotechnology. The two-part scheme of this book properly addresses this fact: Part I is on self-assembly and Part II is on nanotechnology.

I designed this book to be a concept book. My experience is that too many details often hinder underlying principles and logics. Comprehensive delivery of the right concepts is the first step toward successful teaching, especially for a complex subject like nanotechnology. I came up with clear schematic illustrations for almost every section to properly represent the mainstream principles behind each topic. Care has been taken to avoid having the book become an exhausting review, with selective use of specific data. However, those who desire more advanced study will find thorough citations at the end of each chapter.

The book is primarily designed for both undergraduates and graduates who have at least mid-level background in chemistry or chemistry-related fields. Those who have taken basic organic, physical, and/or inorganic chemistry courses should have little difficulty following the streamlined topics of this book. This feature will make this book a good tool when the course objective is to bridge the topics of self-assembly, colloids, and surfaces with nanotechnology. It can also be used as a part of the teaching materials when the courses are joint-efforts across different disciplines or different departments that intend to cover a broader range of nanotechnology. Joint-courses have become increasingly popular these days; in fact, this is an especially effective teaching scheme for nanotechnology.

At the same time, this book is intended for academic/industrial professionals, too. Its whole scope is networked around one stem concept: *force balance*. This is to show that a good deal of the related topics in self-assembly and nanotechnology can be approached with one unified concept, once we expand our view on self-assembly. This feature could provide some useful insights into the research of professionals, especially when they try to understand the seemingly complex self-assembly phenomena behind the nanotechnology issues. Considering the inter- and multidisciplinary natures of nanotechnology, this book should be friendly reading not just for chemistry majors, but for those in chemical engineering, physics, and materials science as well.

My first thanks go to Prof. Sangeeta Bhatia (Massachusetts Institute of Technology), Dr. Jun Liu (Pacific Northwest National Laboratory), and Prof. Todd Emrick (University of Massachusetts, Amherst) for their valuable manuscript reviews. Also, I would like to send my heartfelt thanks to Dr. Oksik Lee at Chemical Abstracts Service for her advice and our discussions throughout the years. I am much indebted to Prof. Kyu Whan Woo (Seoul National University) and Prof. James Rathman (Ohio State University), who have given me a great deal of inspiration about this topic from the very beginning. As always, my deepest thanks go to my family—my wife, Jee-A, my son, Jong-Hyuk, my parents, and my parents-in-law—for their endless support and love.

Yoon Seob Lee
Dublin, Ohio
ylee@cas.org

PART I

SELF-ASSEMBLY

1

UNIFIED APPROACH TO SELF-ASSEMBLY

Traditionally, *self-assembly* has been defined as *spontaneous* association of molecules into defined three-dimensional geometry under a defined condition. It thus refers to a thermodynamics process, and the molecules and the self-assembled aggregates are in equilibrium. Formation of surfactant micelles might be one of the most widely studied systems that fits into this scheme of self-assembly. For this system, thermodynamic description starts from the equilibrium between surfactant molecules (monomer) and surfactant micelles (self-assembled aggregates). An alternative way is to treat the surfactant molecules in bulk (usually aqueous solution) and the surfactant micelles as a different phase (*pseudo*-phase separation) in equilibrium. These two major approaches for the surfactant self-assembly have been well formulated since the 1970s (Clint, 1992), and successfully been applied to a similar type of self-assembly for amphiphilic polymers, such as block copolymers, later in the 1990s (Alexandridis and Lindman, 2000). They are a useful tool to follow the thermodynamics of these self-assembly processes and give a reasonable prediction for the major parameters such as *critical micellar concentration* (*cmc*), aggregation number, counterion binding, micelle size, and micelle size distribution.

The phenomena associated with this scheme of *spontaneous* association are abundant in nature, and its building unit (or association unit) is not limited to

Self-Assembly and Nanotechnology: A Force Balance Approach, by Yoon S. Lee

the surfactant molecules. Association of much bigger colloidal–size objects without involving strong chemical bonds has been known since the 1940s (Verwey and Overbeek, 1948; Overbeek, 1952). Formation of metal and semiconductor nanoparticles through the self-assembly of atoms in bulk has also been well established since the late 1990s (Fendler and Dékány, 1996). The self-assembly of dendric polymers is also now well documented (Emrick and Fréchet, 1999). Thus, the term *self-assembly* actually embraces a wider range of building units. And based on the size/nature of the building units (primary building unit, defined in Section 1.2), they can be viewed mainly as atomic, molecular, and colloidal self-assemblies. Polymeric self-assembly can be classified as molecular self-assembly as the sense of the building unit is polymer molecules.

Spontaneous association phenomena have also been found in biological systems. They are not necessarily limited to the bulk solution, and can also occur at two-dimensional systems such as surfaces and interfaces. The biological system has long been known as a treasure house of intriguing self-assembly processes. Most of the cases are the processes of spontaneous association of biological building units such as lipids and amino acids. There are few covalent bonds involved except for the cases of peptides and thiol bonds. For two-dimensional systems, spontaneous association of metal or semiconductor atoms on a solid surface is now being observed *in situ*. A variety of self-assembly processes at different interfaces have been documented, too. Therefore, in addition to the above classification, self-assembly can be classified as biological or interfacial with the view where the self-assembly occurs. Figure 1.1 shows the schematics. Self-assembly can be classified:

1. By the size/nature of building unit: atomic, molecular, and colloidal
2. By the system where it occurs: biological and interfacial

The classification of self-assembly can be further expanded by the nature of its process: thermodynamic or kinetic. The former includes atomic, molecular, bio-

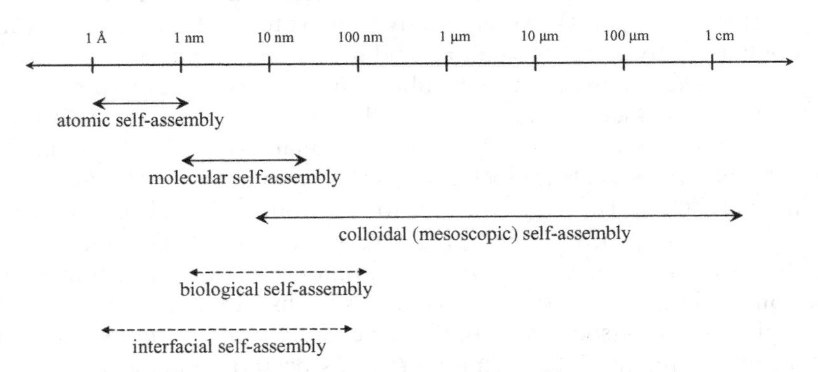

Figure 1.1. Classification of self-assemblies based on the size/nature (atomic, molecular, and colloidal) of building units and on the system where the self-assembly occurs (biological and interfacial); the length scale is also of building units.

logical, and interfacial self-assemblies, while the latter has colloidal and some interfacial self-assemblies. Some of the self-assembly processes are random, while others are directional to some degree. Molecular, colloidal, interfacial self-assemblies are random cases, and some atomic and biological self-assemblies are directional. Self-assembly that is associated with large building units, that is, colloidal self-assembly, can be sensitive to the external stimuli such as electric field, magnetic field, gravity, flow, and so forth.

Thus, the view of spontaneous association covers a broad range-of-length scale from Angström to centimeter, different dimensions, and different sources of origins. The main purpose of this chapter is to propose some unifying approach to this broad range of self-assembly. The very common aspect of these self-assemblies, that is, the interplay of intermolecular and colloidal forces, will be the starting point. It will be discussed for each case of self-assembly process, and then will be followed by the view of the force balance for the formation of self-assembled aggregates. The general scheme of self-assembly and the subsequent formulation will be presented, too. The rest of the chapters in Part I are based on the concept and scheme presented in this chapter. It will be also directly expanded to the implication of the self-assembly for nanotechnology later in Part II.

1.1. SELF-ASSEMBLY THROUGH FORCE BALANCE

Surfactant self-assembly is often called *micellization*: the process for the formation of micelles. With the view of the forces acting on this process, it is actually a process toward the delicate balance between the attractive and repulsive intermolecular forces. Attractive forces directly act on surfactant molecules to bring them close together, while repulsive forces act against the molecules. Hence, the former can be defined as the *driving force* for the micellization, and the latter as the *opposition force*. No strong chemical bond such as a covalent bond is involved during this process. More specifically, the driving force for this process is usually the hydrophobic attraction and the opposition force is the electrostatic repulsion and/or solvation force. First, the long-range hydrophobic force acts as a main force to bring the surfactant molecules together. As the process continues, the opposition forces such as electric double-layer repulsion or hydration forces start to impose. These forces originate from the charge-bearing or hydrated head groups, and are relatively short-range forces compared with the hydrophobic interaction. As will be discussed in Chapter 2, these two types of forces are variable as a function of intermolecular distance, but in opposite ways. Consequently, the attractive and repulsive forces should be balanced at a certain point of the process. Micelles are formed at this point, and the further growth of micelles is prevented. But, since there are no chemical bonds involved, the surfactant monomers in the micelles are free to be exchanged with the monomers in the bulk solution, depending on their molecular dynamic properties. The concentration of this monomer is the concentration that is necessary to form the first micelle (critical micellar concentration). Any additional amounts of surfactant molecules in the bulk solution will follow the same force balance scheme, thereby forming the additional amounts of micelles while keeping the size of the micelles constant. The concentration of surfactant monomer in solution is also kept constant.

Surfactant micelles are not the only system that fits into this picture of self-assembly. Long-studied colloidal suspensions, emulsions, and microemulsions are also systems where the interaction between the similar intermolecular/colloidal attractive and repulsive forces determines the formation of these self-assembled aggregates.

For colloidal suspension, no coagulation will occur while the repulsive forces are dominant between colloidal objects. However, when the attractive forces are dominant, it is coagulated. Now, let us look at this concept of colloidal stability with the notion of the self-assembly discussed above. The van der Waals force is now the self-assembly driving attractive force, whereas the electric double-layer interaction is the self-assembly opposition repulsive force. Then, the situation of the formulation of the DLVO theory (Derjaguin-Landau-Verwey-Overbeek; Chapter 2) can become a useful tool to describe the self-assembly processes of colloidal objects. Self-assembly of nanoparticles with charged surfaces can be one good example. When the potential barrier between nanoparticles is overcome, the coagulation begins as a result of van der Waals attraction. But, since the electric double-layer repulsion is already there along with the van der Waals force (both as a function of the distance between the nanoparticles), any changes that can change the potential curve can change the whole coagulation process. As long as there is a constant supply of nanoparticles that overcome this energy barrier either by change of the electrolyte concentration or by change of pH, the coagulation will continue until it is compensated by the thermal or gravitational force. With the sense of spontaneous association by the interplay of intermolecular/colloidal forces, this coagulation process can be considered as the self-assembly that now occurs with colloidal-size objects. The opposite change of condition that can make the electric double-layer repulsion dominant will reverse the whole process.

Microemulsion is formed based on the surfactant micelle. But the process is somewhat more complex than surfactant micellization. The attractive driving force is hydrophobic interaction between the surfactant molecules. As for the micellization, the surfactant molecules are brought together by this force. Then, the electric double-layer repulsion and/or hydration force is being balanced with the hydrophobic force. The difference is that there is a significant amount of water or oil in the systems, and they are part of the micelle. This situation is usually recognized as the formation of nanometer-sized water droplets in reverse micelles or as swelled normal micelles. They are thermodynamically stable systems and the process is reversible.

Emulsion (or macroemulsion) is formed when two immiscible liquids (usually water and oil phases) are mixed and stabilized by the self-assembled surfactant, polymer, or colloidal particle at the water–oil interface. Since the interfacial tension at this interface can never reach zero, this is a thermodynamically unstable system. The long-term stability is acquired by its extremely slow phase separation kinetics. Besides this difference, the self-assembly process itself for emulsion formation is quite similar to the formation of microemulsion. For the surfactants and polymers, the attractive driving force for the self-assembly is again hydrophobic force, and the opposition repulsive force is electric double-layer and/or hydration force. For the colloidal particles, the DLVO-force mentioned above for the self-assembly of colloidal particles becomes the main mechanism. Table 1.1 represents the typical attractive and repulsive forces that can be found in self-assembly processes.

Biological systems are full of self-assembly processes in this sense. Biological membranes, DNA, RNA, enzymes, and proteins are formed by the delicate force balance between the attractive and repulsive forces. However, the uniqueness of these systems compared with the micelles and colloids is that the biological self-assembled systems, in many cases, are formed with some degree of directionality. And this directionality seems to be closely related with the unique functionality of each self-assembled system and the biological systems in general.

Biological systems are not the only ones that show directionality during self-assembly processes. Many bio-mimetic systems, such as systems with synthetic amino acids, carboxylic acids, and dendric polymers, and even nonbiological graphitic supermolecules, show a unique directionality during the self-assembly processes. This directionality is closely related with a unique functionality such as transport, conductivity, and catalytic activity. Helical structure is among the

TABLE 1.1. Representative intermolecular/colloidal attractive and repulsive forces for self-assembly.

Attractive Force	Repulsive Force
Van der waals[a]	Electric double-layer[b]
Solvation	Solvation
Depletion	Hydration
Bridging	Steric
Hydrophobic	
π–π stacking	
Hydrogen bond	
Coordination bond[c]	

[a] Some cases of interaction between dissimilar colloidal objects can be repulsive (Figure 2.2).

[b] This force sometimes can be attractive when (1) interaction occurs between molecules or colloids with different charges, (2) with the same charge but at very small separation, and (3) between zwitterionic molecules and colloids.

[c] Coordination bond is a strong chemical bond compared with the rest of the forces, but serves as a unique attractive force for some of the supramolecular self-assembly systems.

common self-assembled structures, but others such as tube, rod, and ring structures are also being found.

For these directional self-assembly processes, the attractive driving forces and repulsive opposition forces always function as those in the nondirectional self-assembly ones. But there is another class of forces in these directional self-assembly systems that is directly responsible for the directionality. These forces act uniquely as a *functional force*. Hydrogen bond and coordination bond are among the most commonly found functional forces. But much weaker forces, like steric repulsion, are also commonly found functional forces. These forces can be a part of a driving or opposition force during the self-assembly process, but sometimes act almost exclusively as directional force.

1.2. GENERAL SCHEME FOR THE FORMATION OF SELF-ASSEMBLED AGGREGATES

Based on the above discussion, the general scheme for the self-assembly process that can encompass the length scale from atomic to colloidal can be drawn. Figure 1.2 shows the schematics. Self-assembly is the force balance process between three classes of forces: attractive driving, repulsive opposition, and directional force. Directional force can be considered functional force in the sense that it is also responsible for the functionality. When only the first two classes of forces are in action, the self-assembly process is a random and usually one-step process. The self-assembled aggregates show nonhierarchical structure. Most of the molecular self-assembly processes such as micellization and most of the colloidal systems belong to this category of self-assembly. When the third class of force is involved with the first two classes of force, the self-assembly processes are now directional, and in many cases, they occur as multi-stepwise processes. The self-assembled aggregates usually show hierarchical structure. Most of the biological and bio-mimetic systems belong to this category of self-assembly.

This picture also can be applied to more complex two-dimensional self-assembly systems. Spontaneous association of metal or semiconductor atoms on solid substrates forms a unique self-assembled aggregate, such as quantum dots.

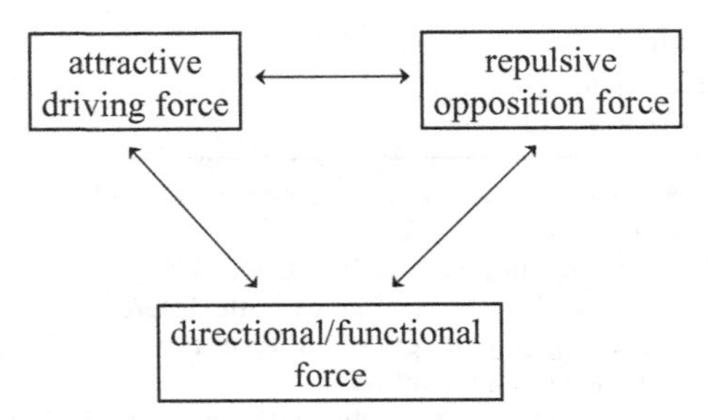

Figure 1.2. Self-assembly *in general* can be defined as the cooperative interaction and balance between three classes of distinctive forces.

Their size can range 1–10 nm and the shapes can be spherical or pyramidal. This is the result of the force balance mainly between the attractive van der Waals force and repulsive electrostatic force. The unique atom–substrate interaction for this system can be considered as directional force, because this force is responsible for the two-dimensionality of this self-assembly. Epitaxial film growth is a good example. The atom–substrate interaction and the epitaxy of the substrate strongly determine the direction of the patterning of quantum dots. This self-assembly process is directional and one-step, and the self-assembled aggregates have nonhierarchical structure.

The same principle can be deduced from interfacial self-assembly processes. Three liquid phase–based solid–liquid, liquid–gas, and liquid–liquid interfaces can be a *confined* substrate for two-dimensional interfacial self-assembly. Surfactant, polymer, and colloidal particles can be self-assembled in this two-dimensional space. The picture for the attractive and repulsive forces is similar to the self-assembly process in macroemulsion systems. The characteristics of the interfaces such as interfacial energy, mechanical force, or interaction of building units with interfaces now act as functional force. Thus, this self-assembly process is also directional and one-step, and the self-assembled aggregates have mostly nonhierarchical structure. Table 1.2 summarizes these aspects. It also shows the

TABLE 1.2. Five classes of self-assemblies, typical building units, examples of self-assembled systems, and characteristics of assembly process.

Classification	Building Units	Self-assembled Systems	Characteristics
Atomic	Metal atom	Epitaxial film, quantum dot	Directional, one-step, nonhierarchical
Molecular	Surfactant, polymer	Micelle, bilayer microemulsion, emulsion	Random, one-step, nonhierarchical
Colloidal	Nanoparticle, nanotube, fullerene colloidal object	Suspension, dispersion, sol, colloidal crystal	Random, one-step, nonhierarchical
Biological	Amino acid, lipid biopolymer	DNA, RNA, protein enzyme, membrane	Directional, stepwise, hierarchical,
Interfacial	Surfactant, polymer, lipid	Surface micelle, Langmuir monolayer, Langmuir-Blodgett film, self-assembled monolayer	Directional, one-step, nonhierarchical

five classes of self-assemblies defined in the first section, the typical building units of each system, and examples of self-assembled aggregates.

As the scheme of Figure 1.2 can predict, if the system is in the right condition, that is, when the attractive and repulsive forces are balanced, even the colloidal systems that usually show kinetic self-assembly process can experience the thermodynamical self-assembly phenomena. These thermodynamically stable self-assembled colloidal aggregates were recently discovered experimentally (Buitenhuis et al., 1994) and confirmed theoretically (van der Schoot, 1992; Groenewold and Kegel, 2001; Likos, 2001; Muratov, 2002; Sciortino et al., 2004). Sterically stabilized or partially charged colloidal objects can be in a condition of delicate balance between the attractive force (van der Waals or depletion) and the repulsive force (electrostatic) at a certain volume fraction. Much like the micellization of surfactant molecules, a certain number of colloidal objects in this condition can self-assemble into the colloidal aggregates with ~20–~1,000 of finite aggregation number. The individual colloidal particles (monomer) are in equilibrium with the self-assembled aggregate, and the whole process is dependent on physicochemical parameters such as temperature and solvent. There is also the exchange of free monomer with self-assembled aggregates. And the change in the shape of the self-assembled aggregates can be induced from spherical, to disk, and to rod as the force balance changes. This force balance change can be induced by the change in the shape/size of colloidal object (monomer), surface charge density of colloid, and dielectric constant of solvent. By rough analogy, for the case of surfactant micellization, the main factors for the change of force balance between attractive and repulsive forces are the shape/length of surfactant molecule (monomer), charge density (for ionic surfactant) or degree of hydration (for nonionic surfactant) on the micelle surface, and the solvent properties such as dielectric constant or pH. While the concept of DLVO describes the irreversible kinetical self-assembly of colloidal objects, again this case represents the reversible equilibrium self-assembly of colloidal objects.

1.3. GENERAL SCHEME FOR SELF-ASSEMBLY PROCESS

In the previous section, the balance between the distinctive but cooperative three classes of forces has been proposed for the formation of self-assembled aggregates. This general scheme can encompass a variety of self-assembly building units with the length scale ranging from atomic to colloidal. Since the self-assembly process can occur in such a wide range of length scale (10^7 difference of order from Angström to centimeter) and the same types of forces are governing the process, the self-assembled aggregates formed by the initial self-assembly step can in many cases become another building unit for the subsequent self-assembly processes at given conditions. That is, self-assembly in fact is not always a single-step process; it can occur in a double-, triple-, and multi-stepwise pattern.

A typical example can be found in the formation of surfactant micelle and its subsequent transition to mesophase structures (Clint, 1992). First, the most common spherical micelles are formed by the typical self-assembly of surfactant molecules. As the solution condition is changed into the subsequent favorable self-assembly, such as increased surfactant concentration, change of pH, or increased concentration of counterion, these micelles begin to interact with each other and can self-assemble together. This process is governed by the intermicellar colloidal forces. Thus, the surfactant molecule can be defined as the *primary building unit* in this sense and the micelle as the *secondary building unit*. And the micelle can be viewed as the *primary self-aggregate* and mesophase (the self-assembled micelle) as the *secondary self-aggregate*. Amphiphilic polymers such as a block copolymer can in many cases follow a similar scheme and form similar polymer mesophases (Alexandridis and Lindman, 2000).

Another example can be found in the consecutive self-assembly of atoms to colloidal-size objects. Certain numbers of metal or semiconductor atoms (<1 nm diameter) (known as the *magic number of aggregation*) can self-assemble into quantum dots or nanoparticles in bulk (2–5 nm diameter) (primary self-aggregates), and the subsequent self-assembly (again associated with the magic number of aggregation) brings those quantum dots or nanoparticles into giant quantum dots or giant nanoparticles of 20–50 nm diameter (secondary self-aggregates) (Rao et al., 2000; Rao, 2001). Van der Waals attraction is the primary driving force for both processes, while some degree of structural constraints seems to be the opposition force. The second process is different from the self-assembly of surfactant- or alkyl chain–modified nanoparticles at the surface or in bulk that occurs by van der Waals and electric double-layer forces. A similar process can occur during the epitaxial film growth of metal or semiconductor at solid surfaces, which can be considered as interfacial self-assembly with a multi-stepwise process from atomics to colloidal-length scale.

Formation of large-scale aggregates of colloidal particles with a centimeter-length scale such as fractals (secondary self-aggregate) occurs in many cases through the assembly of clusters that are formed by the self-assembly of individual colloidal objects (primary self-aggregate).

For biological self-assembly processes, a typical example can be found in the formation of proteins. First, DNA is formed by the self-assembly of amino acids. Thus, the amino acids are the primary building units for this initial self-assembly, and the DNA is the primary self-aggregate. DNA is then self-assembled into a primary structure of protein via the secondary self-assembly process. DNA is now the secondary building unit and the protein is the secondary self-aggregate. Further self-assembly (tertiary, quaternary, etc.) is abundant in biological systems and often is involved with the hetero-building units such as membrane and bioinorganics. Tertiary and quaternary structures

> Along with the intermolecular forces, peptide bond formation is greatly involved. This issue will be discussed in detail in Chapter 7.

of proteins are the result of the intra-self-assembly of the secondary self-aggregates.

The general scheme for the self-assembly process that can occur as a single- to multiple-step and hierarchical-wise pattern can be proposed as shown in Figure 1.3. Any of the atoms, molecules, polymers, colloidal objects, or biological molecules can be a primary building unit for the initial step of the self-assembly. This primary self-assembly is governed by the balance between the intermolecular/colloidal forces as shown in Figure 1.2. The self-assembled aggregate formed is the primary self-aggregate, and can be the building unit (secondary building unit) of the subsequent self-assembly (secondary self-assembly). This process forms the secondary self-aggregate, which can be the building unit of the next self-assembly process. As long as the major forces for this process are the intermolecular/colloidal forces, further assembly is possible as tertiary and quaternary self-assemblies. And the general scheme in Figure 1.2 governs each of the processes.

We now consider the general formulation of this scheme that includes Figures 1.2 and 1.3. Figure 1.4 represents the schematics of some of the intermolecular/colloidal forces that will be discussed in Chapter 2. As a function of the distances between either molecules or colloidal objects, it shows quite complex features both in magnitudes and the ranges of length scale. For example, the curve that represents the case when the van der Waals attractive force has relatively comparable magnitude to the electric double-layer repulsive force shows both the attractive and repulsive nature of the total force as the distances between the two colloidal objects are changed. This is a typical situation for *kinetically* stable colloidal suspension. As long as the maximum energy barrier is high enough and the energy minimum after that (at longer distance; secondary minimum) is deep enough, the colloidal objects keep the constant distance in average. But, this energy barrier is not ultimate, so there are colloidal subjects that can overcome this barrier at any time. By the terms of self-assembly we discussed above, this situation means that the explicit formulation such as for the micellization of surfactant is not quite possible for the kinetical self-assembly of colloidal objects. Also, the range in which each of the forces is exerted is different. While the van

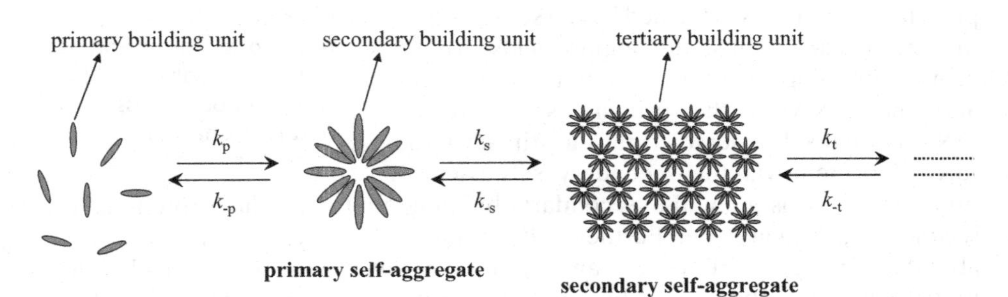

Figure 1.3. General scheme for self-assembly.

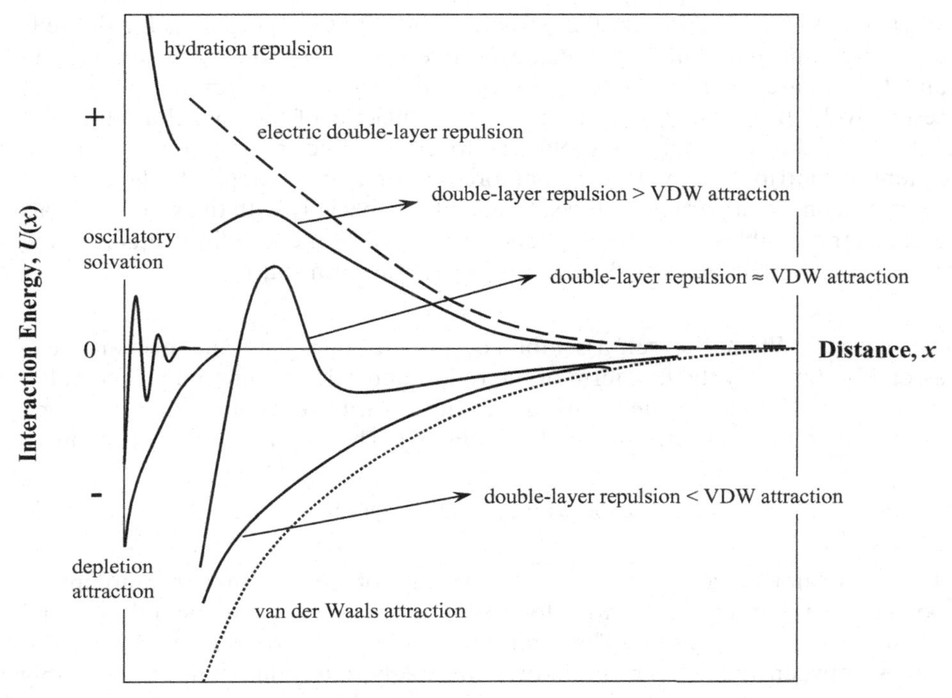

Figure 1.4. Schematic representation of the intermolecular and colloidal potential energies as a function of the distance between two objects (molecule or colloid). Range and magnitude are relative scales.

der Waals and electric double-layer forces are exerted at long and wide ranges, the hydration, solvation, and depletion forces have short- and narrow-range characters. This creates the notion that a *unique* expression that can cover the entire general scheme of the self-assembly might not be possible.

The following formulation is an expression with which we can overview the general self-assembly scheme. Suppose that the total net potential of the entire self-assembly processes with a given building unit is $U_{\text{total}}(x)$. Then, the $U_{\text{total}}(x)$ can be described as the net total of all of the attractive and repulsive potentials involved in each step of the self-assembly as follows:

$$U_{total}(x) = f_P \cdot \lfloor U_{A,P}(x) + U_{R,P}(x) \rfloor + f_S \cdot \lfloor U_{A,S}(x) + U_{R,S}(x) \rfloor + f_T \cdot \lfloor U_{A,T}(x) + U_{R,T}(x) \rfloor + \cdots + U_{ext}(x) \quad (1.1)$$

with

$$\sum f_P + f_S + f_T + \cdots = 1 \quad (1.2)$$

$U_{A,P}(x)$ and $U_{R,P}(x)$ represent the attractive and repulsive potentials for the self-assembly of primary building units, respectively. $U_{A,S}(x)$ and $U_{R,S}(x)$, and $U_{A,T}(x)$ and $U_{R,T}(x)$ are for the self-assembly of secondary and tertiary building units, respectively. f_P, f_S, and f_T are the fraction coefficients of the contribution of the net potential of each self-assembly step to the total net potential. $U_{ext}(x)$ is the potential contribution by the external forces when they are applied. The external forces become comparable to the self-assembly especially with the van der Waals and electric double-layer forces whenever the size of the building units is in the range of colloidal size. The details of this general form follow.

Type I. When the self-assembly occurs through only the primary self-assembly step, only the first term of the right-hand side of equation (1.1) is valid, with $f_P = 1$. The rest of the terms are not necessarily zero, but should be much smaller than the first term, so can be negligible. Thus, equation (1.1) becomes

$$U_{total}(x) = U_{A,P}(x) + U_{R,P}(x) \tag{1.3}$$

Typical examples are the micelle formation of surfactants or amphiphilic polymers at low concentration, formation of vesicle or microemulsion, and stable colloidal suspension. The interplay between the attractive and repulsive forces between the primary building units solely determines the self-assembly process.

For the case of colloidal suspension, equation (1.3) becomes the DLVO force with $U_{A,P}(x)$ and $U_{R,P}(x)$ as van der Waals force and electric double-layer force, respectively. The primary building unit is the colloidal objects.

For the cases of surfactant or polymer micelles, the primary building units are the surfactant or polymer molecules. Intermolecular hydrophobic force is now the major component of $U_{A,P}(x)$, and intermolecular steric, hydration (or solvation), and electric double-layer forces are of $U_{R,P}(x)$. The exact solution for equation (1.3) for this case is not known. But the semiempirical dimensionless thermodynamic solution of the packing parameter (or g-factor) (Chapter 17 of Israelachvili, 1992) provides an excellent tool for the formation and structural transition of the surfactant and polymer micelles. As long as the monomer concentration is kept low enough to minimize the intermicellar interaction (colloidal interactions between self-aggregates), this relation is also valid for the formation of vesicle, bilayer, and microemulsion.

Type II. The cases of self-assembly with both primary and secondary processes can be found in the formation of surfactant and polymer mesophases such as liquid crystals and the formation of secondary structures, such as tube and ring, of certain bio-mimetic systems (Chapter 7). Equation (1.1) now becomes

$$U_{total}(x) = f_P \cdot [U_{A,P}(x) + U_{R,P}(x)] + f_S \cdot [U_{A,S}(x) + U_{R,S}(x)] \tag{1.4}$$

As in the type I cases, surfactant, polymer, and bio-mimetic molecules are the primary building units in this scenario. But the secondary self-assembly is being induced either by interaggregates or by specific functional forces. For the formation of mesophases that can be induced above a certain concentration of surfactant or polymer, the micelles that have been formed via the primary self-assembly now face strong intermicellar colloidal interactions due to the increased concentration of the micelles. This interaction can be either attractive or repulsive. When it is mainly attractive, the micelles are directly assembled together. Thus, the micelles are the secondary building unit for this secondary self-assembly. The formed mesophases are the secondary self-aggregates. Due to the fact that these building units are the self-aggregates of molecules where the molecular rearrangement is obeyed by the energetics of each case, the mesophase can be in different forms. This includes typical liquid crystal structures such as hexagonal, cubic, and lamellar. When the interaction is mainly repulsive, the primary self-aggregates experience the structural transition that is involved with mainly additional monomer in the micelle (increased aggregation number) rather than the direct assembly between the micelles. This results in the formation of ellipsoidal, rodlike, or wormlike micelles.

For bio-mimetic molecules, the functional force, like hydrogen bond, induces the specific assembly of the primary self-aggregates. Thus, the secondary self-aggregates show characteristic directionality.

The fraction coefficients f_P and f_S in equation (1.4) should be correlated at some degree, and $f_P + f_S$ is unity. When $f_P > f_S$, the primary self-assembly should be dominant with the fraction of the secondary self-assembly. When $f_P \cong f_S$, the primary and secondary self-aggregates should be coexisting and thermodynamically favorable. The case of $f_P < f_S$ represents the favorable proceeding to the secondary self-assembly, and the dominance of the secondary self-aggregates. The exact solutions for any of these cases will require the exact knowledge of each form of the potentials as a function of the distances between each of the building units. This is a formidable task. Thus, this equation is not going to be able to provide preknowledge on the self-assembly processes. However, by acknowledging the individual forces functioning on the self-assembly processes along with their functioning range, this concept can provide the qualitative route to predict the entire self-assembly process with quite reasonable accuracy.

Type III. The third type of self-assembly is the self-assembly with the higher order of primary, secondary, tertiary, and above. Many of the self-assembly processes from biological systems show these types of characteristics. The abovementioned case of the formation of proteins that started from the self-assembly of amino acids is a typical example. Formation of a typical extracellular protein such as collagen that is being formed via multilevel hierarchical self-assembly is an excellent example of the self-assembly that occurs well beyond the tertiary self-assembly. The abovementioned formation of giant nanoparticles via magic number of aggregation is an example of the self-assembly up to the tertiary step. Equation (1.1) becomes

$$U_{total}(x) = f_P \cdot [U_{A,P}(x) + U_{R,P}(x)] + f_S \cdot [U_{A,S}(x) + U_{R,S}(x)] + f_T \cdot [U_{A,T}(x) + U_{R,T}(x)] + \cdots \tag{1.5}$$

The summation of all fraction coefficients should be $\sum f_P + f_S + f_T + \cdots = 1$. As with type II above, the exact solution of equation (1.5) can be obtained only by identifying the forces involved in each of the self-assembly steps and with the exact forms of the potentials. But the relation between the fraction coefficients can provide some qualitative picture of this type of self-assembly. For example, when any one of the coefficients is dominant over the rest of them, the self-aggregate represented by that coefficient is dominant with the coexistence of the rest of the two types of self-aggregates with minor amounts. When $f_P > (f_S, f_T)$, the primary self-aggregates are the major forms of the system, while $(f_P, f_S) < f_T$ represents the case of the dominance of the tertiary self-aggregates such as proteins. $f_S > (f_P, f_T)$ may be the case of the high concentration of nanoparticle assembly with magic number. Part of the giant nanoparticles can form clusters such as fractals that can be considered as the tertiary self-aggregates. When $f_P \cong f_S \cong f_T$, all three types of self-aggregates coexist.

Type IV. Since self-assembly is a process of force balancing between building units without the intervening of strong chemical bonds, it can be affected by the external forces that can have an influence on the intermolecular/colloidal forces during the process. Thus, for cases with the influence of external forces, equation (1.1) generally represents the entire self-assembly. The summation of fraction coefficients should be fitted with equation (1.2). For example, when this is the case with the primary self-assembly only, equation (1.1) becomes

$$U_{total}(x) = f_P \cdot [U_{A,P}(x) + U_{R,P}(x)] + U_{ext}(x) \tag{1.6}$$

When the secondary self-assembly process is also involved, it becomes

$$U_{total}(x) = f_P \cdot [U_{A,P}(x) + U_{R,P}(x)] + f_S \cdot [U_{A,S}(x) + U_{R,S}(x)] + U_{ext}(x) \tag{1.7}$$

Typical examples of the external forces include magnetic force, electric force, flow stress, capillary force, gravity, and interaction with substrate in a confined space. These forces actually can be present at all times in a real situation of the self-assembly process. But, regardless of the magnitude difference between the fraction coefficients, whenever the external potential begins to be dominant (or can compete) over the summation of intrinsic intermolecular/colloidal forces, the whole self-assembly process is affected.

For example, when the sterically modified polymer colloidal spheres that are given as a glass state under gravity (on Earth) are placed under microgravity (in space), they are rapidly crystallized. On Earth, gravity is comparative (or dominant) to the particle diffusion and intercolloidal interaction at this given condition, and thus acts on them to settle on the bottom of the container as a glass state. This intervening gravity effect is minimized in space; the self-assembly of this colloidal system now is solely controlled by the intercolloidal forces. Colloidal spheres that can have enough diffusion time to be balanced by the attractive and repulsive forces are crystallized into the regular lattice. The first experimental observation of this phenomenon was made on the Space Shuttle (Zhu et al., 1997) and was proved by theoretical calculation later (Simeonova and Kegel, 2004).

Another example that has become an interesting issue is the self-assembly of surfactant molecules or colloidal objects in a confined space. As the space between the self-assembly building units and the substrate is decreased below a certain range, the interaction of the building units with the surface of the substrate becomes comparable to the intermolecular/colloidal forces. Thus, this situation can considerably affect the whole self-assembly process. This interaction can be considered as the external force of this type of system. Recent examples include the self-assembly of mixed ionic micelles in a confined space of two parallel charged spaces (Yuet, 2004) and the dramatic effect of a geometrical confinement on the shear-induced self-assembly of colloidal polymer spheres (Cohen et al., 2004).

The general picture can be summarized as follows:

1. When $U_{total}(x)$ is equal to or close to zero with zero of $U_{ext}(x)$, the self-assembly is thermodynamically driven. The building units of each of the self-assembly steps are in equilibrium with the self-aggregates. The self-aggregates have finite sizes and defined shapes. Examples of this category include surfactant or polymer micelles, vesicles, proteins, and microemulsions.
2. When $U_{total}(x)$ is negative with zero of $U_{ext}(x)$, the self-assembly is kinetically driven. The self-assembly, in most cases, occurs until most of the building units are exhausted. The self-aggregates have indefinite sizes and less-defined shapes. Examples are coagulated colloidal or nanoparticle precipitates, bilayers, gels, some types of liquid crystals, and macroemulsions.
3. When $U_{total}(x)$ is positive with zero of $U_{ext}(x)$, the self-assembly is not possible in most cases. If this condition is exerted on self-assembled systems or during the self-assembly process, disassembly will be the most likely scenario.

1.4. CONCLUDING REMARKS

It would be fair to say that the schemes proposed in this chapter are nowhere near perfection, nor can then bring the exact solution for the exact prediction of a variety of self-assembly processes. But it would also be fair to say that by accepting the concept of force balance for self-assembly and the general concept of multistep self-assembly processes, we can benefit in the following ways:

1. A variety of processes in nature that are mainly governed by the intermolecular and/or colloidal forces can be integrated into the picture of the self-assembly, which includes a variety of building units with different-length scales and different origins.
2. A variety of self-assembly processes we acknowledged earlier in the chapter can be understood as one unified concept.
3. Each of those self-assembly processes and the physical properties of the self-assembled aggregates can be qualitatively explained with reasonable accuracy.

These issues will be examined throughout the rest of Part I with detailed explanations and examples. Also, it will be shown that they can be directly correlated with the self-assemblies in nanotechnology and provide useful tools to address a variety of nanotechnology issues. The general outline for this will be presented in Chapter 8 followed by the details in the rest of Part II.

REFERENCES

Alexandridis, P., Lindman, B., eds. *Amphiphilic Block Copolymers: Self-assembly and Applications* (Elsevier: 2000).

Buitenhuis, J., Dhont, J. K. G., Lekkerkerker, H. N. W. "Static and Dynamic Light Scattering by Concentrated Colloidal Suspensions of Polydisperse Sterically Stabilized Boehmite Rods," *Macromolecules* **27**, 7267 (1994).

Clint, J. H. *Surfactant Aggregation* (Blackie: 1992).

Cohen, I., Mason, T. G., Weitz, D. A. "Shear-Induced Configurations of Confined Colloidal Suspensions," *Phys. Rev. Lett.* **93**, 046001/1 (2004).

Emrick, T., Fréchet, J. M. J. "Self-Assembly of Dendritic Structures," *Curr. Opin. Colloid Interface Sci.* **4**, 15 (1999).

Fendler, J. H., Dékány, I. *Nanoparticles in Solids and Solutions* (Kluwer Academic Publishers: 1996).

Groenewold, J., Kegel, W. K. "Anomalously Large Equilibrium Clusters of Colloids," *J. Phys. Chem. B* **105**, 11702 (2001).

Israelachvili, J. N. *Intermolecular and Surface Forces*, 2nd ed. (Academic Press: 1992).

Likos, C. N. "Effective Interactions in Soft Condensed Matter Physics," *Phys. Rep.* **348**, 267 (2001).

Muratov, C. B. "Theory of Domain Patterns in Systems with Long-Range Interactions of Coulomb Type," *Phys. Rev. E* **66**, 066108 (2002).

Overbeek, J. Th. G. *Colloid Science*, Kruyt, H. R. ed. (Elsevier: 1952).

Rao, C. N. R. "Universal Aspects of Self-Assembly: The Wide Domain of Weak Interaction," *Curr. Sci.* **81**, 1030 (2001).

Rao, C. N. R., Kulkarni, G. U., Thomas, P. J., Edwards, P. P. "Metal Nanoparticles and Their Assemblies," *Chem. Soc. Rev.* **29**, 27 (2000).

Sciortino, F., Mossa, S., Zaccarelli, E., Tartaglia, P. "Equilibrium Cluster Phases and Low-Density Arrested Disordered States: The Role of Short-Range Attraction and Long-Range Repulsion," *Phys. Rev. Lett.* **93**, 055701/1 (2004).

Simeonova, N. B., Kegel, W. K. "Gravity-Induced Aging in Glasses of Colloidal Hard Spheres," *Phys. Rev. Lett.* **93**, 035701/1 (2004).

van der Schoot, P. "Remarks on the Association of Rodlike Macromolecules in Dilute Solution," *J. Phys. Chem.* **96**, 6083 (1992).

Verwey, E. J. W., Overbeek, J. Th. G. *Theory of the Stability of Lyophobic Colloids* (Elsevier: 1948).

Yuet, P. K. "A Simulation Study of Electrostatic Effects on Mixed Ionic Micelles Confined between Two Parallel Charged Plates," *Langmuir* **20**, 7960 (2004).

Zhu, J., Li, M., Rogers, R., Meyer, W., Ottewill, R. H., STS-73 Space Shuttle Crew, Russel, W. B., Chaikin, P. M. "Crystallization of Hard-Sphere Colloids in Microgravity," *Nature* **387**, 883 (1997).

2

INTERMOLECULAR AND COLLOIDAL FORCES

Five representative forces that govern the binding/interaction of atoms or molecules are ionic, metallic, covalent, hydrogen, and van der Waals forces. When two or more atoms come close together to form a new organic molecule, the force that binds them together is covalent force. It is mostly strong (200–800 kJ/mol) and acts at a very short range of distance (less than a few Angström). The metallic bond operates in a similar way as the covalent bond but mostly among metals that have an *electron sea*. Each atom or molecule that participates with these bonds loses its identity after the bonding. The electron density distribution of each atom or molecule is completely changed after the bonding. In this sense, these forces can be called *chemical forces that give chemical bonding*. Hydrogen and van der Waals forces are usually weak and act on a relatively long range of distance. The results of these forces do not change the identity of each atom or molecule involved. As opposed to the chemical forces, these forces can be called *physical forces that give physical bonding*.

In the self-assembly process, whether it occurs at an atomic-, molecular-, or even colloidal-length scale, rather weak and much longer-range forces compared with the chemical forces take important roles. As described in Chapter 1, they can be viewed as three representative groups: (1) the driving forces that bring the self-assembly units together, (2) the opposing forces that balance with the

Self-Assembly and Nanotechnology: A Force Balance Approach, by Yoon S. Lee

driving forces, and (3) the functional forces that determine the directionality and functionality of self-assembled aggregates.

The purpose of this chapter is to deal with the molecular origin of these weak long-range forces (intermolecular forces) and their relation with a rather large-length scale of colloidal forces. Five distinctive forces will be presented first: van der Waals force, electrostatic force (electric double-layer force), steric/depletion forces, solvation/hydration forces, and hydrophobic force. Each section begins with the basic concept of molecular origin followed by intermolecular force and its expansion to colloidal force. Colloidal forces basically originate from the intermolecular forces usually in a cumulative way but with intervention and/or correction by their geometry and length scale. This inevitably brings numerical approaches or risky assumptions, in many cases, for their formulation.

I have tried to get to the point directly in each section, so as to avoid confusion by lengthy description. References at the end of the chapter will provide more details for interested readers. The final section is for the hydrogen bond. It is a relatively strong and short-range force compared with typical intermolecular forces but has important roles as a functional force in many self-assembly processes.

The coordination bond can be a key driving force for the structuring of certain coordination compounds into unique supramolecular self-assembled structures (Lehn and Ball, 2000). It also serves as a functional force in certain biological systems. This aspect will be revisited in Chapter 7. But the details of this force are beyond the scope of this book. Typical inorganic chemistry textbooks share the great volume of this issue.

Thanks to the development of new measurement devices over the last few decades, such as the surface force measurement apparatus, and scanning probe microscopy (SPM) and its derivative versions, intermolecular and colloidal forces are now measured with great accuracy. This aspect will be mentioned throughout this book whenever it is necessary to provide a clear explanation for a given issue.

2.1. VAN DER WAALS FORCE

Van der Waals force is originated by dipole or induced-dipole interactions at the atomic and molecular level. Thus, there can be three different types of van der Waals forces:

1. Keesom interaction: permanent dipole–permanent dipole interaction
2. Debye interaction: permanent dipole–induced dipole interaction
3. London (or dispersion) interaction: induced dipole–induced dipole interaction

Figure 2.1 shows the schematic representation. These three types are collectively called *van der Waals interactions*. They are proportional to molecular parameters

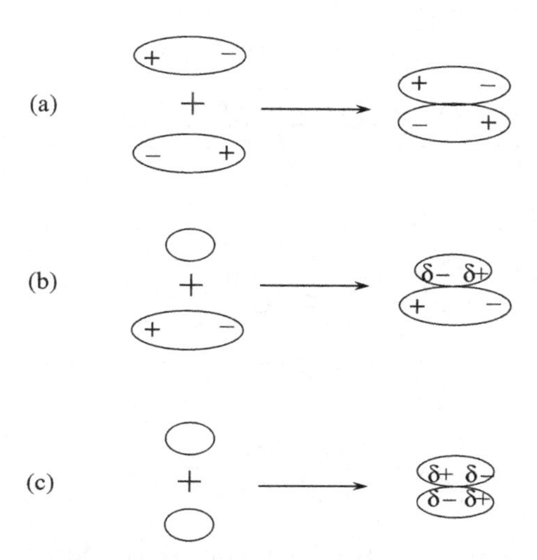

Figure 2.1. Schematic representation of the three types of van der Waals interactions between atoms or molecules: (a) permanent dipole–permanent dipole interaction, (b) permanent dipole–induced dipole interaction, and (c) induced dipole–induced dipole interaction.

that represent the polarization of molecules (polarizability, dipole moment of molecule), and have inverse sixth-power dependence on the separation between the nearest two molecules. There is a portion of interparticle repulsion at the molecular level that is also a function of interparticle separation. This force has inverse twelfth-power dependence.

The combined expressions of van der Waals attraction and the repulsive force are called the *12–6 power law* or the *Lennard-Jones potential*, which appears in most of the physical chemistry textbooks. The London interaction is always present because it does not require the presence of a permanent dipole or charge. A change of symmetry in electron clouds by any means such as charge, permanent dipole, or induced dipole can originate a London interaction. For most of the molecules, London interaction takes the largest contribution of van der Waals attraction, except for a highly polar molecule such as water.

Overall, van der Waals interactions are always attractive at atomic and molecular levels. But, between colloidal objects under certain conditions, they can be repulsive. Self-assembly is an event not only in the range of the molecular level but in the range of the submicroscopic level as well. Thus, it is important for us to understand the nature of the van der Waals interaction both as an expression in the molecular level and in its relation with supermolecular to macroscopic objects.

For the system of a pair of molecules, Debye, Keesom, and London interactions are expressed as follows:

Debye equation:

$$U(x) = -\frac{(\alpha_1\mu_2^2 + \alpha_2\mu_1^2)}{(4\pi\varepsilon_0)^2}x^{-6} \tag{2.1}$$

Keesom equation:

$$U(x) = -\frac{2}{3}\frac{\mu_1^2\mu_2^2}{(4\pi\varepsilon_0)^2 k_B T}x^{-6} \tag{2.2}$$

London equation:

$$U(x) = -\frac{3}{2}h\frac{\alpha_1\alpha_2}{(4\pi\varepsilon_0)^2}\frac{\nu_1\nu_2}{(\nu_1 + \nu_2)}x^{-6} \tag{2.3}$$

where x is the distance between two interacting molecules, α_1 and α_2 is the polarizability of each molecule, μ_1 and μ_2 is the dipole moment of each molecule, k_B is the Boltzmann's constant, T is temperature, h is the Plank's constant, ε_0 is the dielectric permittivity of vacuum, and ν_1 and ν_2 is the vibrational frequency of electron of each molecule. These three equations can be combined into one equation that can represent the net (or total) contribution to van der Waals attraction. For a simple example, that is, for a pair of identical molecules:

$$U_t(x) = -\frac{1}{(4\pi\varepsilon_0)^2}\left(2\alpha\mu^2 + \frac{2}{3}\frac{\mu^4}{k_B T} + \frac{3}{4}h\nu\alpha^2\right)x^{-6} = -\beta x^{-6} \tag{2.4}$$

where the subscript t on potential means total potential and β is the van der Waals interaction parameter.

For a thermodynamics point of view, β can be well correlated with van der Waals parameters a and b in the van der Waals equation for state of gas through equations

$$a = \frac{2}{3}\frac{\pi N_A^2 \beta}{\sigma^3} \tag{2.5}$$

and

$$b = \frac{2}{3}\pi\sigma^3 N_A \tag{2.6}$$

where N_A is Avogadro's number and σ is the diameter of atom or molecule. These relations provide one way to estimate van der Waals interaction parameter β between molecules from known values of a and b.

The scaleup of van der Waals interaction to a colloidal-length scale is strongly dependent on the geometry and composition of the colloidal objects involved. All molecules from one object interact with all molecules from the other object pairwisely.

Note: This does not suggest that van der Waals interaction is pairwise additive like gravitational and Coulomb forces.

The summation of these attractive forces will be the interaction force between the two colloidal objects considered. This issue is well described in Hiemenz and Rajagopalan (1997).

For two spherical colloidal objects with the same size and composition, the same interactions (whether it is each of Debye, Keesom, or London interaction, or the total interaction) that can induce the association of individual molecules are responsible for the aggregation of colloidal-scale objects up to a certain length scale. However, the sixth-power dependence potential decay on the separation of two objects turns into seventh-power dependence decay at the separation in the range of 10–100 nm. This means the van der Waals interaction (attraction) between colloidal bodies decays more rapidly as the separation between them increases.

The degree of this decay as a function of the distance between the colloidal objects is dependent on the geometry of the objects. The more complex geometry brings the more complicated form of expression. Hamaker constant A is a well-established parameter that can provide the relative picture and magnitude of the van der Waals interaction between colloidal objects (Hamaker, 1937). It typically has the value between 10^{-19} and 10^{-20} J. For example, it can be expressed as follows for two identical interacting colloidal blocks (semi-infinite parallel plates):

$$U(x) = -\left(\frac{\rho N_A \pi}{M}\right)^2 \beta \frac{1}{12\pi} x^{-2} = -A \frac{1}{12\pi} x^{-2} \tag{2.7}$$

where β is the van der Waals interaction parameter between the molecules, ρ is the density of molecules in the block, M is the molecular weight, and x is the distance between the two blocks.

The intrinsic assumption made for this expression does not count on the important facts of a real situation including surface heterogeneity, screening of inside molecule by surface molecule, and interaction with a medium such as solvent.

Note: This is the reason that the van der Waals interaction is not pairwise additive.

For the colloidal objects interacting through a vacuum, the Hamaker constant should be always positive, meaning the interaction is always attractive (Table 2.1). The Hamaker constant is also always positive when identical colloidal objects are interacting through solvents. The force is usually much reduced by the action of solvent. But, when nonidentical colloidal objects are interacting through solvent, the Hamaker constant can be positive or negative, meaning the van der Waals interaction can be attractive or repulsive (Table 2.2).

Repulsive van der Waals interaction occurs when the value of Hamaker constant of solvent is in between the one of one colloidal object and the one of the other colloidal object. In molecular property terms, when the dielectric prop-

TABLE 2.1. Hamaker constants calculated from Lifshitz theory for typical inorganic, organic, polymeric, metallic, and biological materials in vacuum and in water.

	Hamaker Constant, 10^{-20} J		
Material	Vacuum (air)	Water	Ref.
$BaTiO_3$	18	8	Bergström, 1997
α-Al_2O_3	15.2	3.67	Bergström, 1997
Mica	9.86	1.34	Bergström, 1997
SiO_2 (amorphous)	6.50	0.46	Bergström, 1997
TiO_2	15.3	5.35	Bergström, 1997
Dodecane	5.04	0.502	Hough and White, 1980
Tetradecane	5.10	0.514	Hough and White, 1980
Hexadecane	5.23	0.540	Hough and White, 1980
Stearic acid	—	0.079–0.368	Visser, 1972
Benzene	—	0.04	Visser, 1972
Poly(methylmethacrylate)	7.11	1.05	Hough and White, 1980
Poly(vinylchloride)	7.78	1.30	Hough and White, 1980
Polystyrene	6.37	0.911	Hough and White, 1980
Pt	—	8–47.7	Visser, 1972
Au	29.6–45.5	5.5–140	Visser, 1972
Ag	16.4–40.0	3–28.2	Visser, 1972
Bovine serum albumin	1.04–2.56	1.40–2.00	Visser, 1972
Biological cells	—	0.02–0.25	Visser, 1972
Leucocytes	—	0.01–0.1	Visser, 1972

TABLE 2.2. Hamaker constants between two different inorganic materials that can show negative value in water. (Data from Bergström, 1997.)

Material 2 / Material 1	Hamaker Constant, 10^{-20} J			
	PbS	α-Al_2O_3	Mica	SiO_2 (amorphous)
PbS	4.98	−0.20	−0.03	−0.08
α-Al_2O_3	—	3.67	2.15	—
Mica	—	—	1.34	0.69
SiO_2 (amorphous)	—	—	—	0.46

erty of the solvent is in between those of the two colloidal objects, the Hamaker constant will be negative. Figure 2.2 represents the schematics for each case. Electrostatic effects such as counterion binding with the same charge on the surface of colloidal objects or steric effects by polymer brush adsorbed on the surface of a colloidal object can also induce repulsive van der Waals interaction in the solvent, which has important implications for the self-assembly of the colloidal-length scale.

Lifshitz theory (Lifshitz, 1956; Dzyaloshinskii et al., 1961) deals with the problem inherited from the basic assumption of pairwise additivity. Equation (2.7) is the expression for the Hamaker constant for two colloidal blocks derived based on the assumption of simple pairwise additivity of the molecular property. But this completely ignores the influence of neighboring molecules. For example, the molecules at/near the surface screen those buried deep in the materials. And, if there is a molecule that has permanent dipole or charge, these may affect the effective polarizability of neighboring molecules or they may themselves experience some orientational effect. The surface heterogeneity that is always a possibility in the real system can also induce significant error.

Lifshitz theory is derived entirely based on measurable bulk properties such as dielectric constant and refractive index. It completely ignores the molecular properties. The formulation itself is quite complicated and beyond the scope of this book. For example, for the two colloidal blocks 1 and 2 interacting through medium 3, the Hamaker constant A_{132} can be written

$$A_{132} = \frac{3}{8\pi^2} h \int_0^\infty \left[\frac{\varepsilon_1(i\nu) - \varepsilon_3(i\nu)}{\varepsilon_1(i\nu) + \varepsilon_3(i\nu)} \right] \left[\frac{\varepsilon_2(i\nu) - \varepsilon_3(i\nu)}{\varepsilon_2(i\nu) + \varepsilon_3(i\nu)} \right] d\nu \tag{2.8}$$

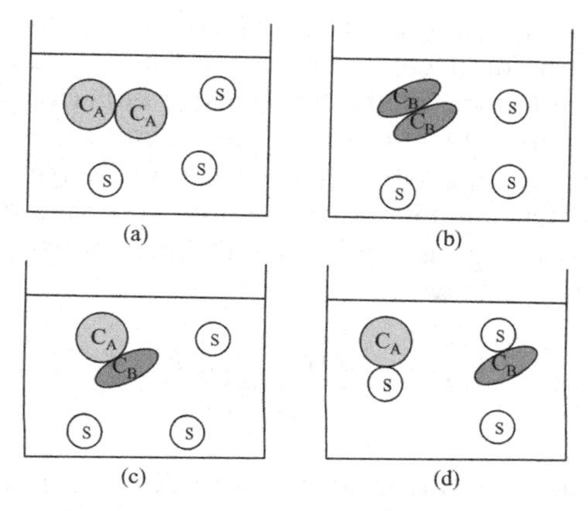

Figure 2.2. Van der Waals interaction between colloidal particles in solvent: (a) and (b) interaction between identical particles that is always attractive, (c) attractive interaction between dissimilar particles, and (d) repulsive interaction between dissimilar particles. C_A and C_B represent colloidal particles, and S represents solvent molecule.

where $\varepsilon(i\nu)$ is the dielectric constant at imaginary frequency and h is the Planck's constant. The value of $\varepsilon(i\nu)$ can be measured from the dissipative part of the dielectric constant spectrum for each material. The complexity of original Lifshitz theory has limited its use in many real systems. Simpler, more useful versions of equations have been derived (Mahanty and Ninham, 1976). Hamaker constants for common materials derived from these equations show relatively good consistency with full Lifshitz calculation (Visser, 1972; Hough and White, 1980; Bergström, 1997). In many cases, they also show a fairly good match with experimental values. But, these Hamaker constants in some cases show a wide range of discrepancy, especially when the medium is other than vacuum (Gregory, 1969; Ackler et al., 1996). This again points to the complexity of the role of the intervening medium in the van der Waals interaction. Tables 2.1 and 2.2 provide some sense of the Hamaker constant for typical materials calculated based on Lifshitz theory in vacuum and in water.

2.2. ELECTROSTATIC FORCE: ELECTRIC DOUBLE-LAYER

The forces originating from electrostatic interaction are another example of common forces that have a strong effect on many self-assembly processes. They can be either balanced with hydrophobic interaction, which results in the finite size of the self-aggregates, or sometimes be added onto during self-assembly. Atomic-scale self-assembly processes are strongly involved with the electrostatic interaction in vacuum or in air. Most of the molecular and colloidal- (meso-) scale self-assembly processes are mainly involved in solution. To stay on track in this book and not be distracted by a lengthy description of the theories and models, I will stick to the relevance of electrostatic interaction in the self-assembly processes in this section; mainly in solution (electric double-layer and its interaction), with a brief introduction of electrostatic interaction in vacuum first.

Whenever two or more charged atoms, ions, or molecules interact in vacuum or in medium, Coulomb interaction occurs between them. Compared with the covalent bond, which is a somewhat strong and short-range force, the Coulomb interaction is a very strong (500–1,000 kJ/mol) and long-range force (up to ~50 nm). Its potential energy is the inverse-distance dependence between two charges of Q_1 and Q_2 as follows:

$$U(x) = \frac{Q_1 Q_2}{4\pi\varepsilon_0 x} = \frac{z_1 z_2 e^2}{4\pi\varepsilon_0 x} \tag{2.9}$$

where z_1 and z_2 are the ionic valence of each charge respectively, e is the elementary charge in vacuum (1.602×10^{-19} C), ε_0 is the dielectric permittivity of vacuum ($8.854 \times 10^{-12}\, C^2J^{-1}m^{-1}$), and x is the distance between the two charges. When the interaction happens in medium, ε_0 becomes $\varepsilon_0\varepsilon$ with ε as the relative permittivity or dielectric constant of medium.

A surface, whether it is soft or hard, can acquire charges in a number of ways when contacted with or immersed into aqueous solutions. The most feasible mechanisms are adsorption (physisorption and chemisorption) of ions from solution or as a result of dissociation, ionization, and surface reaction on surfaces. Figure 2.3 represents the schematics of this situation. Assuming the surface is now fully charged with positive ions, the near surface area will be mostly attracted by negatively charged ions from solution. As the distance goes farther from the surface, the positively charged ions will be diffused into the region while the density of negatively charged ions will be gradually decreased. The mathematical models developed to address this situation reveal two hypothetical layers as shown in Figure 2.3: Stern layer (Stern, 1924; or Helmholtz layer) and Gouy-Chapman double-layer (Gouy, 1910; Chapman, 1913; or diffuse layer, electric double-layer, diffuse electric double-layer). The Stern layer is defined from the boundary where the saturation limit of ions adsorbed (or ionized) on the surface is drawn. The Gouy-Chapman double-layer is the layer where the gradual diffusion of positively charged ions can be recognized. Outside this layer, the ion density can be assumed to be neutral by the compensation between two oppositely charged ions.

There is no experimental technique that can directly measure the actual thickness of these layers and the potential changes as a function of the distance from the surface. The Debye-Hückel approximation addresses this issue, using the classical Poisson-Boltzmann equation. The Gouy-Chapman model acknowledges the potential limitation of the Debye-Hückel approximation. But, under the limitation of low potential that is the case for most of the colloidal systems in aqueous solution, both theories show the same qualitative representation for the picture of the electric double-layer. That is,

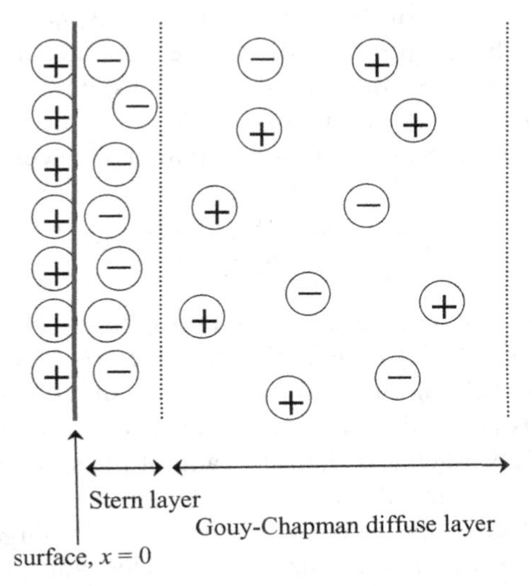

Figure 2.3. Schematic representation of Stern layer and Gouy-Chapman diffuse layer.

$$U = U_0 \cdot \exp(-\kappa x) \tag{2.10}$$

where U is the potential at $x = x$, U_0 is the potential at $x \to 0$, and x is the distance from the surface. This is known as the *Debye-Hückel equation*. For ordinary systems that have their concentration below 0.1 M with mono-, bi-, or triple-valence electrolytes, κ^{-1}, which is defined as the *thickness* of the double layer, can be reached up to ~10 nm.

Equation (2.10) also tells us that, in the diffuse layer, the potential decreases exponentially as the distance goes away from the surface. When two surfaces with the same type of charge are approaching together, the first contact area is obviously the outermost region of the diffuse layers. As two surface are coming closer, the diffuse layers (the potentials) from each surface come to overlap. This generates excess pressure that originates the repulsive force between the two surfaces: *electrostatic double-layer repulsion* (Verwey and Overbeek, 1948). The potential in the diffuse layer is mainly changed by the geometry of the surface, and the concentration and type of electrolyte. So does the repulsive force. The effect of surface roughness and nonuniform surface charge density on the electrostatic interaction between electric double-layers has been well formulated by Duval et al. (2004).

In colloidal solutions or biological systems, interaction between the curved or more likely spherical double-layers is evident. Formulation of the forces between two curved surfaces in solution requires solving the Poisson-Boltzmann equation for the change of potential energy in a curved double-layer. This is a complex process and requires numerical approaches that induce a large margin of error. Derjaguin approximation (Derjaguin, 1934) has been widely accepted as a useful solution to address this issue. It relates the force between two curved surfaces with the potential energy between two planar surfaces through the assumption that the surface of the curved objects or spheres may be considered as the summation of the infinitesimal size of the stepped surface of planar geometry. For instance, for the two interacting colloidal spheres with radius of R_1 and R_2, respectively, the force between two spheres, *F(x)*, can be correlated with potential energy per unit area, *U(x)*, of two planar surfaces by

$$F(x) \approx 2\pi \left(\frac{R_1 \cdot R_2}{R_1 + R_2} \right) \cdot U(x) \tag{2.11}$$

where x is the closets distance between the two spheres. This is called *Derjaguin approximation* and is valid as long as the distance x is much less than the radii of the spheres or the radii of the curvatures of the curved surfaces. More specifically in solution, it is applicable when the thickness of the electric double-layer or Debye length, κ^{-1}, is much less than the radii of the spheres or the radii of the curvatures of the curved surfaces.

This much-used equation is relevant generally for large colloidal systems such as suspensions of micrometer sizes of particles, and provides an excellent estimation for both attractive and repulsive forces. However, when the size of the colloids approaches to the length scale of the interaction, a violation can occur. Derjaguin approximation often overestimates the forces between the small colloids with longer Debye length (Todd and Eppell, 2004). This is usually the case for many of the polymer colloids and biomolecules in biological systems, and for most of the nanoparticles and nanotubes. Because they usually have a size range of ~1–~50 nm, their Debye lengths can easily reach over a couple of nanometers and even larger than their sizes. There have been recent efforts to address this issue. The surface element integration approach (Bhattacharjee and Elimelech, 1997) predicts much closer values of the interaction force between the small spherical colloid and planar surface than the calculation based on the Derjaguin approximation. Much progress on this issue is needed for the deeper quantitative understanding of especially self-assembly processes for nanoparticles, biological systems, and colloidal- (meso-) scale systems, which should include the development of reliable solutions for the interactions between a variety of geometries.

In reality, the total force acting on the colloidal objects in solution is always the sum of the electric double-layer force and the van der Waals force. Depending on the geometry of the colloidal subject and the physical/chemical composition of the surface of the objects, other forces such as steric force, hydrophobic interaction, and solvation force should be included as well. As long as we focus on the former case (the case with only electric double-layer force and van der Waals force), the legendary DLVO (Derjaguin-Landau-Verwey-Overbeek) (Derjaguin and Landau, 1941; Verwey and Overbeek, 1948) theory well addresses this total force acting on colloidal objects. It also beautifully predicts the experimental values for a wide range of the separation between two colloidal objects in water and also in a variety of electrolyte solutions. This range covers from well below the Debye length to over 200 nm.

For the mathematical expression, DLVO theory is the sum of the expression for the electric double-layer interaction and that of the van der Waals interaction. Thus, the net interaction potential between two planar colloidal surfaces becomes

$$U(x) = 64k_BTn_\infty\gamma^2 \exp(-\kappa x) - (A/12\pi)/x^2 \tag{2.12}$$

where k_B is the Boltzmann's constant, T is temperature (K), n is the concentration of ion in bulk, $1/\kappa$ is the Debye length, A is the Hamaker constant, and x is the distance between two surfaces.

By using the Derjaguin approximation of equation (2.11), the expression for the force between two spherical colloidal objects with the same radius R can be written as

$$F(x) = \pi R \cdot U(x) \tag{2.13}$$

Thus, the net DLVO interaction potential between two spherical colloidal objects becomes

$$U(x) = 64\pi R k_B T n_\infty \gamma^2 \kappa^{-2} \exp(-\kappa x) - AR/12x \tag{2.14}$$

For equations (2.12) and (2.14),

$$\gamma = \tanh(ze\psi_0/4k_BT) \tag{2.15}$$

where γ is the surface tension (or surface energy) and ψ_0 is the electrostatic surface potential of ion with its valence of z.

For equations (2.12) and (2.14), the first terms on the right-hand side of the equations are the electric double-layer repulsion terms and the second terms are the van der Waals attraction terms. Details on the expression for the van der Waals energy of colloidal objects with different geometries can be found in Israelachvili (p. 177, 1992).

This DLVO theory well describes the issues of colloidal stability (thermodynamically and kinetically), coagulation, and critical coagulation concentration of colloidal suspensions. The details, including the shapes of potential curves as functions of a variety of parameters in the equations, can be found in Israelachvili (1992) and Hiemenz and Rajagopalan (1997). But here let me mention a couple of important and common implications of this theory.

Equations (2.12) and (2.14) say that the colloidal suspension will be coagulated as the pair potential becomes attractive by

1. The higher Hamaker constant A
2. The smaller surface potential ψ_0
3. The larger κ that means smaller Debye length that can be obtained from higher electrolyte concentration or from higher valence of ions

By simply solving equations (2.12) and (2.14) (details are provided by Hiemenz and Rajagopalan, 1997) by setting both the potential and its first derivative at zero, we can deduce the Schulze-Hardy rule from this DLVO theory. It states that the colloidal stability primarily is affected by the valence of the ions of the opposite charge to the colloid rather than the characteristics of the ions. DLVO theory predicts that the critical coagulation concentration of electrolytes $[C]_{elec}$ follows the inverse sixth power of the valence of the ion in solution. That is,

$$[C]_{elec} \propto 1/z^6 \tag{2.16}$$

Classical theory on the kinetics of coagulation views the coagulation process as mainly of two classes. They are *perikinetic* coagulation, where the Brownian diffusional motion is the main force for the contact between the colloidal particles, and *orthokinetic*, where the contact is mainly caused by the velocity gradient, that is, when the potential energy curve has an energy barrier larger than ~kT. DLVO theory predicts relatively well the rate of coagulation for both processes.

The ratio of the rate functions for these two processes is well known as the *stability ratio*, and is well correlated with the value of critical coagulation concentration. It also connects in a nice relation with the fractal dimension of colloidal self-aggregates. Limitation of this theory comes from its basic assumption during the formulation that does not account for the polydispersity of colloidal particles in the system and nonspherical morphology of the colloidal particles.

2.3. STERIC AND DEPLETION FORCES

Since the ancient civilizations of Egypt, Greece, and China, people have used a variety of natural polymeric materials as stabilizers for their paints, inks, and foods. Increased colloidal stability in these systems by the action of the polymers, mainly by steric and depletion forces, was one of the keys of their success. This has been an important issue for decades whenever there has been a great need to control coagulation, emulsification, and flocculation, which widely appears in traditional colloid science and related industries, such as food, petroleum, cosmetics, and pharmaceuticals, as well as in the recent development of environmental issues and nanoparticle- and nanomaterial-related processing and manufacturing.

Steric (or overlap) force is mainly induced by polymers, polyelectrolytes, or biomacromolecules that are adsorbed or grafted onto the surface of colloidal objects. It is a long-range force whose interaction can reach up to ~$10 \cdot R_g$ (where R_g is the radius of the gyration of the polymer chain) in aqueous solution. Every possible interaction, including polymer–polymer, polymer–solvent, and polymer–colloid interaction, and the physical/chemical conditions of the system, including solvent (good or poor in a given polymer), temperature, and nature of charge (especially in the case of polyelectrolytes) involved can affect this force. Thus, the quantitative formulation has encountered quite a bit of limitation. In this section, I briefly mention the qualitative approach, which has provided quite a good insight on this issue.

Polymer-induced forces can be either repulsive or attractive. As shown in Figure 2.4, *repulsive* force arises when:

1. Solvent in the system is a good solvent to the polymer. Good solvent molecules have affinity with the polymer segment. Thus, when the two colloidal objects coated with polymer chains contact each other, the solvent molecules are constrained near chains. This reduces the dielectric constant of medium, which results in larger repulsive force than predicted by the conventional Gouy-Chapman theory (Huang and Ruckenstein, 2004).
2. The number density of the polymer chains per unit area of the surface is in the range where the interaction between the polymer chains restricts the molecular motion or orientational freedom. This causes the loss of

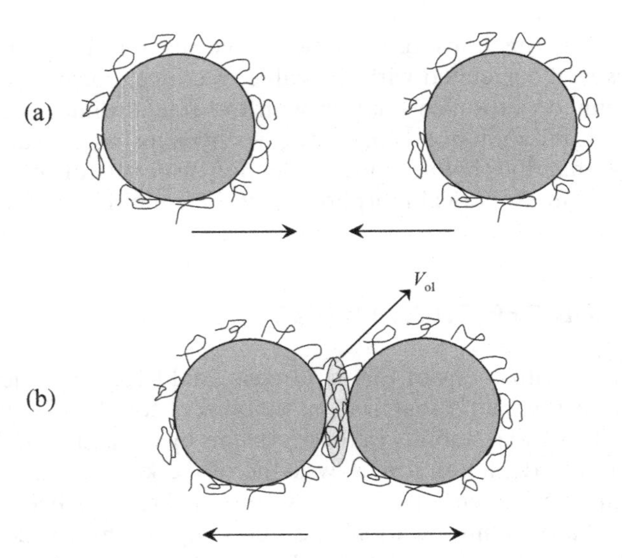

Figure 2.4. Schematic representation of steric repulsion between polymer-decorated colloidal spheres: (a) two colloidal spheres approach each other, and (b) repulsive interaction potential increases as polymer chains overlap each other. V_{ol} is the volume where chain-overlap occurs, so the polymer concentration is nearly doubled.

entropy due to the chain confinement, and thus causes the repulsive force.

3. The elasticity of polymer coils is large enough to oppose the compression by the approaching of two surfaces. While case 2 is the entropically driven interaction induced by contact or interdigitation of polymer chains, this is the case where the change is in conformational entropy between stretched and compressed polymer chains.

Attractive force, on the other hand, arises when:

1. Solvent is a poor solvent. As the two surfaces with polymer on the surface approach each other, the polymer–polymer attraction is being preferred to polymer–solvent attraction, which will result in the net attractive force. The theta (θ) temperature becomes an important factor in this case.
2. The polymer chain has enough affinity with the surfaces. As shown in Figure 2.5, *bridging attraction* can be possible in this case. Two colloidal surfaces are attracted by attraction with polymer chains. Along with theta (θ) temperature, the number density of the polymer chains on the surface that determines the available adsorption (physisorption or chemisorption) sites for the approaching polymer chain is an important factor in this case.

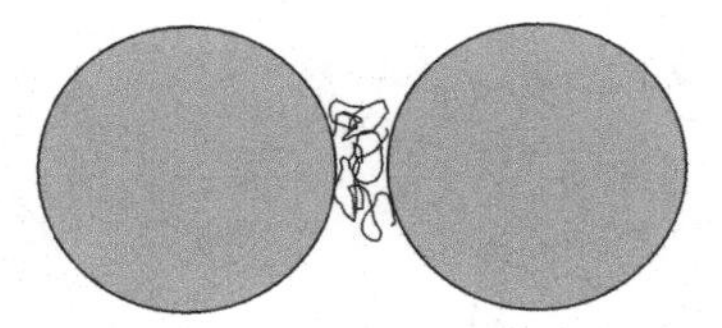

Figure 2.5. Schematic representation of attractive interaction between colloidal spheres via polymer bridging.

3. When the colloidal objects are interacting in the solution with nonadsorbing polymer, attractive depletion interaction can occur. This is discussed in more detail in the next three paragraphs.

Whenever the colloidal objects are interacting in the solution with nonadsorbing polymer (or nonadsorbing micelle or particle), there will be evenly distributed polymer chains between colloidal particles when the distance between the colloids is long enough. When the distance between the approaching colloids becomes smaller than the size of polymer ($\sim 2 \cdot R_g$, where R_g is the radius of gyration of polymer; or diameter of micelle or particle), there likely is a depletion of polymer chains at the contact region (depletion volume) as the polymer chains are being squeezed out of this region. Thus, the osmotic pressure force that is exerted by the molecules on the outside region (outside of depletion volume region) of the colloid exceeds that on the inside region (depletion region). This will induce net attractive force between the colloids (Figure 2.6; Asakura and Oosawa, 1958).

Strength of depletion force is mainly dependent on the concentration and molecular weight of polymers. It is weaker than van der Waals, electrostatic, and even solvation forces, but can be an important force for colloidal stability when there are no other significant attractive forces. But, due to this nature, depletion force is often difficult to distinguish from bridging force. The first successful measurement was made in 1988 (Evans and Needham, 1988) by using the synthetic lipid bilayers in an aqueous solution of dextran (polyglucose). This result verified the above scenario of the origin of depletion force and confirmed its relation with polymer concentration, size of polymer chain, and temperature. When the colloidal object and/or polymer chain (or micelle, particle) is charged (either the same or different), which is the case in many real systems, electric double-layer repulsion and depletion attraction compete as a function of the distance. At a very small distance, repulsive electric double-layer force dominates the net interaction. But, as the distance between colloidal objects increases up to the size of polymer ($2 \cdot R_g$) (or diameter of micelle or particle), the repulsive force begins to decrease as a function of the distance. But, as the attractive depletion force remains almost constant, the net interaction can become attractive. As the distance becomes larger than the size of polymer, the electric double-layer repulsion and depletion attraction become comparable, and the net interaction becomes sharply small or close to zero (Huang and Ruckenstein, 2004).

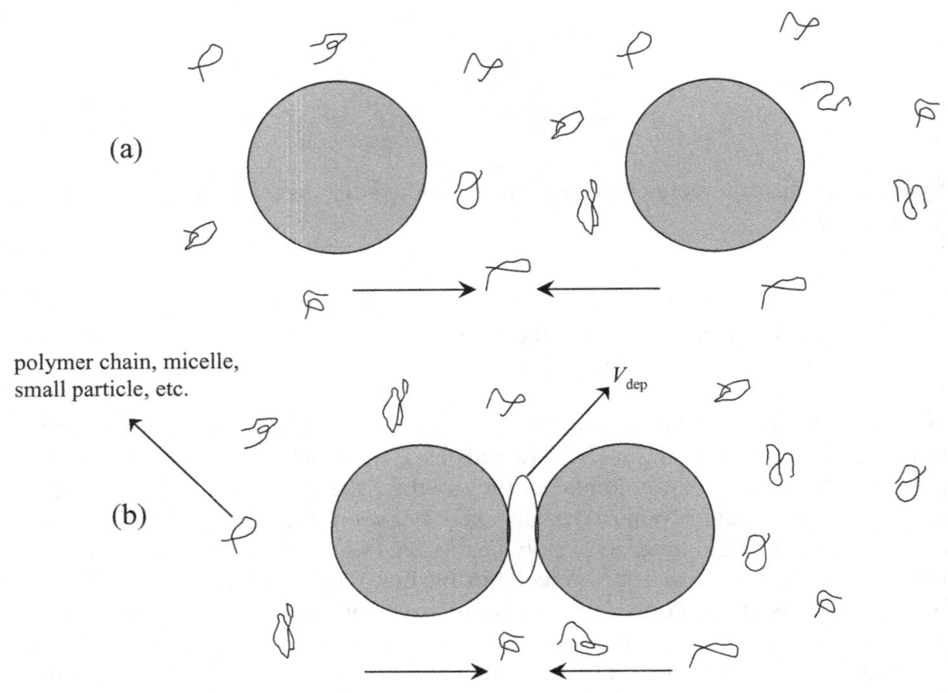

Figure 2.6. Schematic representation of depletion attraction between colloidal spheres in polymer solution: (a) two colloidal spheres approach each other, and (b) attractive interaction potential increases as depletion of polymer chains occurs at V_{dep}.

Depletion force occurs in virtually every size or shape of colloidal particles. For example, the system of spherical colloidal particles with small rods as a depletion agent has been studied recently. While the attraction on the system of sphere (colloids that experience attraction)–sphere (depletion agent such as polymer or small colloid) purely arises by the entropy gain from the translational degree of freedom of the depletion agent, the depletion attraction from a sphere–rod system arises by the entropy gain from both translational and rotational degrees of freedom. Thus, the latter system shows stronger attraction force both in a pure entropic system (Helden et al., 2003) and in the system coupled with electric double-layer repulsion (Helden et al., 2004).

An often neglected but important aspect of polymer-induced forces is their nonequilibrium nature. Due to the fact that there always exists the confinement of polymer chains and also of solvent molecules involved between the nanometer ranges of colloidal surfaces, the dynamics of polymer chains, the dynamics of polymer chains from surfaces, and even the dynamics of solvent molecules can take many hours or even days to reach their true thermodynamic equilibrium. This is often the main reason for the discrepancy between the theory and the measurement. Furthermore, for the cases of soft and rough surfaces, such as those in many biological systems or micelles, the undulation force and peristaltic force can become involved. The former is related to the bending modulus and the latter to the area expansion modulus. Thus, in addition to the well-formulated van der Waals force, electric double-layer force, and hydrophobic force, all these geometrically and dynamically induced forces become a part of the net interaction of soft self-assembly systems such as biological systems and surfactant micelles.

2.4. SOLVATION AND HYDRATION FORCES

When two colloidal surfaces or objects are closer than a few nanometers, the DLVO calculation, which was based on the van der Waals attractive force and the electric double-layer repulsive force, often shows a large discrepancy from experimental results. This rather short-range interaction (compared with van der Waals and electric double-layer interactions) arises from the layering (or ordering) of solvent molecules on/near the surface of colloidal objects. This is, in general, called *solvation force* (or structural force), and when water is the solvent, it is called *hydration force*. It often shows an oscillatory pattern as a function of the distance between two surfaces, and the pattern decreases exponentially as the distance increases.

2.4.1. Solvation Force

As shown in Figure 2.7, with the assumption of an atomically smooth surface, when the separation between the two surfaces is close enough to the molecular size (diameter for the spherical molecule and size of the short axis for the

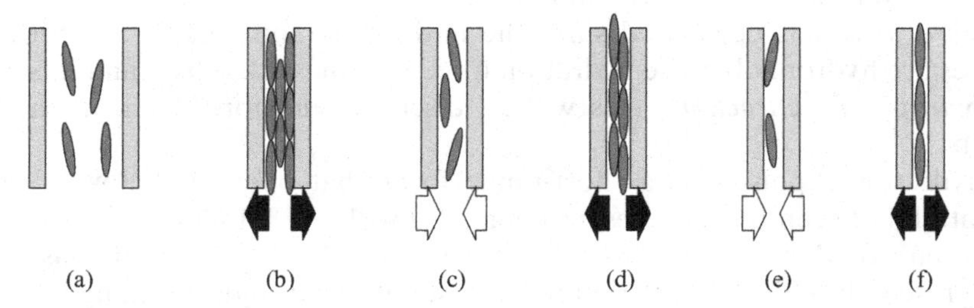

Figure 2.7. Schematic diagram of solvation force. The interaction is negligible when the two surfaces are at large separation (a). As the separation between the two surfaces is decreased, attractive force arises where the solvent molecules are less ordered, such as in (c) and (e). Repulsive force arises where the solvent molecules are well ordered, so entropy can be gained as a result of the repulsive act, such as in (b), (d), and (f). Magnitude of the force is increased monotonically as the separation is decreased. (Schematic representation is in Figure 1.4 of Chapter 1.)

rod-shaped one) of the solvent, the solvent molecules become ordered along/between the two surfaces. As this situation increases the molecular density in this region compared with the bulk, this region becomes entropically unfavorable, so the two surfaces experience a repulsive force. At distances for the close-packing of the molecules (also at double, triple, and so on of the molecular size of the solvent), the same situation causes a repulsive force between the two surfaces, but with the magnitude usually decreasing monotonically. At distances in between, the molecules are disordered. This produces less molecular density between the surfaces, and hydrodynamic pressure from the other side of the surfaces causes the attractive interaction between the surfaces. This attractive force also diminishes exponentially as the distance between the surfaces increases.

Solvation force depends on the physical factors both of solvent molecule and of colloidal surface. From the solvent side, it includes shape, size, and polarity of solvent molecule. From the surface side, it includes surface properties such as hydrophobicity, hydrophilicity, surface roughness, and surface homogeneity. The Angström scale of layering (or structuring) of solvent molecules along the surface of colloidal objects over multiple layers of length scale is the key to this oscillatory solvation force. Thus, it requires a high degree of order or symmetry from both sides. For example, measured results from linear alkane and branched alkane show a marked difference on their solvation force curves (Gee and Israelachvili, 1990). Also, the oscillatory curves are often totally disrupted by the slight surface roughness of a few Angström (Gee and Israelachvili, 1990).

2.4.2. Hydration Force

When the interaction occurs solely in water, the interaction between the colloidal objects described above differs significantly. It is always repulsive whenever the surfaces that are interacting are hydrophilic in nature. The concept of hydration was proposed long ago (Kruyt and Bungenberg de Jong, 1928), and the term *hydration force* has been used since the mid-1970s (Jordine, 1973). When the surfaces are hydrophobic, the hydration force becomes attractive, and it is called the *hydrophobic attraction*. This will be discussed in more detail in the next section.

Hydration repulsion is a short-range force that operates below ~3 nm of separation between the surfaces, as compared with the DLVO force or double-layer repulsion that operates over the range up to ~100 nm. Its origin has been a strong research subject for a long time, but still its clear understanding is controversial. However, one common feature among the research data is that hydration force seems to appear whenever the water molecules strongly bind to the hydrophilic groups on the surfaces such as hydroxyl, ammonium, phosphate, or sugar, especially in the case of biological systems. Thus, the binding nature of water molecules on the surface such as orientational degree and density, and the nature of the surface, such as softness, hardness, roughness, and homogeneity, and the density/nature of functional groups on the surfaces are important determining factors for hydration force.

Depending on the nature of surfaces, such as softness or hardness, the hydration force shows two typical patterns. For the hydrophobic soft surfaces, for example, across soap/bubble film, between emulsion droplets, or between amphiphile bilayers, the hydration repulsion increases monotonically as the distance between the surfaces decreases. This is believed to be an entropic effect just as in steric repulsion. As the two surfaces approach each other, the protruded head-groups of the amphiphiles or, in many cases, parts of the hydrocarbon chains usually thermally excited begin to overlap each other. This will, of course, cost the entropy of the system, so repulsive force will be executed. Again, DLVO theory does not a satisfactory explanation in this region.

When the surfaces are hard, such as silica, mica, glass, or clays (montmorillonite, bentonite, liponite, kaolinite, etc.), the hydration force is believed to arise due to the relatively wide range of the hydrogenbonding network or multilayer structuring of water molecules on these strongly hydrophilic surfaces. The surface functional groups that are either the nature of surfaces or obtained by adsorption are responsible for the hydrogen bonding. The strong repulsive hydration force appears at the range below double-layer repulsion, and DLVO theory does not fit in this region. One logical explanation is that when the two surfaces approach each other in this region, it is likely that the energy is required to dehydrate the ions that bind with water molecules. Sometimes, this hydration force shows oscillatory features with its mean periodicity roughly equal to the diameter of the water molecule. The minima can reach negative energy (attractive force), which again reflects the complexity of the origin of the hydration force (Pashley and Israelachvili, 1984).

Hydration force is important in many industrial processes such as froth flotation, clay processing, and clay-related nanocomposites preparation/processing. Also, along with hydrophobic interaction and electric double-layer repulsion, hydration force plays an important role in especially the self-assembly of many biological systems such as membranes, protein folding/unfolding, and enzymes.

2.5. HYDROPHOBIC EFFECT

Hydrophobic effect plays the central role in the understanding of molecular self-assembly phenomena in a wide-length scale. It is mainly an entropic effect, but for those who remember the thermodynamic principle, $\Delta G = \Delta H - T \cdot \Delta S$, the hydrophobic effect is also affected by enthalpy contribution along with its entropy term. And each contribution is changing as a function of temperature and pressure.

When a substance that is sparingly soluble in water, such as paraffin-chain alcohols, long-chain amines, hydrocarbons, and rare gases, contacts with water, the tetrahedral bonding sites of water molecules are disrupted around the solute–water interface or at the surface of solute. Such molecules are known as *hydrophobic substances*. The most favorable way to reorganize energetically for water molecules is giving up one bonding site and forming a water molecular network around the solute molecules. This is called *iceberg cluster* or *iceberg formation* (Frank and Evans, 1945; Shinoda, 1977). The iceberg formation itself is not an entropic effect. The molecular motion (both translational and reorientational motions) of the water molecules at this iceberg structure is measured at ~10^3 times slower than those in bulk water. The most recent published evidence is presented by Yamaguchi et al. (2004).

> Yamaguchi et al. (2004) provides excellent simulation evidence of its own, and comprehensively reviews the experimental, theoretical, and computer-modeling results.

This is actually a loss of entropy of the system, and it is compensated for and overcome by the enthalpy term. When the solute molecules with water clusters come close together, this iceberg structure starts to break into individual water molecules. As they go back into the bulk water, they regain their original molecular motion. Now, the solute molecules that come close together obviously lose their entropy, but the overall system gains a significant amount of entropy that overcomes the entropy loss from the aggregation of solute molecules. This is why hydrophobic substances often show an unusually stronger interaction in water than in a gas state, that is, mainly van der Waals interaction. Figure 2.8 shows the schematics for this process.

Table 2.3 represents the thermodynamics parameters (Gibbs free energy, enthalpy, and entropy) for bringing two methane molecules to the distance of 1.533 Å at 10 °C. Among the different solvents shown, water shows the largest magnitude of Gibbs free energy (1.99 kcal/mol) compared with the rest of the polar and nonpolar solvents, which range 1.28–1.49 kcal/mol. The sign for all solvents tested is negative, which means the process is spontaneous. The enthalpy,

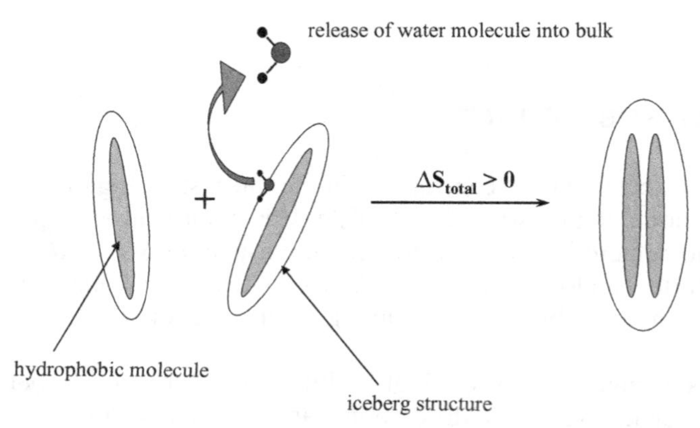

Figure 2.8. Schematic representation of hydrophobic effect.

TABLE 2.3. Thermodynamic parameters for bringing two methane molecules to the separation of 1.533 Å at 10 °C. HI denotes hydrophobic interaction. (Data from Yaacobi and Ben-Naim, 1974, and Ben-Naim, 1980.)

Solvent	δG^{HI} (kcal/mol)	δS^{HI} (cal/mol·deg)	δH^{HI} (kcal/mol)
Water	−1.99	11	1.5
Methanol	−1.28	0	−1.4
Ethanol	−1.34	0	−1.3
1-propanol	−1.39	2	−0.6
1-butanol	−1.44	−1	−1.9
1-pentanol	−1.49	0	−1.5
Cyclohexane	−1.36	1	−1.2

on the contrary, shows positive value for water and negative values for the rest of the solvents, which implies that this process in water is unfavorable at least enthalpy-wise, while enthalpy is the major contribution term for the rest of the solvents. The unfavorable contribution from the enthalpy term in water is compensated for by the large value of entropy. The entropy contribution from the rest of the solvents is negligible.

Detailed thermodynamic formulation for the hydrophobic interaction can be found in Ben-Naim (1980). This book is devoted solely to the hydrophobic interaction. It also provides the effect of temperature and pressure on the hydrophobic interaction, and the role of hydrophobic interaction in many solute particle systems such as micelles or biological systems. The lack of data on the water structuring (iceberg) around the hydrophobic solutes back then render the argument on water structuring somewhat ambiguous in parts of this book. But, by combining it with the recent progress on this issue, readers will find that the author's judgment and the subsequent formulations were quite correct.

Here, let me introduce both sides of a story that is becoming a subject of fierce debate. The model that was explained above is based on the classical model proposed by Frank and Evans (Frank and Evans, 1945) more than a half-century ago. It has long been the subject of debate and studies, not only due to pure scientific curiosity, but because of the increasing consensus that the hydrophobic effect is at the heart of our understanding of the many crucial biological systems and their functions as well. Numerous studies conducted for the last several decades back up the general picture of Frank and Evans's model (Yamaguchi et al., 2004; and reference therein). They found evidence of an iceberglike structure and confirmed the decreased mobility of water molecules almost 10^3 times. But, so far they have not provided the exact form of this iceberglike structure and compelling evidence as to how the iceberg structure might be related to (or affects) the slowing down of the mobility of the water molecules.

The other side of the story starts with recent observation of nanobubbles on the surface of hydrophobic molecules that might be strongly related to the strangely long-range nature of the hydrophobic interaction. This feature can reach up to 100 nm. And the unusual strength of the hydrophobic interaction is the result of the capillary action between the nanobubbles. The findings related to the nanobubbles have been pouring into the scientific journals since the late 1990s, mostly after their observation with AFM (atomic force microscopy) and other experimental methods such as x-reflectivity or neutron scattering. There have been some simulational supports for this picture, too. A recent article by Ball well contrasts both sides of this story (Ball, 2003). It is also argued that a partial drying (or partial de-wetting) of water that induces the gaslike water, rather than the icelike water, on the surface of hydrophobic molecules is a more realistic case of the hydrophobic effect than the water structuring (Jensen et al., 2004; and references therein). I believe this healthy debate will go on. If the second version of the story turns out to be correct, the thermodynamic schematic that has been used to explain the entropy gain should be corrected. But for now I will accept the icelike-structure story.

2.6. HYDROGEN BOND

A hydrogen bond is a strong and directional intermolecular interaction. It is one of the most important forces not only to get the entire picture of intermolecular interactions, but especially to understand the variety of events in biological systems. It is directly responsible for all those unexpected but interesting and vital physical properties of water molecules. Some examples include its unusually high melting and boiling points when compared with those molecules with similar molecular weight, its density minimum at 4 °C, and the increasing of the dielectric constant when liquid water is transformed into ice, unlike most typical polar molecules.

Water is not the only molecule that can form a hydrogen bond. When highly electronegative atoms such as oxygen, nitrogen, and fluorine are covalently bonded with hydrogen, their high electronegativity strongly attracts an electron cloud of hydrogen atoms, which makes the hydrogen atom electron deficient, meaning positively polarized. This positively polarized hydrogen atom can then interact strongly with the nearby electronegative atoms. The bond length of a typical hydrogen bond is less than that of the combined van der Waals radii of the two atoms involved, but larger than that of the covalent bond of the two atoms involved. This is in line with the general order of bond strength: covalent bond (~500 kJ/mol) > hydrogen bond (10–40 kJ/mol) > van der Waals interaction (~1 kJ/mol). Water–water interaction from a hydrogen bond is about 12–20 kJ/mol.

What makes the hydrogen bond directional is its unique bonding sites. For example, even though the van der Waals interaction can sometimes make molecules align along their axis of dipole moment, it does not occur on the specific interaction site, meaning its direction of interaction is always random. On the other hand, the hydrogen bond–capable molecules always interact only through specific bonding sites that are polarized hydrogen atom and lone-pair electron. Depending on the number of these bonding sites on the molecules, different types of hydrogen-bonded structures can be formed. This includes the linear chain, ring, dimer, layered sheet, and three-dimensional network. For certain types of molecules, intramolecular hydrogen bonding can also occur (Figure 2.9). It is especially important for biological molecules such as DNA, RNA, proteins, and membranes. For instance, it is greatly responsible for the formation and

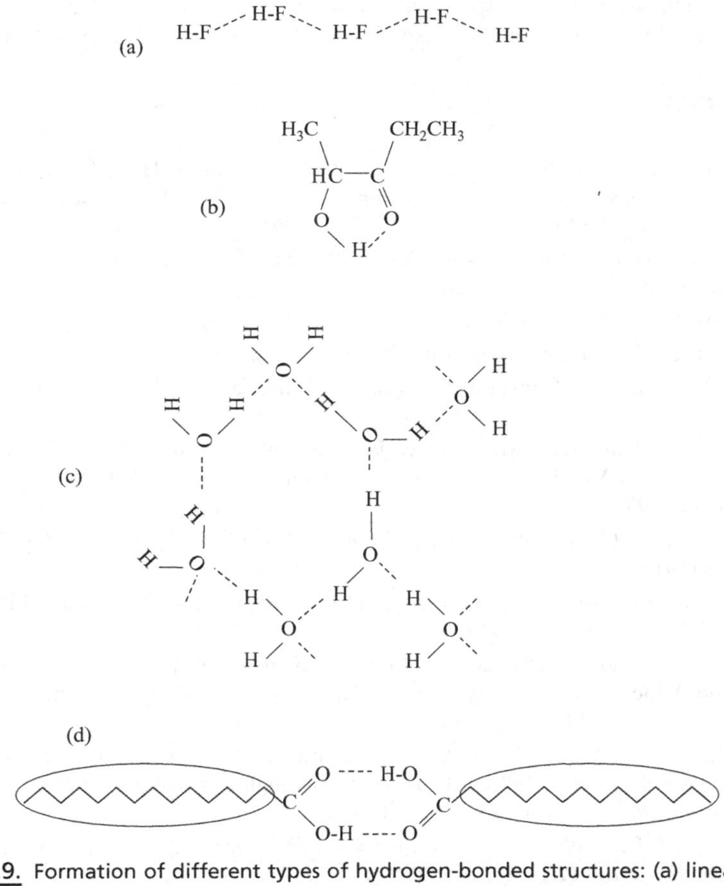

Figure 2.9. Formation of different types of hydrogen-bonded structures: (a) linear chain of hydrofluoric acid, (b) intramolecular ring of 2-hydroxy-3-pentanone, (c) six-membered ring network of water, and (d) dimer of pentadecanoic acid. This dimer geometry can limit the direct hydrophobic interaction between the two pentadecyl chains; rather it can provide the possibility of stacking-type directional self-assembly via hydrophobic interaction between the dimers.

stability of the DNA double-helical structure, integrity and function of biological membranes, and cell functions such as cellular transport. Both the strength of the hydrogen bond and its directionality are critical for these events.

One of the unique features of the hydrogen bond in molecular self-assembly is that it provides the stability and the directionality to the molecular self-assembly. This causes the self-assembled structures to be formed with a rich range of morphology and important functionality. The hydrogen bond is also a key driving force for the gelation of many organic molecules, macromolecules, and biological molecules and systems such as cells. Pollack (2001) introduces the revolutionary view on biological cells and their functions based on the gelation process in the biological systems. This concept is based on the hydrogen bond between water and biological molecules, and its cascade-like formation and destruction.

REFERENCES

Ackler, H. D., French, R. H., Chiang, Y.-M. "Comparisons of Hamaker Constants for Ceramic Systems with Intervening Vacuum or Water: From Force Laws and Physical Properties," *J. Colloid Interface Sci.* **179**, 460 (1996).

Asakura, S., Oosawa, F. "Interaction between Particles Suspended in Solutions of Macromolecules," *J. Polym. Sci.* **33**, 183 (1958).

Ball, P. "How to Keep Dry in Water," *Nature* **423**, 25 (2003).

Ben-Naim, A. *Hydrophobic Interactions* (Plenum Press: 1980).

Bergström, L. "Hamaker Constants of Inorganic Materials," *Adv. Colloid Interface Sci.* **70**, 125 (1997).

Bhattacharjee, S., Elimelech, M. "Surface Element Integration: A Novel Technique for Evaluation of DLVO Interaction between a Particle and a Flat Plate," *J. Colloid Interface Sci.* **193**, 273 (1997).

Chapman, D. L. "A Contribution to the Theory of Electrocapillarity," *Philos. Mag.* Ser. 6, **25**, 475 (1913).

Derjaguin B. V. "Friction and Adhesion. IV: The Theory of Adhesion of Small Particles," *Kolloid-Zeitschrift* **69**, 155 (1934).

Derjaguin, B. V., Landau, L. "Theory of the Stability of Strongly Charged Lyophobic Sols and of the Adhesion of Strongly Charged Particles in Solutions of Electrolytes," *Acta Physicochim. URSS* **14**, 633 (1941).

Duval, J. F. L., Leermakers, F. A. M., van Leeuwen, H. P. "Electrostatic Interactions between Double Layers: Influence of Surface Roughness, Regulation, and Chemical Heterogeneities," *Langmuir* **20**, 5052 (2004).

Dzyaloshinskii, I. E., Lifshitz, E. M., Pitaevskii, L. P. "The General Theory of Van der Waals Forces," *Adv. Phys.* **10**, 165 (1961).

Evans, E., Needham, D. "Attraction between Lipid Bilayer Membranes in Concentrated Solutions of Nonadsorbing Polymers: Comparison of Mean-Field Theory with Measurements of Adhesion Energy," *Macromolecules* **21**, 1822 (1988).

Frank, H. S., Evans, M. W. "Free Volume and Entropy in Condensed Systems. III: Entropy in Binary Liquid Mixtures; Partial Molal Entropy in Dilute Solutions; Structure and Thermodynamics in Aqueous Electrolytes," *J. Chem. Phys.* **13**, 507 (1945).

Gee, M. L., Israelachvili, J. N. "Interactions of Surfactant Monolayers across Hydrocarbon Liquids," *J. Chem. Soc. Faraday Trans.* **86**, 4049 (1990).

Gouy, G. "Constitution of the Electric Charge at the Surface of an Electrolyte," *J. Physique (Paris)* **9**, 457 (1910).

Gregory, J. "The Calculation of Hamaker Constants," *Adv. Colloid Interface Sci.* **2**, 396 (1969).

Hamaker, H. C. "The London–Van der Waals Attraction between Spherical Particles," *Physica (The Hague)* **4**, 1058 (1937).

Helden, L., Koenderink, G. H., Leiderer, P., Bechinger, C. "Depletion Potentials Induced by Charged Colloidal Rods," *Langmuir* **20**, 5662 (2004).

Helden, L., Roth, R., Koenderink, G. H., Leiderer, P., Bechinger, C. "Direct Measurement of Entropic Forces Induced by Rigid Rods," *Phys. Rev. Lett.* **90**, 048301/1 (2003).

Hiemenz, P. C., Rajagopalan, R. *Principles of Colloid and Surface Chemistry*, 3rd ed. (Marcel Dekker: 1997).

Hough, D. B., White, L. R. "The Calculation of Hamaker Constants from Lifshitz Theory with Applications to Wetting Phenomena," *Adv. Colloid Interface Sci.* **14**, 3 (1980).

Huang, H., Ruckenstein, E. "Double-Layer Interaction between Two Plates with Hairy Surfaces," *J. Colloid Interface Sci.* **273**, 181 (2004).

Huang, H., Ruckenstein, E. "Interaction Force between Two Charged Plates Immersed in a Solution of Charged Particles: Coupling between Double Layer and Depletion Forces," *Langmuir* **20**, 5412 (2004).

Israelachvili, J. N. *Intermolecular and Surface Forces*, 2nd ed. (Academic Press: 1992).

Jensen, M. Ø., Mouritsen, O. G., Peters, G. H. "The Hydrophobic Effect: Molecular Dynamics Simulations of Water Confined between Extended Hydrophobic and Hydrophilic Surfaces," *J. Chem. Phys.* **120**, 9729 (2004).

Jordine, E. St. A. "Specific Interactions which Need Nonequilibrium Particle/Particle Models," *J. Colloid Interface Sci.* **45**, 435 (1973).

Kruyt, H. R., Bungenberg de Jong, H. G. "The Lyophilic Colloids I: General Introduction: Agar Sol," *Kolloidchem. Beihefte* **28**, 1 (1928).

Lehn, J.-M., Ball, P. "Supramolecular Chemistry," *The New Chemistry*, Hall, N., ed., pp. 300–351. (Cambridge University Press: 2000).

Lifshitz, E. M. "The Theory of Molecular Attractive Forces between Solids," *Soviet Phys. JETP* **2**, 73 (1956).

Mahanty, J., Ninham, B. W. *Dispersion Forces* (Academic Press: 1976).

Pashley, R. M., Israelachvili, J. N. "Molecular Layering of Water in Thin Films between Mica Surfaces and Its Relation to Hydration Forces," *J. Colloid Interface Sic.* **101**, 511 (1984).

Pollack, G. H. *Cells, Gels and the Engines of Life* (Ebner and Sons: 2001).

Shinoda, K. " 'Iceberg' formation and solubility," *J. Phys. Chem.* **81**, 1300 (1977).

Stern, O. "The Theory of the Electrolytic Double-Layer," *Zeits. Elektrochem. Angew, Phys. Chem.* **30**, 508 (1924).

Todd, B. A., Eppell, S. J. "Probing the Limits of the Derjaguin Approximation with Scanning Force Microscopy," *Langmuir* **20**, 4892 (2004).

Verwey, E. J. W., Overbeek, J. Th. G. *Theory of Stability of Lyophobic Colloids* (Elsevier: 1948).

Visser, J. "On Hamaker Constants: A Comparison between Hamaker Constants and Lifshitz—Van der Waals Constants," *Adv. Colloid Interface Sci.* **3**, 331 (1972).

Yaacobi, M., Ben-Naim, A. "Solvophobic Interaction," *J. Phys. Chem.* **78**, 175 (1974).

Yamaguchi, T., Matsuoka, T., Koda, S. "Mode-Coupling Study on the Dynamics of Hydrophobic Hydration," *J. Chem. Phys.* **120**, 7590 (2004).

3

MOLECULAR SELF-ASSEMBLY IN SOLUTION I: MICELLES

This chapter describes the first part of the molecular self-assembly in solution. The formation of micelles from surfactants (and amphiphilic polymers) will be presented. The next chapter (Chapter 4) will cover the second part of the molecular self-assembly in solution, including bilayers, liquid crystals, and emulsions. Micelle formation is a process of thermodynamic random self-assembly with the length scale of molecules. The formation of bilayers, liquid crystals, and emulsions often involves a tricky, kinetically driven process, and ends up with much larger self-assembled aggregates than the usual micelles. By following this format of two separate chapters, hopefully readers will acquire a better picture of the similarities, differences, and relationships between these two systems. It will also benefit readers to go ahead to the implications of these systems on nanotechnology issues in Part II.

This chapter is structured as follows:

1. Introduce a clear picture of micelles and their formation process.
2. Describe the critical physical properties of micelles and define their basic parameters.

Self-Assembly and Nanotechnology: A Force Balance Approach, by Yoon S. Lee

3. Describe the structural change of micelles and introduce the surfactant packing parameter, *g*, as a useful tool to follow it.
4. Define the concept of micellar catalysis and relate it with critical applications including nanostructured inorganic materials.

3.1. SURFACTANTS AND MICELLES

A micelle is a colloidal-size object with 2–20 nm diameter that is formed by spontaneous association of surfactant (or amphiphilic polymer) molecules. It is formed mainly in aqueous solution, but a variety of micelles also are formed in nonaqueous solutions and aqueous/nonaqueous solvent mixtures (Tanford, 1980; Rosen, 2004). Figure 3.1 shows the representative scheme of a spherical micelle of surfactant molecule in aqueous solution. Typically, the ionic (or hydrophilic) head groups are exposed to the bulk aqueous solution, while the hydrophobic hydrocarbon tail groups form the interior of the micelle.

As the name *surfactant* (short for *surf*ace *act*ive *a*gen*t*) implies, surfactant molecules are the molecules that are active at a variety of surfaces, especially at air–liquid and liquid–liquid interfaces. Not all surfactants form micelles, but they are always surface active. Thus, they have tendency to change their interfacial properties dramatically, such as surface tension, interfacial tension, surface diffusion, and so forth. Some examples and structures of the micelle-forming surfactants and polymeric amphiphiles are given in Figure 3.2. All of them are composed of two main parts: the head group and the tail group. Ionic surfactants have a third part: the counterion. A widely accepted classification of surfactants is based on the class of head group. When the head group bears a cationic group, it is called a *cationic surfactant*. The same stands for the anionic, nonionic, and

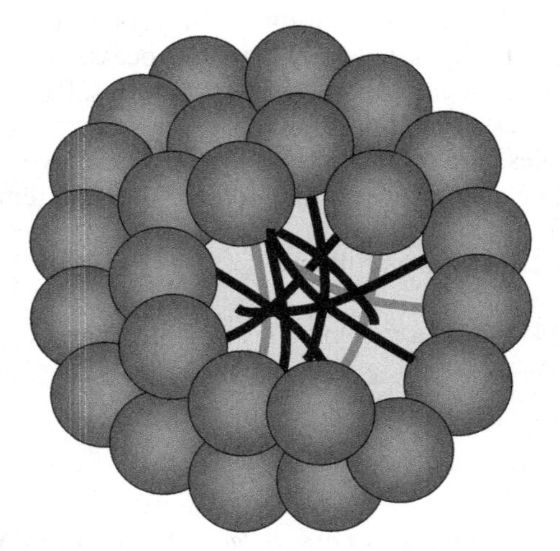

Figure 3.1. Schematic structure of a spherical micelle.

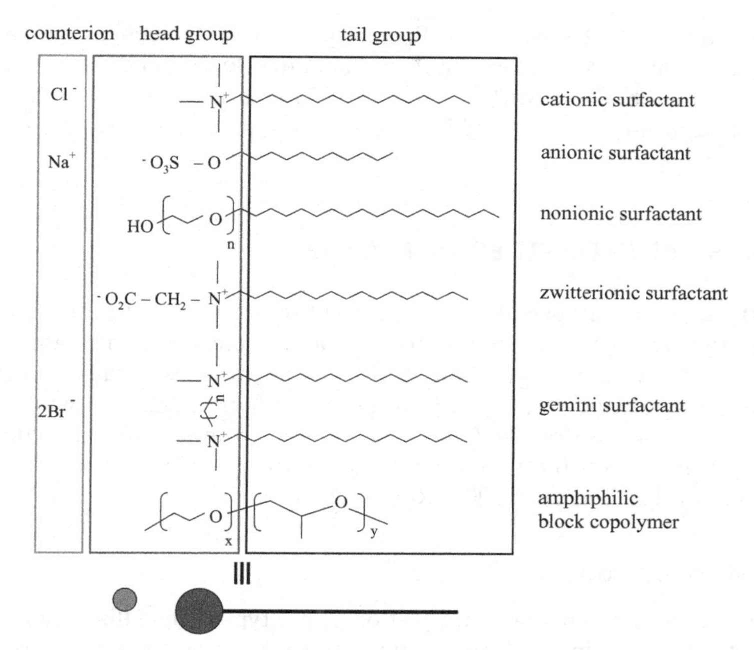

Figure 3.2. Examples of typical surfactants and amphiphilic polymer.

zwitterionic surfactant. Gemini surfactant is a family of synthetic amphiphiles whose two long hydrocarbon chains with ionic head groups are covalently bonded with a *spacer* (Menger and Littau, 1993). The spacer is usually another hydrocarbon chain. A typical micelle-forming amphiphilic block copolymer is of the Pluronic series. It has two basic composing units of ethylene oxide and propylene oxide. The former unit shows a hydrophilic character in aqueous solution, while the latter has hydrophobicity.

The number of hydrocarbon chains of surfactants can be 1–3, and each chain has 8–20 (up to ~30 for cases of nonionic surfactants) CH_2 units, including terminal CH_3. It can be saturated or nonsaturated. Molecules with four hydrocarbon chains are too bulky to show surfactancy. Hydrocarbon chains with less than seven CH_2 units have too little hydrophobicity, which cannot be an attractive enough driving force for the self-assembly. Hydrocarbon chains with more than 20 CH_2 units have too much hydrophobicity. They thus show an oily character; phase separation with water.

The types of amphiphiles are also classified by their origins, such as synthetic, polymeric, and biological. Details and lists of surfactants are well described by Porter (1994). Surfactants for industrial purposes have a wide variety of trade names and are commonly used as mixtures. Good material is available that covers this information (Ash and Ash, 1993). A wealth of information on synthetic amphiphatic polymers can be obtained from the literature (Alexandridis and Lindman, 2000; Riess, 2003). Specialty surfactants such as siloxane surfactant, fluorocarbon surfactant, labile surfactant, bile salt, and so on will not be presented in this book. However, Robb (1997) is an excellent source for these details. Biological-origin amphiphiles (mostly biological lipids) will be presented in detail in Chapter 7.

3.2. PHYSICAL PROPERTIES OF MICELLES

Again, the common aspects of the self-assembly of surfactant (and amphiphilic polymer) systems are that it is a thermodynamic, random, and nonhierarchical process with molecular-length scale. This section provides details on this issue. The definitions of three important parameters for micelles also will be given: critical micelle concentration (*cmc*), aggregation number (n), and degree of counterion binding on the surface of micelles (α) (Tanford, 1980; Zana, 1991; Nagarajan and Ruckenstein, 2000; Rosen, 2004).

3.2.1. Micellization

Figure 3.3 shows the schematic illustration of the typical micellization process of surfactant molecules in aqueous solution. The surfactant molecules are soluble in aqueous solution. Part of the initially added molecules is adsorbed onto the air–liquid interface and forms an adsorbed monolayer. Since the space for the monolayer formation is limited at the surface, the rest of the molecules remain in the solution as a free form of molecules (monomers). For most of the cases, the main attractive driving force for the micellization is hydrophobic interaction. As discussed in Chapter 2, it is mainly of entropic origin. As the concentration of surfactant is increased, the monomers come close together by this interaction. This also brings closely together the head groups, which can be either ionic (for ionic surfactants) or hydrated (for nonionic surfactants and amphiphilic polymers). Thus, the repulsive force between the head groups begins to arise on the

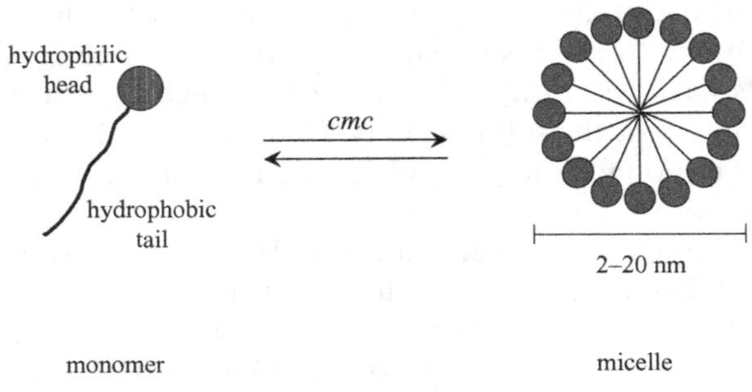

Figure 3.3. Micelle formation: micellization.

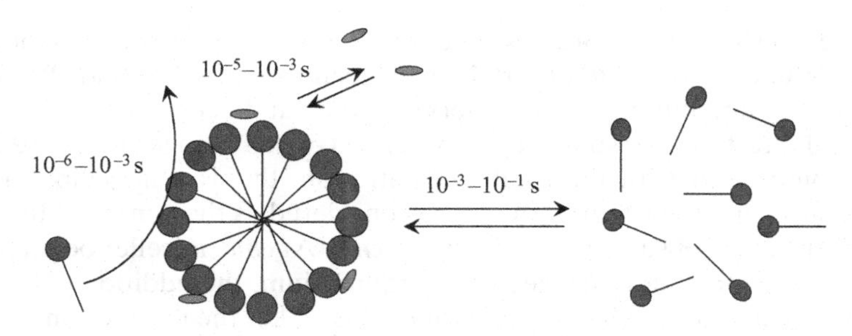

Figure 3.4. Micelle is a dynamic structure.

surface of the self-assembled aggregate (micelle). The transition of monomers into this self-assembled aggregate form is balanced when these two opposite forces are balanced. The micelle is at equilibrium with monomers.

For most of the surfactants, the initially formed micelle has spherical or near-spherical shape (close to ellipsoidal in the case of longer chain length). There are no covalent bonds involved in the entire process. The free monomers are free to be exchanged with the one within micelles with the time scale of 10^{-6}–10^{-3} s. Figure 3.4 shows this dynamic aspect of micelles. For ionic surfactants, counterions that are bound on the surface of micelles are free to be exchanged with the free one in the solution with the time scale of 10^{-5}–10^{-3} s. The whole micelle itself is a dynamic entity that experiences a constant disassemble–assemble process (constant formation and breakup) with the time scale of 10^{-3}–10^{-1} s.

3.2.2. Critical Micellar Concentration and Aggregation Number

The monomer concentration where the first micelle begins to appear is defined as *cmc*. Since pre-micellar aggregates are formed in many cases, this point for the *cmc* is not always clear cut. Thus, the *cmc* is often determined by the empirical average of the narrow ranges of concentration. Typical *cmc* ranges are from 10^{-5} to 10^{-2} mole/liter for most of the single-chain surfactants and the amphiphilic polymers, such as the Pluronic series. The size (diameter) is determined by the length of the surfactant or polymer molecule and its dynamic structure, such as *cis–trans* conformation of a hydrocarbon chain or folding of a polymer chain. It usually is 2–20 nm. The aggregation number is the number of surfactant or polymer molecules within a micelle. It ranges 50–10,000. No micelle has a clear-cut aggregation number. It always shows a Gaussian-type distribution. Thus, the aggregation numbers from the references are the average number. Naturally, this is quite dependent on the measurement method and calculation process. So is the *cmc*. Studies using various techniques show that the state of the micelle core region is close to that of the bulk hydrocarbon. The microviscosity of the micelle core of typical surfactants ranges ~10—~50 cP at room temperature.

Figure 3.5 shows the typical changes of the concentrations of monomer and micelle as a function of the total surfactant concentration before and after micelle formation. The amount of the monomers adsorbed at the air–liquid interface and liquid–solid interface (for some of the systems) typically ranges below 10^{-8} mole/liter. In comparison with the total concentration, this is a negligible amount. Thus, the monomer concentration can be considered as the concentration of the total concentration of surfactants below the *cmc*. At *cmc*, micelles begin to form. Since the monomers and micelles are at equilibrium, the additional amount of surfactants forms the micelles after the *cmc*. This makes the micelle concentration increase linearly with the total concentration, while the monomer concentration remains almost constant. It continues until the structure of micelles is changed by the constraint from the increased monomer concentration. The concentration of micelles is corrected at the point where the spherically shaped micelles are transformed into higher-order structures such as rod-shape or wormlike micelles. Since these types of micelles have a much higher aggregation number than the spherical micelle, the micelle concentration can be lower than that of spherical micelles after this point. The monomer concentration usually remains unchanged.

When the total concentration exceeds the point of the phase transition into liquid crystal, the concentration of monomers is significantly changed. The concentration where the liquid crystal phase begins to form varies widely, depending on the nature of surfactant. For ionic surfactants, it usually ranges ~20–~40 wt. % of surfactant at room temperature (Laughlin, 1991). For nonionic surfactants, it decreases to ~5–~20 wt. % (Laughlin, 1996; Chernik, 2000). The details of the structural evolution of micelles will be covered in Section 3.4.

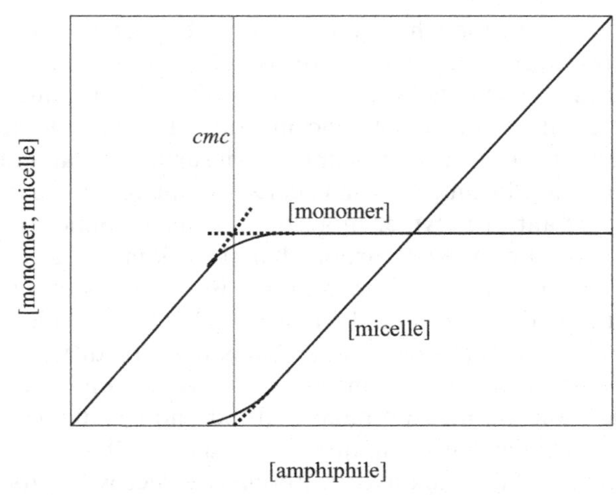

Figure 3.5. Schematic representation of the changes of monomer and micelle concentrations against the total amphiphile concentration.

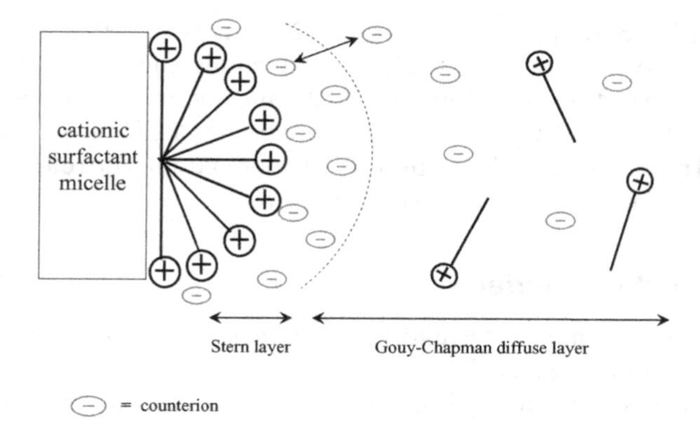

Figure 3.6. Schematic representation of the concepts of double-layer and counterion binding for ionic micelles.

3.2.3. Counterion Binding

Figure 3.6 shows the schematic representation of the concept of double-layer and counterion binding for the case of ionic micelles. When the ionic micelle is formed, its surface becomes either cationic or anionic, depending on the nature of surfactant molecules. Some of the counterions are free in the solution, but some are bound on the surface of the micelle. The ratio of counterion bound on the micelle surface to the whole concentration of counterion in the system is defined as the degree of counterion binding (α). α ranges 0.2–0.8. $1 - \alpha$, which is usually defined as β, is the degree of counterion dissociation. There is an excellent review in Grieser and Drummond (1988) for *cmc*, aggregation number, and the degree of counterion binding for a variety of surfactants and their measurement techniques.

Besides its dynamic nature, the general picture of ionic micelles in Figure 3.6 is the same as the spherical colloidal suspension with ionized species. The sizes are at the same range of nanometer scale, and the disassemble–assemble process of micelles is two orders of magnitude on average slower than the exchange rate of counterion. Thus, the concept of Stern and Gouy-Chapman double-layer for the planar and curved colloidal surfaces (Chapter 2) is also valid for the surface of micelles. The region where the counterion is bound can be considered as a Stern layer and the region where the concentration gradient can be recognized is a Gouy-Chapman diffuse layer. The adoption of this double-layer concept of the ionic micelle has been successful for the development of the thermodynamic understanding of the surface of micelles (Tanford, 1980).

3.3. THERMODYNAMICS OF MICELLIZATION

The theory of micellization has been well formulated through thermodynamics and kinetics, and also with a statistical thermodynamics approach (Tanford, 1980;

Puvvada and Blankschtein, 1991; Nagarajan, 2003; Rosen, 2004). This section will go through the two most common thermodynamic models: mass-action and pseudo-phase separation. Two long-debated issues on micellization are whether this process should be viewed as a reaction or whether a micelle should be viewed as a separated phase of the bulk solution. The former adopts the reaction approach, while the latter takes the phase-separation view.

3.3.1. Mass-Action Model

This approach treats micelle formation as a *reaction* of n (aggregation number) monomers to form a micelle.

For nonionic surfactant, zwitterionic surfactant, and amphiphilic polymer, there is no counterion present in solution. Thus, the scheme of Figure 3.3 can be expressed as

$$nD \xleftrightarrow{K} D_n \tag{A}$$

where K is an equilibrium constant, and D and D_n represent monomer and micelle, respectively.

For ionic surfactants, counterion takes account. For anionic surfactants, the following relation can be set:

$$nD^- + mC^+ \xleftrightarrow{K} [D_nC_m]^{Z-} \tag{B}$$

where K is an equilibrium constant, D^- is the monomer that is anionic, C^+ is the counterion that is cationic, and $[D_nC_m]^{Z-}$ represents the micelle that is composed of n monomer and m counterion. Z is the net charge of the micelle, which becomes $n - m$. For ionic micelles, there is no completely *naked* micelle (no counterion bound) or completely *dressed* micelle (all counterion bound). n is always larger than m, and the ratio m/n is the degree of counterion binding.

For both [A] and [B], the reality is that there can be an infinite number of possible reactions in the system. The monomer reacts with another monomer to form a dimer, then this dimer can react with the third monomer to form a trimer, and so forth to tetramer, and so forth to n-mer, that is, the micelle. $(n + z)$-mer (z is an integer with ≥ 1), that is, the micelle with an aggregation number larger than the average number n, is possible, too. The random reactions between $(n - x)$-mer and $(n - y)$-mer (x and y are integers with $\leq n - 1$) are also possible. Each of the possible reactions can have a different reaction rate and equilibrium constant. Thus, assuming a single apparent K for the entire process is quite an extreme simplification, and can cause some deviation for the systems with larger aggregation numbers such as nonionic surfactants and ionic surfactants with long hydrocarbon chains.

Let us take reaction [B] as a general type of micellization; then K becomes

$$K = a_{mic}/a_D^n \cdot a_c^m \tag{3.1}$$

where a_D, a_C, and a_{mic} is the activity for monomer, counterion, and micelle, respectively.

The standard free energy change per mole of monomer for micellization can be obtained by

$$\Delta G^o_{mic} = -\frac{RT}{n} \cdot \ln K \tag{3.2}$$

For most of the surfactants, *cmc* ranges at a very low concentration of 10^{-5}–10^{-2} mole/liter. Thus, at *cmc*, the assumption for dilute solution stands valid, which can yield the relation of $a_D \cong a_C \cong a_{cmc} \cong cmc$. By taking this after taking equation (3.1) into equation (3.2), one can obtain

$$\Delta G^o_{mic} = RT\left[\left(1+\frac{m}{n}\right) \cdot \ln cmc - \frac{1}{n}\ln a_{mic}\right] \tag{3.3}$$

By considering the usual aggregation number of 50–10,000, the second term on the right-hand side can be negligible compared with the first one. This makes

$$\Delta G^o_{mic} \cong RT[(1+\alpha) \cdot \ln cmc] \tag{3.4}$$

with $\alpha = m/n$, which is the degree of counterion binding. Both α and *cmc* can be measured by typical experiments. Thus, ΔG^o_{mic} can be readily obtained from equation (3.4). For nonionic surfactant, α can be taken as zero, which gives

$$\Delta G^o_{mic} \cong RT \cdot \ln cmc \tag{3.5}$$

Both equations (3.4) and (3.5) are widely used, and provide quite reliable results for a wide range of surfactants and polymers.

3.3.2. Pseudo-phase Separation Model

This approach treats micellization as a *phase transition* between two separate thermodynamic monomeric and micellar phases. It is useful to explain the concentration changes of monomer and micelle as a function of total surfactant concentration. Figure 3.5 is the plot based on this model.

The standard free energy of micellization, ΔG^o_{mic}, is the free energy difference between the final standard state of micellization and the initial standard state of micellization; it is the difference per mole between the free energy of surfactant in a micelle and in water. Thus, it should be viewed as three processes: of (1) *from* the solution with surfactant monomer at the initial standard state concentration *to* solution with surfactant monomer at *cmc*, (2) *to* surfactant in micelle at *cmc*, and (3) *to* surfactant in micelle at the final standard state. This makes the chemical potential difference for each step as follows:

For step (1):

$$\Delta\mu_1 = RT \ln cmc - RT \ln [D]^o_{mon} \tag{3.6}$$

For step (2):

$$\Delta\mu_2 = \Delta G_{mic} \tag{3.7}$$

For step (3):

$$\Delta\mu_3 = \frac{RT}{n} \ln \left[\frac{[D]_{mic}}{n}\right]^o - \frac{RT}{n} \ln \frac{[D]_{mic}}{n} \tag{3.8}$$

where $[D]^o_{mon}$ is the monomer concentration at the initial standard state, ΔG_{mic} is the free energy change for micellization, $[D]_{mic}$ is the surfactant concentration in the micelle, $\frac{[D]_{mic}}{n}$ is the concentration of the micelle, and $\left[\frac{[D]_{mic}}{n}\right]^o$ is the concentration of the micelle at the final standard state.

ΔG^o_{mic} is the sum of the chemical potential differences of these three processes. The *standard state of surfactant* refers to the infinitely dilute solution of surfactant, thus the second term on the right-hand side of equation (3.6) is zero. The monomer and micelle are at equilibrium at step 2; this means ΔG_{mic} is zero. And the standard state for the micelle means a pure micelle, which makes the first term of the right-hand side of equation (3.8) zero. The sum of equations (3.6), (3.7), and (3.8) becomes

$$\Delta G^o_{mic} \cong RT \ln cmc - \frac{RT}{n} \ln \frac{[D]_{mic}}{n} \tag{3.9}$$

The ordinary aggregation number n of 50–10,000 is a large enough number so that the second term on the right-hand side can be assumed to be much smaller than the first one, and can be neglected. This yields

$$\Delta G^o_{mic} \cong RT \ln cmc \tag{3.10}$$

This is in the same form as equation (3.5). Neither model (mass-action nor pseudo-phase separation) fully describes the *true* process or state of micellization. This is largely because the surface of the micelle (the interface between the micelle and water) is not well defined. In most cases, water molecules penetrate into the approximately 1.5–2.5 carbon atom position from the head groups. This makes the surface of the micelle much rougher than the interface of ordinary water–hydrocarbon.

Both models have their advantages and limitations, but they are complementary to each other. Similarly developed theories for more complicated self-assembly systems do not exist. But this view of *process* vs. *state* (*phase*) is one of the basic issues for every self-assembly. It will certainly yield better insight for more complicated self-assembly systems.

3.3.3. Hydrophobic Effect and Enthalpy–Entropy Compensation

The standard enthalpy change for micellization, ΔH^o_{mic}, can be easily obtained by applying Gibbs-Helmholtz equation to equation (3.4) or (3.5). It can be also measured by the common solution calorimetric method. And the standard entropy change for micellization, ΔS^o_{mic}, can be obtained by the famous thermodynamic relation:

$$\Delta G^o_{mic} = \Delta H^o_{mic} - T \cdot \Delta S^o_{mic} \tag{3.11}$$

cmc of 10^{-5}–10^{-2} mole/liter brings the value of ΔG^o_{mic} negative, which means the spontaneous formation of micelles. ΔH^o_{mic} can be either negative or positive, which means that micellization can be either favorable or unfavorable enthalpy-wise. But, the ΔS^o_{mic} usually shows a positive value at the temperature range of ~10—~40°C, which means the micellization is entropically favorable at this condition. For a positive ΔH^o_{mic} value, the entropy contribution to micellization is always greater than the contribution from enthalpy: $|\Delta H^o_{mic}| < |T \cdot \Delta S^o_{mic}|$. This entropy-dominant contribution to micellization is called the *hydrophobic effect* for micellization.

The entropy contribution comes from three sources:

1. Change of the state of surfactant molecule (transfer of surfactant molecule from monomer to micelle state)
2. Change of the state of solvent (water) during micellization
3. Change of the molecular dynamics of surfactant molecule (mainly of the hydrocarbon chain, because the head group is always exposed to solvent during micellization)

Thus, ΔS^o_{mic} is

$$\Delta S^o_{mic} = \Delta S^o_{surfactant} + \Delta S^o_{water} + \Delta S^o_{hydrocarbon} \tag{3.12}$$

where $\Delta S^o_{surfactant}$, ΔS^o_{water}, and $\Delta S^o_{hydrocarbon}$ is the contribution from term 1, 2, and 3, respectively. The magnitude of $\Delta S^o_{hydrocarbon}$ is typically much smaller than the first two terms, and can be neglected. This yields

$$\Delta S^o_{mic} \cong \Delta S^o_{surfactant} + \Delta S^o_{water} \tag{3.13}$$

The transfer of 50–10,000 monomers into a single micelle is a decrease of the entropy, which means that $\Delta S^o_{surfactant}$ is always negative. However, as described in Chapter 2, the release of structured water around the hydrocarbon chains (iceberg structure) during micellization is an increase of the entropy, which gives ΔS^o_{water} always a positive value.

For most of the typical surfactants, $|\Delta S^o_{surfactant}| < |\Delta S^o_{water}|$ is fulfilled usually at the temperature range of ~10—~40 °C. So, ΔS^o_{mic} is always >0, and as stated above, its contribution is always dominant over the enthalpy contribution. But, as the temperature is increased, the condition of $|\Delta S^o_{surfactant}| > |\Delta S^o_{water}|$, thus $\Delta S^o_{mic} < 0$ often becomes the reality, possibly because of the lack of enough structured water. This is the situation where micellization becomes entropically unfavorable. The micellization can occur only when the enthalpy contribution in the system is enough to overcome this negative entropy contribution.

Hopefully, the description in this section clears up the commonly found misstatement in the literature that "micellization is always an entropy-driven process." Micellization *can* be entropically unfavorable; it depends on the system conditions. This usually occurs at high temperature (Evans and Wightman, 1982) and also at high pressure. It can also happen for some cases of micellization in an aqueous-based mixed solvent system when the dielectric constant is not high enough, such as for aqueous binary solvent systems with dimethylformamide and dimethylsulfoxide. This is possibly because the hydrogen bonds are already disrupted; there is not enough formation of hydrogen bond–networked structured water around the hydrocarbon chains. For these, micellization is an enthalpy-driven process. This is called *entropy–enthalpy compensation* for micellization.

3.4. MICELLIZATION VS. GENERAL SCHEME OF SELF-ASSEMBLY

The main topic of this section is the change of micellar structure in solution. Critical factors for the evolution of micellar structures will be discussed. The concept of surfactant (or micellar) packing parameter or, simply, *g*-factor, will be introduced, which can be used to explain this structural change with the concept of force balance (Israelachvili et al., 1976; 1992).

3.4.1. Change of Micelle Structures

As stated earlier, micelle formation is a process of force balance and no strong covalent bonding is involved. Any changes that can influence either the attractive or repulsive force components (or both) can change this force balance. For example, the change of the binding degree of anionic counterion on the cationic micelle of Figure 3.6 will affect the effective density of positive head groups on the micellar surface. This will alter the repulsive force between the head groups, which is the major repulsive force component. The hydrophobic force, that is, the major attractive force will respond toward the new force balance. For micelles, this usually results in the change of micellar structures. Other factors can include molecular structure of surfactant, concentration of surfactant, solvent polarity, dielectric constant of solvent, pH, size and balance of counterion, type and concentration of salt, temperature, pressure, and so on.

Figure 3.7 shows the various commonly found structures of micelles in solution. For comparison, some of the two-dimensional micelles are also presented. The typical evolution of micelle structures in aqueous solution is

$$\text{spherical} \rightarrow \text{rod-shaped} \rightarrow \text{wormlike} \rightarrow \text{liquid crystal (or mesophase)}$$

This change usually occurs as the increase of the surfactant concentration or as the increase of the concentration of counterions. The former is the typical direction of the increasing of attractive force and the latter is that of the decreasing of repulsive force. The rod-shaped micelle has almost the same diameter as the spherical micelle and has length 2–5 times its diameter. The wormlike micelle has the same cross-sectional structure as the rodlike micelle, but its apparent length can reach up to the μm-scale and often has a branched or randomly networked structure (Cates and Candau, 1990).

Hexagonal, cubic, and lamellar are the typical structures of the surfactant (and of amphiphilic polymer) liquid crystals. Hexagonal is the two-dimensional array of long micellar rods, cubic is the three-dimensional structure with a given unit cell, and lamellar is the one-dimensional layered structure of a planar sheet. Usually, hexagonal and lamellar have a wider range of phase diagram than cubic.

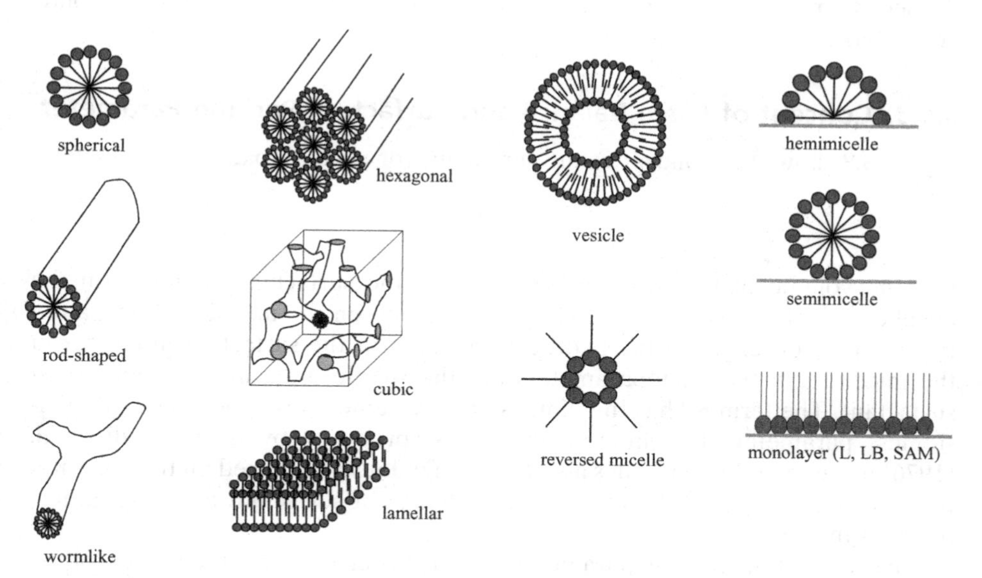

Figure 3.7. Structures of different micelles in solutions and at surfaces. L, LB, and SAM represent Langmuir monolayer, Langmuir-Blodgett film, and Self-Assembled Monolayer, respectively.

The vesicle and reverse micelles will be discussed in detail in Chapter 4. Surface micelles and monolayers will be discussed in later chapters.

3.4.2. General Scheme of Micellization

Here, we will look at micellization and change of micelle structures based on the general scheme of self-assembly proposed in Chapter 1 (Figure 1.3). The surfactant monomer in Figure 3.3 corresponds to the primary building unit of micellization, and the spherical micelle corresponds to the primary self-aggregate. Rod-shaped and wormlike micelles can be understood as forms derived from the spherical micelle due to the shift of force balance inside the micelle. Strong intermicellar repulsive force between these micelles prevents further self-assembly. However, as the condition shifts in favor of more attractive force and/or less repulsive force (as listed in the above section), these micelles begin to self-assemble into liquid crystal structures (Hassan et al., 2002; Kato, 2003). Hexagonal and tetragonal structures can be considered as the result of direct self-assembly of rod-shaped or wormlike micelles into a given geometry. Direct arrangement of spherical micelles can form cubic with the geometry of *Fd*3*m*. Some liquid crystal structures (e.g., cubic with *Ia*3*d*, *Pm*3*n*, *Im*3*m* geometries, and lamellar) do not pertain to the exact track of the spherical, rod-shaped, or wormlike geometries. But they are formed by the small shift in the force balance inside the self-aggregate of spherical, rod-shaped, or wormlike micelles or inside other liquid crystals. Thus, the spherical, rod-shaped, and wormlike micelles are the secondary building units of micellization, and liquid crystals are the secondary self-aggregate.

3.4.3. Concept of Force Balance and Surfactant Packing Parameter

Figure 3.8 shows the concept of the surfactant (or micelle) packing parameter

$$g = v/a_o l \tag{3.14}$$

v is the volume of the hydrocarbon chain of the surfactant monomer in the micelle, a_o is the area occupied by the head group on the micelle surface, and l is the apparent length of the hydrocarbon chain of the surfactant monomer in the micelle. All these subparameters are the simple molecular parameters of surfactant. The term $a_o l$ has the dimension of volume; thus, g becomes a dimensionless parameter. This monumental work, proposed by Israelachvili et al. (1976) more than three decades ago, has tremendously impacted on the quantitative understanding of the formation and change of surfactant and amphiphilic polymer micelles.

The surfactant packing parameter is the rule derived from the picture of the force balance between the attractive driving force and repulsive opposition force for the formation of micelles (see Figure 1.2). Here, we consider only the primary

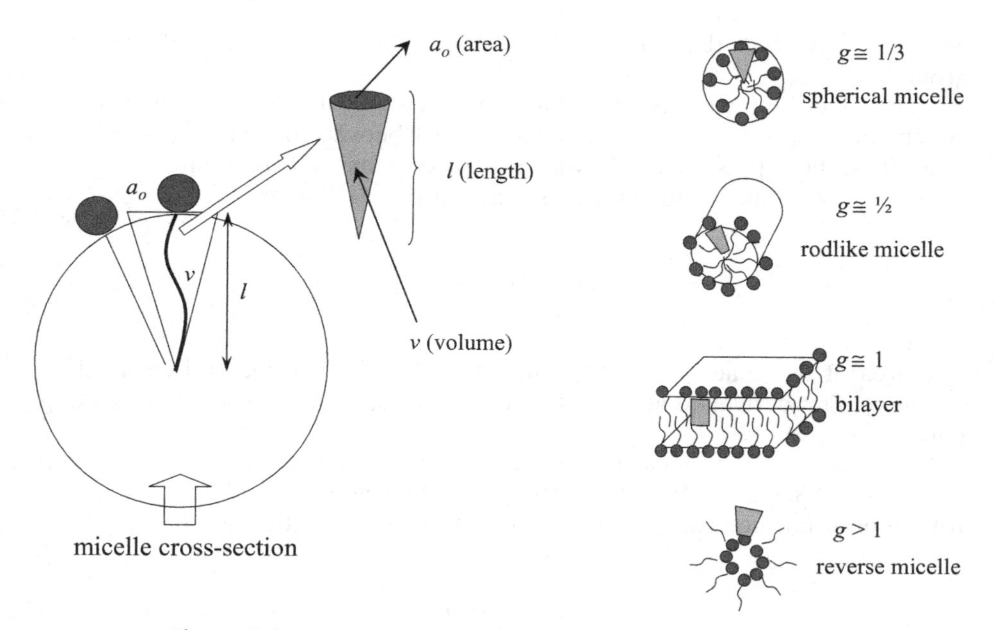

Figure 3.8. Surfactant (or micelle) packing parameter: $g = v/a_o l$.

self-assembly of surfactant. Thus, from equation (1.3), the total free energy, $U_{\text{total}}(x)$, follows:

$$U_{total}(x) = U_{A,P}(x) + U_{R,P}(x) \tag{3.15}$$

> The surface of the micelle is in reality rough due to the penetration of the water molecules inside, and the dynamic nature of micelles and surfactant molecules.

The attractive energy contribution, $U_{A,P}(x)$, that mainly originates from the hydrophobic interaction regardless of the head-group property (ionic or nonionic) can be expressed as $\gamma \cdot A$, where γ is the free energy at the oil–water interface and A is the total area of the micelle surface. We assume here that the micelle surface is absolutely smooth.

Thus, dividing this term by aggregation number n, $\gamma \cdot A/n$, yields the free energy per monomer of the micelle. With the assumption of close packing of the head group, this can be written as $\gamma \cdot a$, where a is the area occupied by each monomer on the surface of the micelle.

> Here we assume that the surface of the micelle is a two-dimensional planar surface.

The repulsive energy contribution, $U_{R,P}(x)$, can originate from many different possible sources such as electrostatic repulsion (for ionic micelles), hydration, and steric interaction. But, since they are all repulsive forces on the surface of micelles, we can expect that they are inversely proportional to the area of the head group, a.

This is the same analogy as that the repulsive forces of one-dimensional systems are inversely proportional to the distance between the interacting objects. Thus, it can be expressed as K/a, where K is a constant. The total interfacial free energy per monomer at the micelle surface can be written as

$$U_{total}^{\text{int}} = \gamma \cdot a + K/a \tag{3.16}$$

The area of the head group at the minimum of U_{total}^{int} can be easily solved by setting its first derivative at zero. This is the a_o in the above surfactant packing parameter g.

Tanford (1980) proposed useful relations to estimate the remaining two subparameters, v and l. With the assumption of a compact micelle (i.e., no penetration of water molecules on the surface of the micelle), they are

$$v = 27.4 + 26.9 \cdot n_{hc} \tag{3.17}$$

and

$$l_{\max} = 1.54 + 1.265 \cdot n_{hc} \tag{3.18}$$

v is in cubic Angström and n_{hc} is the number of carbons in the hydrocarbon chain. l_{max} is the maximum possible extension of the hydrocarbon chain with Angström units (all *trans* C–C bond). Thus, in estimating the l, in reality another assumption is added: $l \leq l_{\max}$. The number of C–C *kink*-configurations in the hydrocarbon chain is one of the main factors to determine this. One kink can reduce the chain length ~1.25 Å, and typical surfactant chains have 1.5–2.5 kinks, depending on their length and dynamic state.

Once the three subparameters v, a_o, and l are estimated, we can determine the most favorable micelle structure by using the surfactant packing parameter (Figure 3.8). It provides a beautiful guide for the micelle structures without requiring knowledge of the complex relations between the forces. For example, when the g-value is close to 1/3, one can predict the formation of a spherical micelle by the simple geometrical consideration of packing of *cone*-type monomer into three-dimensional self-assembled aggregate. Similarly, when the g-value is close to 1/2, one can predict that the rodlike micelle will be formed. When the g-value is close to 1, the unit packing shape becomes the *cylinder* type; thus the bilayer-type of micelle is the most likely structure to be assembled. $g > 1$ results in a reverse micelle.

EXERCISE 3.1

Describe qualitatively the condition that can induce the spherical micelle of DTAB (dodecyltrimethyl ammonium bromide) to change into a rod-shaped micelle based on the packing parameter consideration. Assume that the *g*-value for the spherical micelle is 1/3 and for the rod-shaped micelle is 1/2. Solvent is water and DTAB concentration is fixed.

Solution: Since v and l are fixed, the only way to increase the *g*-value from 1/3 to 1/2 is to decrease a_o. This can be achieved by the reduction of repulsive force on the surface of the micelle. DTAB is a cationic surfactant; thus the density of the positive charge on its micelle surface should be reduced. This can be done by the addition of secondary cationic counterion such as NaBr, increase in the solution pH, or increase in the solution polarity by the addition of a second solvent such as formamide.

This concept of the packing parameter was developed primarily for the surfactant micelles in solution. But its basic principle of *force balance* is also valid for other types of self-assembled aggregates. It provides a reasonable prediction for the structural changes in the systems, including self-assembled monolayer, Langmuir monolayer, Langmuir-Blodgett film, polyelectrolyte multilayer, and biological membrane. Geometrical match and mismatch, which beautifully explains the formation of a nanometer-scale regular pattern on the surface of a mixed phospholipid bilayer (Zasadzinski, 1987) is also a good example of the versatility of this concept.

As one gathers more understanding of the type and role of weak forces on colloidal systems, the packing parameter emerges as a powerful concept to follow self-assembly on a colloidal-length scale, too (Park et al., 2004). Details will be in Chapter 5. In Chapter 9, it will be shown how this concept of force balance can be adopted for the structural changes of many nanostructured materials that are prepared by using self-assembled aggregates as *structure-directing agents* or *templates*.

3.5. MULTICOMPONENT MICELLES

As shown in Figure 3.9, a *multicomponent* or simply *mixed* micelle is a micelle that is formed through the self-assembly of more than two different kinds of surfactant molecules (Abe and Ogino, 1992). In most cases, the composition of monomers between the different surfactant molecules is directly related to the composition in the mixed micelles. Thus, the pseudo-phase separation model that can follow the key thermodynamics quantities is more suitable for the description of the formation of mixed micelles. The basic concepts of micellization, including *cmc*, aggregation number, and counterion binding, are the same as for single-component micelles.

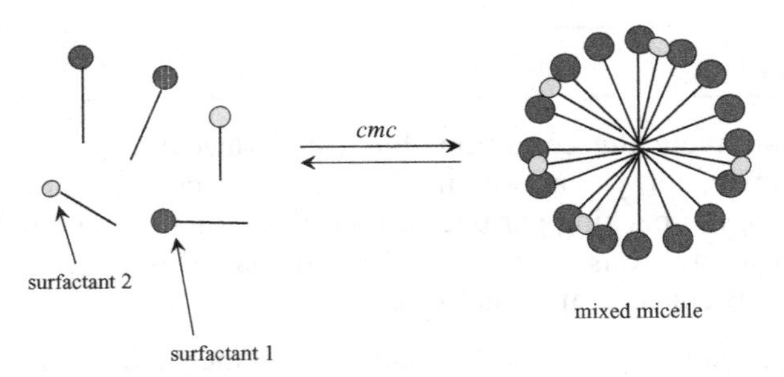

Figure 3.9. Formation of mixed micelles.

For single-component micelle formation, there is only one molecular interaction either between the surfactant molecules 1 or between the surfactant molecules 2. When both surfactant 1 and 2 coexist, there are three molecular interactions between surfactant molecules 1 and 2: between 1 and 1, between 2 and 2, and between 1 and 2. Should all three interactions be the same or similar, the formation of mixed micelle can be ideal or at least close to ideal. *cmc* of surfactant mixture 1 and 2 becomes

$$1/cmc = X_1/cmc_1 + X_2/cmc_2 \tag{3.19}$$

where cmc_1 and cmc_2 are the *cmc* of surfactant 1 and 2, respectively. X_1 and X_2 are the mole fraction of each surfactant with $X_1 + X_2 = 1$ (Clint, 1975). The first case of this ideal mixed micelle formation can be when two surfactants have the same head group. For example, the formation of mixed micelles between the homologs series of surfactants (series of surfactants that have the same head group but different length of hydrocarbon chain) is largely ideal. Cases with very similar head groups can be ideal, too, or a slightly negative deviation from ideality.

For ionic–nonionic surfactant systems, the mixed micelle formation negatively deviates from ideal, which means that there is a net attractive interaction between surfactant molecules. The repulsive force between ionic head groups can be shielded by nonionic groups possibly located between them and there can be an additional attractive force between nonionic and ionic head groups due to charge–dipole interaction. The largest negative deviation from ideality comes from cationic–anionic surfactant systems. Strong electrostatic attraction between head groups is responsible for this.

When the mixed micelle systems show negative deviation from ideality, they usually show synergism in many physicochemical properties. Synergism can be viewed as a property or performance of the mixed micelle systems that is better than any of the single-component micelle systems. *cmc* of the mixed micelle is smaller than the *cmc* of at least one of the single-component micelles. This changes the concentration of monomer in solution, structure of the micelle, aggregation number, and the phase diagram as a whole. For some cases, even a small addition of the second component can dramatically change the micellization process itself. Formation of a vesicle at a relatively low concentration of cationic–anionic systems is one good example (Iampietro et al., 1998).

Synergism has been discovered in many practical properties such as surface tension, interfacial tension, foamability, emulsification, solubilization, detergency, and tertiary oil-recovery. For example, the emulsifying efficiency of the major surfactant can be significantly increased by the addition of a small amount of the right second surfactant component (Huibers and Shah, 1997). For many practical reasons such as efficiency of the synthetic process, byproducts, and the cost–performance relation, the micelle systems that are used in industry inevitably are mixtures. Thus, this issue is important for many industrial applications.

> The phenomena of synergism have been found in a wide field of research areas. When the major component for a given application is mixed with small amounts of the second (or third) component(s) (intentionally or by chance), the performance of this mixture for that given property often shows unexpected enhancement. For example, addition of impuritic amounts of the second component (that actually is a poor catalyst) to the major component of catalyst often reveals a great jump in its overall catalytic activity. It is called *promoter* in the area of the solid catalyst.

When the molecular interactions between 1 and 1, and 2 and 2 in Figure 3.9 are stronger than the interaction between 1 and 2, the mixed micelle formation experiences the positive deviation from the ideality. Contrary to the synergism, *antagonism* is found in this case. *cmc* of the mixed micelle is higher than that of the single-component micelles, and the aggregation number is decreased. For the extreme case of the mixture of hydrocarbon–perfluorocarbon surfactants with the same ionic head group, two different micelles (one hydrocarbon-rich and the other perfluorocarbon-rich) can coexist in solution (Shinoda and Nomura, 1980).

The third type of mixed micelle can be found on amine oxide surfactants. Its general formula is

$$CH_3{-}(CH_2)_n{-}\overset{|}{\underset{|}{{}^{+}N}}{-}OH \quad \longleftrightarrow \quad CH_3{-}(CH_2)_n{-}\overset{|}{\underset{|}{N}}{\rightarrow}O$$

with $n = 7 - 17$. It has a long enough hydrocarbon chain and a right head group that has enough hydrophilicity. It thus forms a micelle with *cmc* range of 1–200 mM in aqueous solution at room temperature. What makes this surfactant unique (regarding formation of mixed micelles) lies in its head group. Depending on the solution pH, it can be either cationic (protonated form) at low pH or nonionic at high pH. Both cationic and nonionic forms coexist at medium pH. Changing the solution pH thus can form cationic micelles, cationic–nonionic mixed micelles, and nonionic micelles. FTIR results on this system (Rathman and Scheuing, 1991) and recent theoretical study (Maeda, 2004) confirm that the thermodynamic nonideality of ionic–nonionic mixed micelles originates from the interaction at the surface of micelles.

As briefly mentioned earlier, the mixed micelle has long been accompanied by a variety of industrial applications. On the nanotechnology side, the concept of mixed micelles provides an important insight for current and future development of nanostructured materials and nanosystems. For the preparation of nanostructured materials, often synergism and antagonism of mixed micelles play an important role in determining/controlling the structure and quality of the final products (Chapter 9). Formation of mixed monolayer and multilayer and their phase behavior can be a critical issue for a variety of nanofabrication or nanopatterning processes (Chapter 13). Also, it provides good insight into the development of the concept of *heterogeneous self-assembly* for further progress in nanodevices (Chapter 14).

EXERCISE 3.2

Suppose you have a mixed micelle of DTAB/n-decanol (mole ratio 8:2; n-decanol in this case is a *co-surfactant*; details in the next section) with spherical shape. Comparing it with the DTAB-only micelle at the same concentration, qualitatively describe the change of its *cmc*, aggregation number, and possible structural change.

Solution: DTAB has 12 carbons of alkyl chain, while n-decanol has only 10. Thus, compared with the DTAB-only micelle, DTAB/n-decanol mixed micelle should have almost the same v and l values. But, since we can expect significantly decreased repulsive forces on the surface of the mixed micelle, a_o for the mixed micelle should be decreased. This will result in the increase of the g-value for the mixed micelle. So, the mixed micelle should show lowered *cmc* value and increased aggregation number, and possibly be changed in its structure in the direction of rod-shape.

3.6. MICELLAR SOLUBILIZATION

It is known that a number of molecules that are insoluble or slightly soluble in water can become soluble in surfactant micellar solution (Nagarajan, 1996) and also in polymer micellar solution (Hurter et al., 1995). For a simple example, if one put a drop of benzene on the surface of water while gently stirring it, the benzene drop would be circling around on the surface of the water or part of it might be dispersed into the water as oil-droplets. If the benzene drop is placed on the surface of aqueous surfactant solution above its *cmc*, one can observe the gentle disappearance of the benzene drop into the surfactant solution. This type of process is commonly called *micellar solubilization*. Figure 3.10 is the schematic representation.

Since the micellar interior provides the hydrocarbon-like environment, lipophilic (oil-friendly) compounds can easily reside in this region of the micelle. However, the actual location of the solubilizates (lipophile) inside the micelle varies due to several factors. First, the polarity inside the micelle is not even. While the core region is more likely hydrocarbon, the region near the micelle surface can be somewhat polar. Thus, the polarity of the solubilizates largely determines the location of solubilization. Highly polar solubilizate can be located near the surface of the micelle, while less polar solubilizate can reside in the core of the micelle. Also, the interaction between the solubilizate and the surfactant can be an important factor. When the less polar solubilizate has a functional group, the possible strong interaction of this functional group with the head group of the micelle can put the solubilizate nearer the surface of the micelle. But, if there were no functional group, the micelle core would be a more favorable place for the solubilizate to reside. When a hydrophobic alkyl chain is part of a less hydrophobic solubilizate body, for example, a benzene ring, the alkyl chain is buried inside the micelle core while the benzene ring is pushed back near the surface of the micelle.

Sometimes, the picture of micellar solubilization and the formation of the mixed micelle is not clear. n-alkanols with a hydrocarbon chain of 6–12 are not soluble in water, so they are not surfactant in water. But, when they are solubilized in the surfactant solution above the *cmc*, they are solubilized in such a way as to be a second surfactant in the micelle. Due to the strong interaction, the alcohol group is forced to be near the micelle surface while the alkyl chain is located inside the micelle core just like the mixed micelle. The alkanols, in this case, are called *co-surfactant* instead of *solubilizate*. n-amines with a hydrocarbon chain of 6–12 can be another example of co-surfactant.

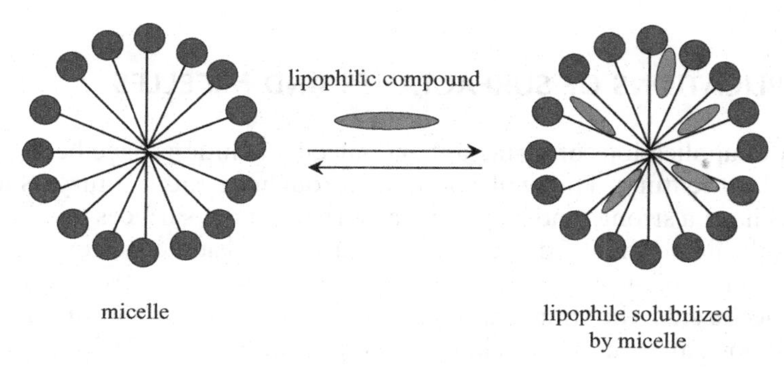

Figure 3.10. Schematic representation of the concept of micellar solubilization.

When a solubilizate begins to be solubilized inside the micelle, the more favorable location begins to be occupied by the solubilizate first. As the concentration of the solubilizate increases, the less favorable places start to be occupied. The point where all possible solubilizates have been taken up by the micelle can be defined as *micellar solubility*. But, in most cases, this point is not clear cut. The solubilized molecule inevitably changes the force balance inside the micelle, which changes the packing status of the micelle. This usually induces the structural change well before the solubilization limit has been reached. Sphere-to-rod and -to-wormlike micelle transitions are common changes. This transition of micellar structure can alter the capacity and amount at the micellar solubilization. A small amount of solubilizate also can change *cmc*.

Above the solubilization limit, the excess amount of solubilizate shows phase separation. The ternary-phase diagram of water–solubilizate–surfactant has long been studied (Laughlin, 1991).

EXERCISE 3.3

Suppose you have rod-shaped DTAB micelle with solubilized benzene (mole ratio 7:3). Comparing it with the DTAB-only micelle at the same concentration, qualitatively describe the change in its *cmc*, aggregation number, and possible structural change.

Solution: A benzene molecule can show the ring current effect inside the micelle. The rough center of mass of the DTAB molecule is about the second carbon from the nitrogen atom of the head group. Thus, a benzene molecule at low fraction mainly resides near this location. v and l values should be the same, compared with the DTAB-only micelle. However, since we can expect the increased value of a_0 due to the inclusion of benzene near the micelle surface, the overall g-value should be decreased. Thus, this micelle should show an elevated *cmc* value and decreased aggregation number, and possibly be changed in its structure in the direction of spherical-shape.

3.7. APPLICATIONS OF SURFACTANTS AND MICELLES

A variety of applications of surfactant and micellar solutions have been explored over the last century. This will continue throughout this century. Surfactant molecules have a strong tendency to be adsorbed on the surfaces and interfaces, which results in a dramatic decrease in the surface tension of water or the interfacial tension at oil–water interfaces. This is the first key to their applications. On the micelle side, the counterion binding and micellar solubilization phenomena are important. Ionic species in surfactant solution can be highly localized on the very narrow region of the micelle surface through the process of counterion binding. Micellar solubilization literally means solubilizing water-insoluble compounds in aqueous solution and localizing them within a small confined space of the micellar interior.

The oldest application of surfactant solution may be its detergency (Culter and Kissa, 1987). First, decreased interfacial tension at the oil–water interface by the adsorption of surfactant will promote the detachment of oil from the solid substrates (glass, fiber, soil, etc.). Then, the oil will be solubilized within the micelle in nearby solution, which will be eventually removed from the solution, leaving the clean solid substrates behind. The well-known *rollup* mechanism and the force balance during the change of contact angle at the three phase contact regions are important. Additional major applications include cosmetics and pharmaceutical formulations, delivery vehicles, micellar liquid chromatography, micellar-enhanced ultrafiltration, rheology modifier, tertiary oil recovery, and so on. Readers will find details in excellent references: Scamehorn and Harwell (1989); Morrow (1991).

3.7.1. Micellar Catalysis

Micelles in solution can dramatically change the reaction rates of a number of different reactions. Also, micelles in solution can make reactions possible that usually are not possible in aqueous solution or that are possible only under very specific condition such as vigorous stirring. This phenomenon is called *micellar catalysis*. While traditional catalysis is involved with the change of activation energy of an intermediate reaction complex, micellar catalysis is the result of the dramatic localization of ionic and non- (or less-) polar compounds on the micelle. This is also differentiated from the biological enzyme catalysis that occurs via molecular-specific bonding. Micellar catalysis is a noncovalent and nonmolecular-specific process. It is rather a zone- (or regional-) specific process at the nanometer scale.

The name *microheterogeneous system* originates here.

Counterion binding and micellar solubilization are again two key points. Also, the extremely high oil–water interfacial area in the surfactant solution greatly contributes to this process. With the size of ordinary micelles and the roughness of their surfaces, the interfacial area of a typical micellar solution can be estimated as ~1,000—~2,000 cm^2 per *g* of surfactant molecule.

Figure 3.11 shows three typical micellar catalysis processes. First, type (a) is involved only with the counterion binding process. When the ionic organic or inorganic compound (or compound with ionic functional group) is solubilized or dispersed in aqueous solution with low concentration, its chance of reaction is very low or the reaction rate is very slow even though it is reactive. However, when it is solubilized or dispersed in the surfactant solution whose micelle head group has opposite charge, strong charge–charge interaction brings it to the

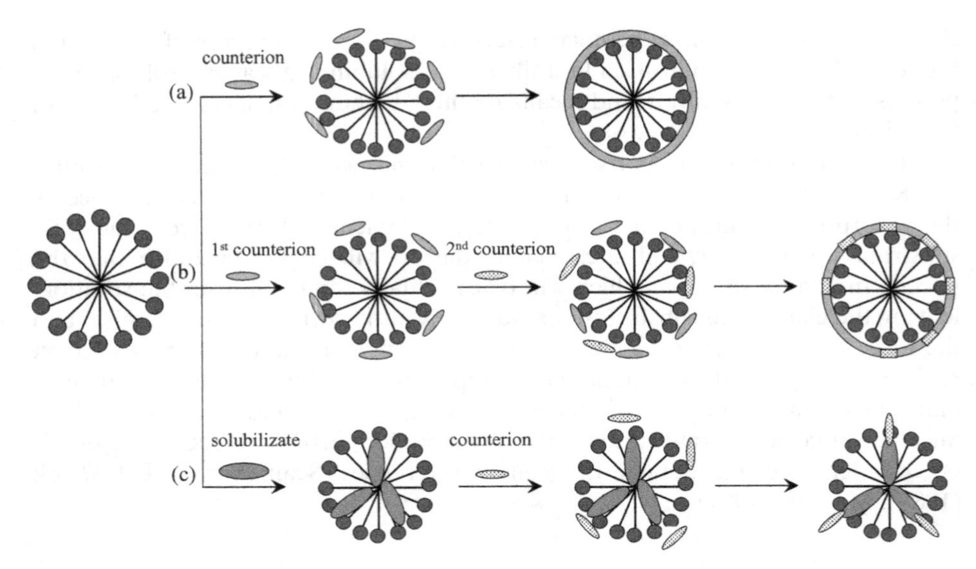

Figure 3.11. Schematic representation of micellar catalysis.

surface of the micelle, and localizes it in a confined space. This can increase the reaction rate up to ~100-fold (the figure represents an example of polymerization or condensation after counterion binding). Second, type (b) is also involved only with the counterion binding process. But the sequential binding of two or more different reactants on the surface of the micelle makes the more diverse reaction possible. The binding competition between the different ionic species and possible saturation by one reactant can limit the catalytic effect. The third type (c) is the combination of micellar solubilization and counterion binding. Reactant that is water insoluble (or less soluble) is first solubilized into the micelle and the second reactant that is ionic is localized on the surface of the micelle. Two reactants that are immiscible can be brought together this way, which makes the reaction possible with a enhanced reaction rate.

Concepts of micellar phase-transfer catalysis and micellar autocatalysis have been well studied based on this scheme (Cordes, 1973; Fendler and Fendler, 1975; Almgren, 1991; Bunton, 1991; Rathman, 1996). Biological systems have evolved some vital functions that are strikingly similar to the scheme of this micellar catalysis. Some examples include transport across the membrane and self-reproduction of lipid vesicles for the model growth of cell membranes (Luisi, Walde, and Oberholzer, 1999). It also provides an excellent model system for the study of enzyme activity enhancement in the human body, such as the role of phospholipase A_2 (PLA_2) in hydrolysis of glycerophospholipids (Tatulian, 2001).

REFERENCES

Abe, M., Ogino, K. "Solution Properties of Anionic-Nonionic Mixed Surfactant Systems," *Surfactant Science Series*, Vol. 46 (Mixed Surfactant Systems), pp. 1–21 (Marcel Dekker: 1992).

Alexandridis, P., Lindman, B., eds. *Amphiphilic Block Copolymers: Self-assembly and Applications* (Elsevier: 2000).

Almgren, M. "Kinetics of Excited State Processes in Micellar Media," *Surfactant Science Series*, Vol. 38 (Kinetics and Catalysis in Microheterogeneous Systems), pp. 63–113 (Marcel Dekker: 1991).

Ash, M., Ash, I. Compiled: *Handbook of Industrial Surfactants: An International Guide to More Than 16,000 Products by Tradename, Application, Composition & Manufacturer*, (Gower: 1993).

Bunton, C. A. "Micellar Rate Effects upon Organic Reactions," *Surfactant Science Series*, Vol. 38 (Kinetics and Catalysis in Microheterogeneous Systems), pp. 13–47 (Marcel Dekker: 1991).

Cates, M. E., Candau, S. J. "Statics and Dynamics of Worm-like Surfactant Micelles," *J. Phys.: Condens. Matter* **2**, 6869 (1990).

Chernik, G. G. "Phase Studies of Surfactant-Water Systems" *Curr. Opin. Coll. Inter. Sci.* **4**, 381 (2000).

Clint, J. H. "Micellization of Mixed Nonionic Surface Active Agents," *J. Chem. Soc. Faraday Trans. 1* **71**, 1327 (1975).

Cordes, E. ed. *Reaction Kinetics in Micelles* (Plenum Press: 1973).

Culter, W. G., Kissa, E., eds. *Surfactant Science Series*, Vol. 20 (Detergency: Theory and Technology) (Marcel Dekker: 1987).

Evans, D. F., Wightman, P. J. "Micelle Formation above 100°C," *J. Colloid Interface Sci.* **86**, 515 (1982).

Fendler, J., Fendler, E. *Catalysis in Micellar and Macromolecular Systems* (Academic Press: 1975).

Grieser, F., Drummond, C. J. "The Physicochemical Properties of Self-Assembled Surfactant Aggregates as Determined by Some Molecular Spectroscopic Probe Techniques," *J. Phys. Chem.* **92**, 5580 (1988).

Hassan, S., Rowe, W., Tiddy, G. J. T. "Surfactant Liquid Crystals," *Handbook of Applied Surface and Colloid Chemistry*, Vol. 1, pp. 465–508 (John Wiley & Sons: 2002).

Huibers, P. D. T., Shah, D. O. "Evidence for Synergism in Nonionic Surfactant Mixtures: Enhancement of Solubilization in Water-in-Oil Microemulsions," *Langmuir* **13**, 5762 (1997).

Hurter, P. N., Alexandridis, P., Hatton, T. A. "Solubilization in Amphiphilic Copolymer Solutions," *Surfactant Science Series*, Vol. 55 (Solubilization in Surfactant Aggregates), pp. 191–235 (Marcel Dekker: 1995).

Iampietro, D. J., Brasher, L. L., Kaler, E. W., Stradner, A., Glatter, O. "Direct Analysis of SANS and SAXS Measurements of Catanionic Surfactant Mixtures by Fourier Transformation," *J. Phys. Chem. B* **102**, 3105 (1998).

Israelachvili, J. N. *Intermolecular & Surface Forces*, 2nd ed. (Academic Press: 1992).

Israelachvili, J. N., Mitchell, D. J., Ninham, B. W. "Theory of Self-Assembly of Hydrocarbon Amphiphiles into Micelles and Bilayers," *J. Chem. Soc. Faraday Trans. 2* **72**, 1525 (1976).

Kato, T. "Microstructures of Nonionic Surfactant-Water Systems: From Dilute Micellar Solution to Liquid Crystal Phase," *Surfactant Science Series*, Vol. 112 (Structure–Performance Relationships in Surfactants), 2nd ed., pp. 485–524 (Marcel Dekker: 2003).

Laughlin, R. G. "Aqueous Phase Science of Cationic Surfactant Salts," *Surfactant Science Series*, Vol. 37 (Cationic Surfactants: Physical Chemistry), pp. 1–40 (Marcel Dekker: 1991).

Laughlin, R. G. "Surfactant Phase Science," *Curr. Opin. Coll. Inter. Sci.* **1**, 384 (1996).

Luisi, P. L., Walde, P., Oberholzer, T. "Lipid Vesicles as Possible Intermediates in the Origin of Life," *Curr. Opin. Coll. Inter. Sci.* **4**, 33 (1999).

Maeda, H. "Electrostatic Contribution to the Stability and the Synergism of Ionic/Nonionic Mixed Micelles in Salt Solutions," *J. Phys. Chem. B* **108**, 6043 (2004).

Menger, F. M., Littau, C. A. "Gemini Surfactants: A New Class of Self-Assembling Molecules," *J. Am. Chem. Soc.* **115**, 10083 (1993).

Morrow, N. R., ed., *Surfactant Science Series*, Vol. 36 (Interfacial Phenomena in Petroleum Recovery) (Marcel Dekker: 1991).

Nagarajan, R., Ruckenstein, E. "Self-Assembled Systems," *Experimental Thermodynamics*, Vol. 5, Issue Pt. 2, pp. 589–749 (Elsevier: 2000).

Nagarajan, R. "Solubilization in Aqueous Soltions of Amphiphiles," *Curr. Opin. Coll. Inter. Sci.* **1**, 391 (1996).

Nagarajan, R. "Theory of Micelle Formation: Quantitative Approach to Predicting Micellar Properties from Surfactant Molecular Structure," *Surfactant Science Series*, Vol. 112 (Structure–Performance Relationship in Surfactants), 2nd ed., pp. 1–109 (Marcel Dekker: 2003).

Park, S. H., Lim, J. H., Chung, S. W., Mirkin, C. A. "Self-Assembly of Macroscopic Metal-Polymer Amphiphiles," *Science* **303**, 348 (2004).

Porter, M. R. *Handbook of Surfactants*, 2nd ed. (Blackie: 1994).

Puvvada, S., Blankschtein, D. "Molecular Modeling of Micellar Solutions," *Surfactants in Solution* **11**, 95 (1991).

Rathman, J. F. "Micellar Catalysis," *Curr. Opin. Coll. Inter. Sci.* **1**, 514 (1996).

Rathman, J. F., Scheuing, D. R. "Alkyldimethylamine Oxide Surfactants," *Fourier Transform Infrared Spectroscopy in Colloid and Interface Science*, ACS Symp. Ser. 447, pp. 123–142 (American Chemical Society: 1991).

Riess, G. "Micellization of Block Copolymers," *Prog. Polym. Sci.* **28**, 1107 (2003).

Robb, I. D., ed. *Specialist Surfactants*, (Blackie: 1997).

Rosen, M. J. *Surfactants and Interfacial Phenomena*; 3rd ed. (John Wiley & Sons: 2004).

Scamehorn, J. F., Harwell, J. H., eds. *Surfactant Science Series*, Vol. 33 (Surfactant-Based Separation Process) (Marcel Dekker: 1989).

Shinoda, K., Nomura, T. "Miscibility of Fluorocarbon and Hydrocarbon Surfactants in Micelles and Liquid Mixtures: Basic Studies of Oil Repellent and Fire Extinguishing Agents," *J. Phys. Chem.* **84**, 365 (1980).

Tanford, C. *The Hydrophobic Effect: Formation of Micelles and Biological Membranes* (John Wiley & Sons: 1980).

Tatulian, S. "Toward Understanding Interfacial Activation of Secretory Phospholipase A_2 (PLA_2): Membrane Surface Properties and Membrane-Induced Structural Changes in the Enzyme Contribute Synergistically to PLA_2 Activation," *Biophys. J.* **80**, 789 (2001).

Zana, R. "Micellization of Cationic Surfactants," *Surfactant Science Series*, Vol. 37 (Cationic Surfactants: Physical Chemistry), pp. 41–85 (Marcel Dekker: 1991).

Zasadzinski, J. A. N. "Ripple Wavelength, Amplitude, and Configuration in Lytropic Liquid Crystals as a Function of Effective Headgroup Size," *J. Physique* (Paris) **48**, 2001 (1987).

4

MOLECULAR SELF-ASSEMBLY IN SOLUTION II: BILAYERS, LIQUID CRYSTALS, AND EMULSIONS

This chapter contains the second part of molecular self-assembly in solution. The formation of bilayers, liquid crystals, and emulsions, and their similarities and differences compared with those of micelles will be covered.

Micelle formation (micellization) is a spontaneous thermodynamically driven process that generates a well-defined *cmc* (critical micelle concentration) and a finite size of self-assembled structures. Many amphiphiles that can form typical micelles usually have an ability to form bilayers, liquid crystals, and emulsions. However, the self-assembly process for their formation often does not show a clear-cut *cmc* and can generate not-well-defined, sometimes infinite-size self-assembled structures. Often, it is highly stepwise (molecule-by-molecule-wise or monomer-by-monomer-wise), highly kinetically driven, or highly dependent on external energy input.

While bilayers and liquid crystals are self-assembled objects that are formed in solutions, emulsions are structures that are stabilized by the results of self-assembly at liquid–liquid (mainly oil–water) interfaces. This may make it reasonable for emulsions to be classified as a surface self-assembly. But, since the process occurs in solutions and their three-dimensional structures have a strong tie with the formation of normal and reverse micelles, it would be logical to discuss them here as part of the self-assembly in solutions. Foam is another type of three-dimensional structure that is formed by self-assembly at interface (liquid–gas). Its formation also has a strong correlation with micelles in solutions. But the process has a stronger tie with adsorption at interface; therefore, we will include it in Chapter 6 as part of surface self-assembly.

This chapter focuses primarily on the formation of bilayers, liquid crystals, and emulsions. It describes the concept of force balance to clarify understanding of their structural transitions (with the packing parameter, *g*-value). Also, it describes the concepts of counterion binding and micellar solubilization and shows that the general concept of micellar catalysis (Chapter 3) stands valid for these systems. Detailed application issues, including the preparation and assembly of nanoparticles, preparation of macrostructured and macroporous inorganic materials, and special applications such as delivery vehicles, will be covered in Part II.

4.1. BILAYERS

4.1.1. Bilayer-Forming Surfactants

Typical micelle-forming surfactants (presented in Figure 3.2) usually have the capability to form bilayers. But it often requires specific conditions that can ensure the increased hydrophobic interaction between surfactant hydrocarbon chains and/or the decreased repulsive interaction between head groups. Thus, typical binary-phase diagrams of those surfactants with water at these conditions show a wide range of micelle region (spherical, rodlike, wormlike, etc.) that is followed by a liquid crystal region at higher concentration. The bilayer area inside this liquid crystal region where cubic and lamellar structures are located appears at even higher concentration.

Figure 4.1 presents typical amphiphiles that can form bilayers and emulsions under the usual conditions of low concentration and at low (or no) concentration of additional counterions. Like the micelle-forming surfactants in Figure 3.2, these amphiphiles also consist of a charged (or hydrated) head group and a hydrophobic hydrocarbon tail group. They are also classified as *cationic, anionic*, or *zwitterionic*, based on the intrinsic charge of their head groups. Most of them have two (or branched) hydrocarbon chains. This structural characteristic helps them have bulky tail groups, which makes them suitable to fulfill the molecular packing requirement for the formation of a bilayer structure.

Dialkyl quaternary ammonium is one of the most common cationic surfactants in this class. The two long alkyl chains typically have 8–20 carbons. Aerosol OT (AOT) is another common but anionic surfactant in this class. It has two branched hydrocarbon chains, which make the tail group even bulkier.

Lipids are amphiphiles in this class with the most abundant variety and structural diversity. Biochemistry uses the term *lipid* to classify a much wider range of biological-origin organic molecules. But, typically, lipids are amphiphiles with two hydrocarbon chains of 8–20 carbons and a single head group. Some

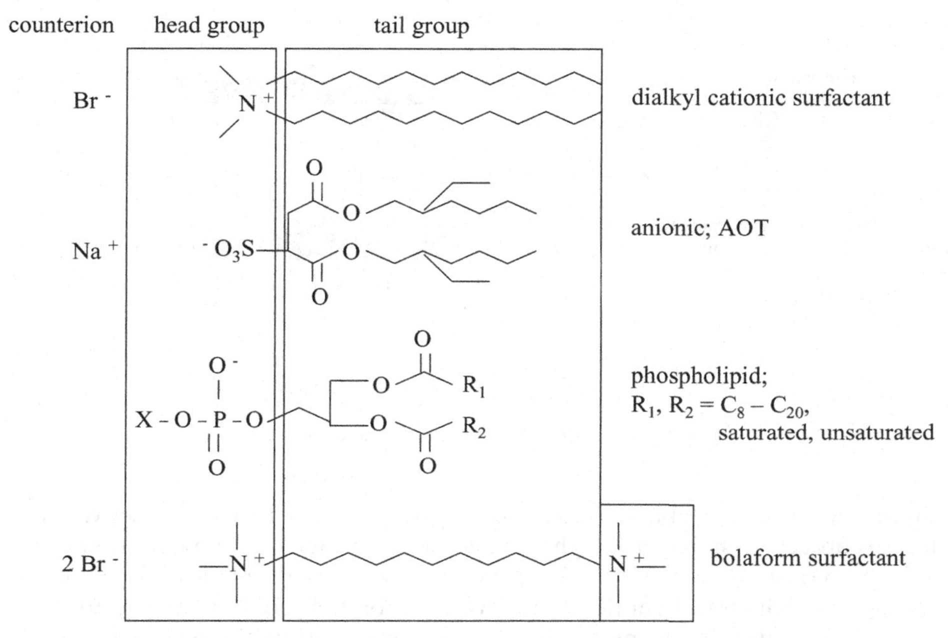

Figure 4.1. Typical amphiphiles that form bilayers and emulsions.

lipids possess one hydrocarbon chain. Major classes of lipids include *phospholipids* (glycerophospholipid and sphingophospholipid) and *glycolipids* (glyceroglycolipid and sphingoglycolipid). The molecular structure in Figure 4.1 represents glycerophospholipids. For all lipids, hydrocarbon chains can be saturated or unsaturated. Unsaturated hydrocarbon chains make the tail group bulkier, which in many cases is the critical molecular packing factor for their structural evolution and physicochemical and even biological functionalities. For most of the phospholipids, the head groups bear either anionic or zwitterionic charge at neutral solution pH. A phospholipid with a positively charged head group is extremely rare.

Bolaform surfactants are amphiphiles with two head groups attached on both ends of a single hydrocarbon chain. Their hydrocarbon chains are not as bulky as those of others in the Figure. But their unique structural features that can induce the head group interactions on both ends give them packing geometry very similar to others.

4.1.2. Bilayerization

Figure 4.2 shows a typical bilayer structure and its formation process. As in the case of micellization, this is a thermodynamically driven process and the self-assembled bilayer is in equilibrium with its monomer. Thus, bilayers are

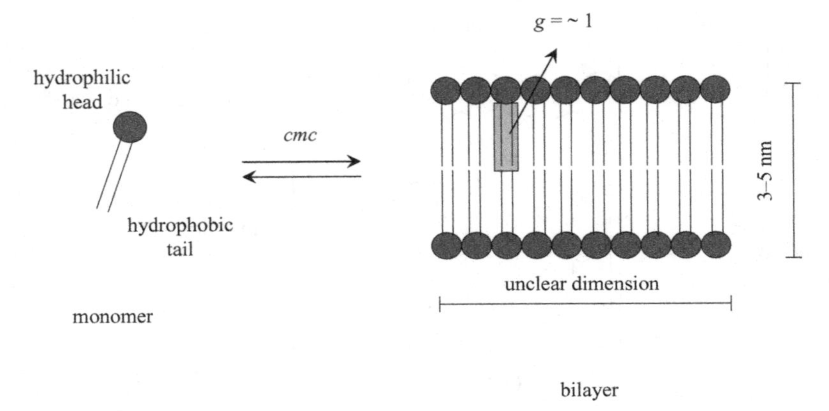

Figure 4.2. Formation of bilayer: bilayerization.

thermodynamically stable self-assembled aggregates. The driving attractive force for this process is mainly hydrophobic interaction, which is originated by entropy gain. Two (also often bulky) hydrocarbon tail groups certainly will contribute to this force much more than the single hydrocarbon tail group of micelle-forming surfactants. This self-assembly driving force is balanced by the repulsive opposition of electrostatic (for ionic or zwitterionic) or hydration (for nonionic) force. No directional force is involved during this process. *Cmc* is determined at the concentration point of this balance. A much stronger attractive force compared with micellization but the same magnitude of repulsive force can make this bilayer-forming process favorable even at low concentration. *Cmc* for bilayers is indeed much lower than for spherical micelles. For typical biological lipids, it spans 10^{-3}–10^{-10} mM and sometimes is too low to be clearly defined.

What determines the unique planar-like structure of a bilayer is the molecular geometry of its monomer, whose two-chain (or bulky) tail group makes its packing geometry g value close to unity. Ideally, this bilayer structure might stretch infinitely. But again this self-assembled structure, like all others, is stabilized by the force balance between intermolecular forces. Thus, it can easily fluctuate due to any factors that can perturb this delicate balance. This can generate defect sites on the surface of the bilayer, which in turn promote the closing up of the open ends of the bilayer structure. This process will minimize the region where the hydrocarbon is directly exposed to water. And it is the main reason why there are so many macroscopic self-assembled structures consisting of a bilayer unit with so much variety and such diverse curvatures (Katsaras and Gutberlet, 2001). On the other hand, the thickness of the bilayer is always determined by the dynamic molecular length of monomers. It mostly ranges 3–5 nm.

As with micelle formation, the bilayer is also formed by a *start–stop* process that is governed by force balance. But, for most of bilayer formation, the actual start and stop points are very hard to detect. First, such a low *cmc* makes most of the experimental techniques not quite useful for the bilayer system. Their macroscopic sizes, with so much structural variety, certainly limit the reliability of the measurements, too. This is one of the main reasons why comprehensive thermodynamic theories that would complement the theories of micellization have not been extensively developed.

4.1.3. Physical Properties of Bilayers

As with micelles (Figure 3.4), a bilayer is also a self-assembled dynamic structure. But the dynamics of bilayers are much slower than for micelles (Figure 4.3). The monomer exchange rate with bilayers is on the order of 10^{-3}–10^{2} second, that is, bilayer dynamics are 2–5 orders of magnitude slower than those of micelles. Thus, the lifetime of bilayers is much longer than that of micelles. Micelles experience constant formation/breakup on the order of 10^{-3}–10^{-1} second. Also, the greater attractive force from much a stronger hydrophobic interaction and the nature of its macroscopic size make the bilayer respond much more slowly to any changes in its surrounding conditions. Constant formation/breakup of bilayers, with a timescale comparable to that of micelles, is unlikely. It takes days to years for bilayers to achieve true equilibrium and to be reequilibrated from any structural fluctuation or perturbation. Since the binding and exchange of counterion occur on the surface of self-assembled aggregates, the dynamics of counterion for bilayers are within the same range as for micelles.

Bilayers serve as a basic repeating or building unit for a variety of macroscopic self-assembled structures. Figure 4.4 shows some examples of different types of bilayers. Lamellar liquid crystals are a stacked form of extended bilayer self-assembled unit (the secondary building unit from Figure 1.3) with one-dimensional symmetry. The other two axes of the lamellar structures are dimensionless. Based on the molecular geometry of the monomers (the primary building unit) within bilayers, they can be classified as L_{α}, L_{β}, $L_{\beta'}$, P_{β}, and $P_{\beta'}$. Bicontinuous cubic liquid crystals consist of a regularly curved bilayer unit (the secondary

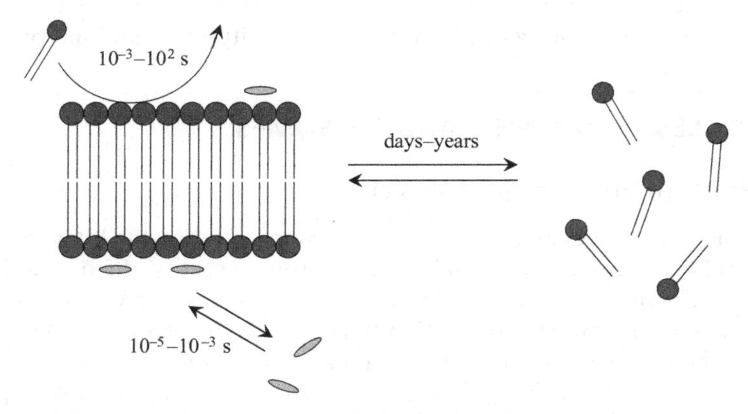

Figure 4.3. Bilayer dynamics are much slower than those of surfactant micelles.

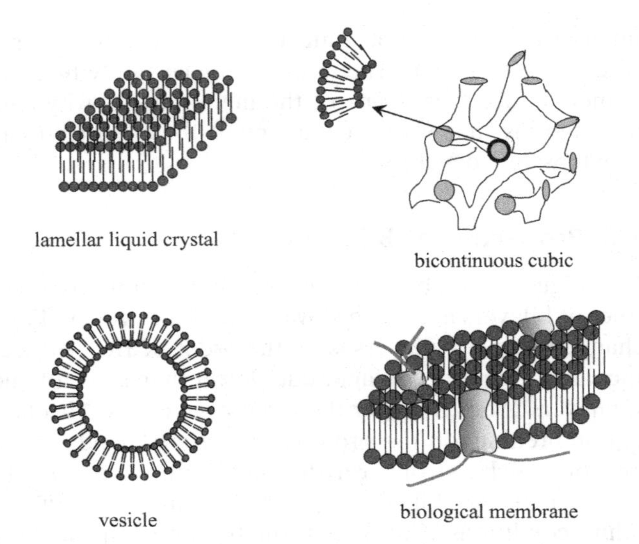

Figure 4.4. Schematic representation of different types of bilayers.

building unit) with clearly defined three-dimensional geometry. Three typical geometries of surfactant bicontinuous cubics are *Ia*3*d*, *Pm*3*n*, and *Im*3*m*. There are some bicontinuous cubics that have indefinitely defined geometry (L_3 or sponge phase). Some of the gels (hydrogel and organogel) are also stabilized by the action of the bilayer unit. A vesicle is a metastable suspension of spherical (globular) enclosed bilayers. A basic structure of biological membranes is a bilayer of various lipids with embedded proteins.

Applications of bilayer systems have long been explored. They include delivery vehicles, catalytic reaction such as self-replicating membranes, surface recognition of biomolecules, nanoscale reactors, model system for biological membranes, and so on. For example, permeability (or diffusion) through model bilayer systems provides an excellent implication of the functions of biological membranes. So does fluidity of bilayers in the activity of membrane proteins.

4.2. VESICLES, LIPOSOMES, AND NIOSOMES

4.2.1. Physical Properties of Vesicles

Suppose the packing geometry of a bilayer is exactly 1; then the bilayer will have an infinite sheet of two-dimensional structure. But, in reality, the bilayer is under a constant pressure of structural fluctuation from a number of factors such as change of environmental condition, thermal diffusion, mechanical instability, and so on. On the other hand, when the packing geometry is close to 1 but slightly different, the bilayer can be curved and eventually form an enclosed spherical shape of self-assembled aggregate (Engberts and Kevelam, 1996). There is the possibility of forming a wavelike infinite structure. But this scenario of enclosed form can eliminate the possibility of unfavorable contact of water-hydrocarbon chains that costs the system more enthalpy, and a much smaller size of self-assembled aggregates than the infinite bilayer will provide additional entropical favorability to the system energetics. Also, the bending modulus that is imposed on the outer layer as a result of this structure can be compensated for by the reduction of modulus in the inner layer. The packing geometry thus becomes slightly less than 1 for the outer layer and slightly more than 1 for the inner layer. There are some literatures that use different nomenclature. But it is safe to accept the term *vesicle* to present this type of enclosed bilayer in general. *Liposome*, which means "fat body" in Greek, refers to vesicles that are composed solely of lipids (both natural and synthetic). *Niosomes* are vesicles formed by solely nonionic surfactants.

The size of vesicles can range ~20 nm–~50 μm in diameter, but the thickness of each single layer is limited to 3–5 nm. Figure 4.5 shows schematic representations of typical types of vesicles. They are usually classified in three groups based on their size and geometry: small unilamellar vesicle (those with radius below ~100 nm), large unilamellar vesicle (those with radius above ~100 nm), and multilamellar vesicle (Rosoff, 1996). A large unilamellar vesicle with different sizes of smaller unilamellar vesicles inside is called a *vesosome*.

Vesicles, in most cases, are thermodynamically unstable but kinetically stable (or metastable) self-assembled aggregates. A vesicle "solution" is a dispersion of those aggregates. Thus, a variety of physicochemical properties of vesicles depend on preparation and post-preparation techniques such as sonication, filtration, extrusion, and so forth. This fact in turn provides a useful means to control the size and shape of vesicles for given applications. Some cases of vesicles that are formed from mixtures of cationic surfactants with anionic surfactants (catanionic) are thermodynamically stable (Iampietro et al., 1998). Thus, the physicochemical properties of those vesicles are determined solely by their intrinsic molecular parameters.

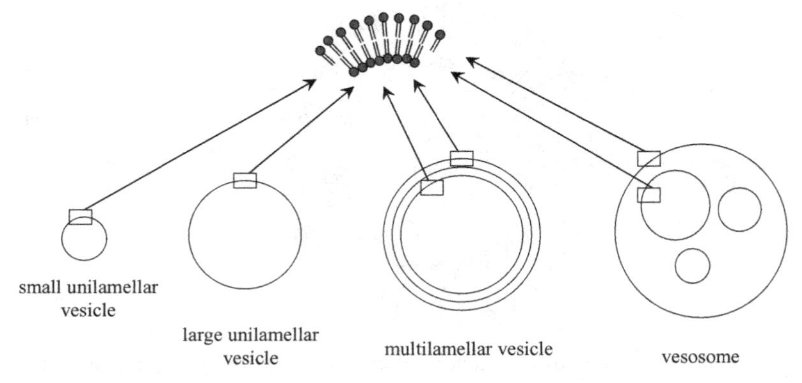

Figure 4.5. Schematic representation of different types of vesicles.

An interesting phenomenon of self-assembly is that there is a counterpart of each of the self-assembled systems. This may be called the *symmetry* of self-assembly. For example, there is a micelle and there is a reverse micelle. There is a liquid crystal and there is a reversed liquid crystal (for hexagonal and bicontinuous cubic). And there is an oil-in-water microemulsion and there is a water-in-oil microemulsion. This type of *mirror* structure (one in a water-rich environment and the other in an oil-rich environment) of self-assembled systems has been identified for many other systems. Indeed, reverse vesicles also have been identified in oil-rich solvent (Kunieda et al., 1991). The unit layer, in this case, is the bilayer whose two head groups contact each other, while the two hydrocarbon chains are apart toward the outside.

4.2.2. Micellar Catalysis on Vesicles

Since the first discovery of the liposome by Alec Bengham over 40 years ago (Bangham and Horne, 1964), its potential applications have been a central issue in many areas (Bangham et al., 1965; Bangham et al., 1973; Lasic, 1993). This section is not intended as an exhaustive review of all these topics. Lasic (1993) provides an especially excellent source for readers who need all the details.

Rather, this section focuses on the underlying issue essential for all applications with vesicles. Two key points for micellar catalysis in Chapter 3 were the enormous energetical and kinetical advantages of counterion binding (Figure 3.6) and micellar solubilization (Figure 3.10). As shown in Figure 4.6, these two processes can be followed for vesicles almost the same way as for micelles. Hydrophobic substances can be solubilized inside the bilayer region of vesicles, and two (inner and outer) surfaces can serve as localized binding sites for counterions. A greater benefit of the structural characteristics of vesicles is the possibility of solubilization of amphiphatic substances that can span through bilayers, and thus can connect the inner and outer sides of aqueous regions.

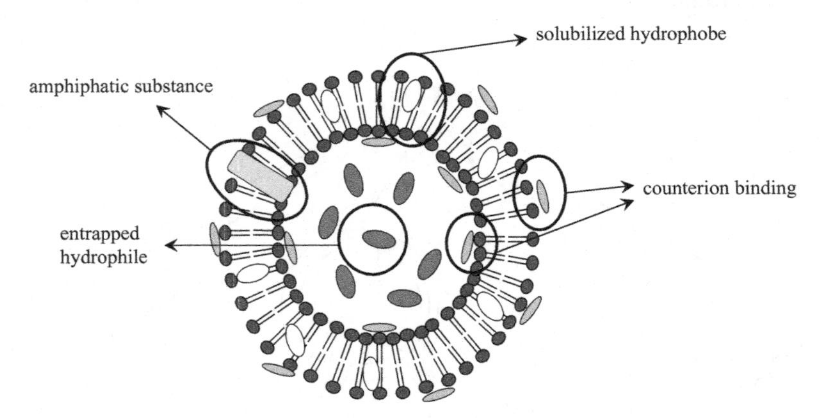

Figure 4.6. Counterion binding, micellar solubilization, and hydrophile entrapment on vesicles.

This type of exercise can provide a route to a biomimic protein channel that might serve as a critical feature for many applications. Also, the inner water pool can hold useful hydrophiles for a significant period of time. Together with amphiphatic substances embedded in bilayers, these are the crucial points for the application of vesicles as controlled delivery vehicles. Control of permeability and diffusion through bilayers is also possible by the control of structural parameters such as size, number of bilayers, molecular composition, monomer structure, and so on. The former two points of micellar solubilization and counterion binding are also basic concepts for applications involved in a variety of reactions using vesicles.

The issue of the origin of life has long been studied through the regeneration of self-replicating vesicles. Production of new amphiphile molecules inside the bilayer region of a vesicle through the process of micellar autocatalysis was demonstrated (Section 3.7.1). This new amphiphile molecule is produced as a new (or additional) component of the vesicles. Like micelles, vesicles can also be used as a structure-directing soft template to produce nano–micro-scale composite or solid materials with well-defined structures (Figure 9.4).

Particularly, vesicles have long been identified as excellent candidates for drug-delivery systems. They can entrap both water-soluble and water-insoluble drugs. They are easy to modify in both their inner and outer surfaces with biologically active ligands. And they can easily hold a variety of physicochemical functional groups. There had been a setback period, especially from the late 1980s to the early 1990s, due to the issue of their vulnerability to the attack of enzyme and immunological systems. But the discovery of the stealth concept of vesicles using the coating of polyethylene glycol has reignited the excitement about them as powerful and selective drug-delivery candidates. The concept of targeting (too minimize the side effect by minimizing the unnecessary amount of drugs) has also been developed. Along with these features, the vesicle itself is biologically benign, nontoxic, and biodegradable. This will be schematically covered with other concepts of delivery vehicles in Chapter 14.

4.3. LIQUID CRYSTALS

All matter exists in a solid, liquid, or gaseous state at ambient temperature and pressure. *Solid* is the state where building units are arranged with long-range positional and orientational orders. This high degree of ordering is achieved by strong bonds such as ionic, metallic, or covalent bonds. Building units of *liquid* are arranged with only short-range order and do not have a long-range correlation as in the solid state. This low degree of ordering is the result of interaction by weak intermolecular forces such as hydrogen, electrostatic, hydrophobic, or van der Waals forces. For the *gas* state, building units show no significant geometrical correlation and have no significant interaction between them.

However, there are groups of organic compounds that show both crystalline solid and isotropic liquid characteristics at ambient temperature and pressure.

They show some degree of geometrical ordering but with some degree of molecular mobility. This state is called the *liquid crystal* or *liquid crystalline* state. Liquid crystals are thermodynamically stable and their phases are also called *mesophase*. No strong bonds such as are involved in the solid state are responsible for the liquid crystalline state. Weak intermolecular forces are the only interaction to retain this state. Their macroscopic appearance often looks like a gellish solid but with some degree of fluidity, and they often show crystalline solid-like properties such as optical anisotropy.

Based on the origin of intermolecular interactions that induce the ordering, liquid crystals are classified as thermotropic and lyotropic liquid crystals. Thermotropic liquid crystals are induced by temperature changes, while lyotropic liquid crystals are formed by the change of concentrations. Thus, lyotropic liquid crystals are always present with solvent.

4.3.1. Thermotropic Liquid Crystals

Thermotropic liquid crystals can be formed from single- or multicomponent systems (Gennes and Prost, 1993). But no solvent is involved in their formation process. In this sense, the formation of thermotropic liquid crystals is not exactly the self-assembly process in solution, as the title of this chapter claims. However, their thermodynamic properties are essentially the same as for lyotropic liquid crystals and many of their applications come from the same principles.

Typically, thermotropic liquid crystals are formed as the temperature changes between the crystalline solid state and isotropic liquid state (Figure 4.7). Compounds that exhibit liquid crystal during this process are called *mesogene*. The rigid rodlike mesogene (calamitic) is the most common one. But the rigid disklike one (discotic) is also well known. 5CB (4-cyano-4′-pentyl-1,1′-biphenyl) and 8CB (4-cyano-4′-octyl-1,1′-biphenyl) are among the most commonly used calamitics. TC 3 (*trans*-4-propyl-cyclohexanecarboxylic acid-4-ethoxyphenyl ester mixture

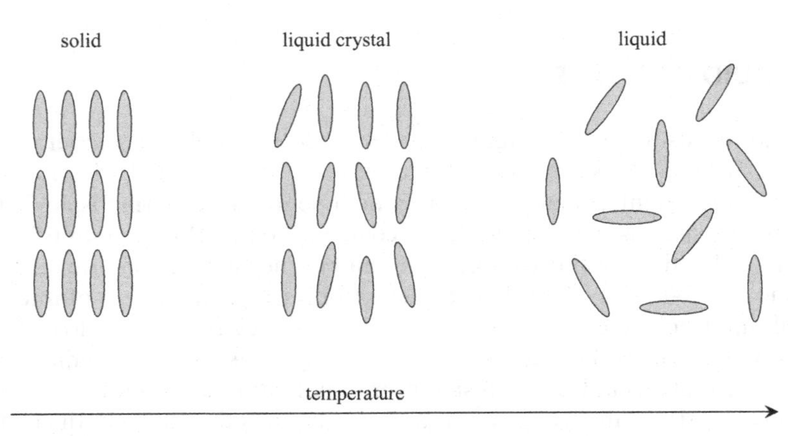

Figure 4.7. Pictorial diagram of the change of state through liquid crystal states.

with 1-ethoxy-4-(*trans*-4-propylcyclohexyl) benzene and *trans*-4-pentylphenyl-4-propylcyclohexanecarboxylate) is a typical example of a multicomponent system that forms thermotropic liquid crystal. There are approximately 250 liquid crystal–forming compounds and compound mixtures registered in the CAS (Chemical Abstracts Service) registry file as of early 2005. Also, polymers, colloidal particles, and even viruses that have anisotropic geometry (clay platelet, tobacco mosaic virus, etc.) can form liquid crystalline phases.

Figure 4.8 shows the different phases of liquid crystals. Smectic liquid crystals are the phases whose molecules are arranged with both long-range macroscopic

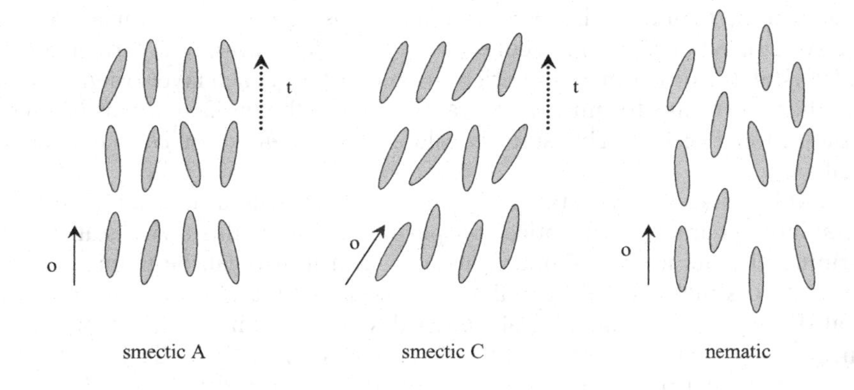

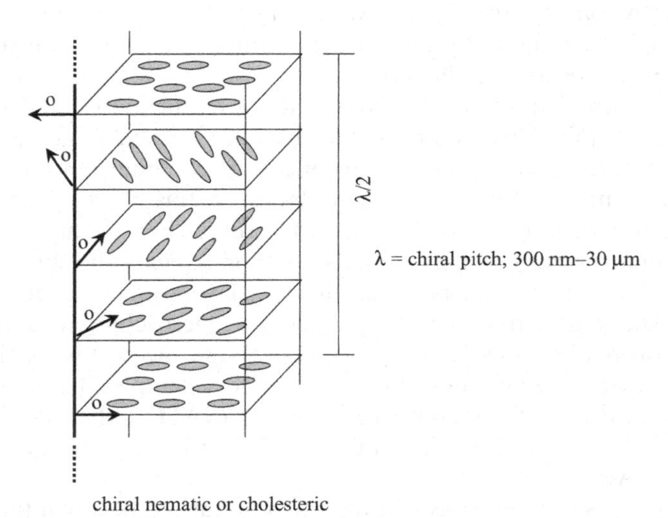

Figure 4.8. Schematic representation of different types of liquid crystal states.

orientational order (the one of molecule) and long-range macroscopic translational order (layering). When these two directions are the same, it is called the *smectic A* phase. There are phases whose two directions are not the same. These are called *smectic C*. Smectic C is a representative name. Depending on the degree of the tilt of molecules (with respect to the layer plane), smectic F, smectic G, and so forth are also known. More than 12 different smectic phases have been identified.

When smectic liquid crystals are further heated or isotropic liquids are cooled, they often lose their translational order but still possess their orientational order. This state is called *nematic* liquid crystal. This is the most common type of liquid crystal. Nonchiral (achiral) molecules form nematic liquid crystals as shown in the figure. Their orientational order is homogeneous throughout the entire system. When chiral molecules (but not a racemic mixture) form a nematic liquid crystal, the direction of the orientational order of each layer orients slightly each other layer, thus forming the spiral twisting of the molecular axis (direction of orientational order). This state is called *chiral nematic* or *cholesteric* liquid crystal.

Liquid crystals always retain a certain degree of structural order, so they can always show characteristic optical properties. At first, this structural order is determined by the nature of unit molecules. But it also can be manipulated by external forces including electric field, magnetic field, and temperature. Liquid crystal display (LCD) and optical switch devices are among the most popular applications of liquid crystals. Both of their basic principles lie in the use of changes of optical property that are controlled by an external electric field. A liquid crystal thermometer is a device that uses the changes of color induced by changes of temperature. Another example can be found from the chiral pitch (length for 360° rotation of the spiral twisting) for cholesteric liquid crystals, which usually spans 300 nm–30 μm. When the pitch matches with the wavelength of visible light, the liquid crystal can show dramatic changes of color, which can open up diverse optical applications.

Thermotropic liquid crystal is one state of matter between crystalline solid and isotropic liquid. But, as for most of the self-assembled aggregates, its structure is maintained by a very delicate balance between intermolecular forces. And it is not uncommon that even a slight change in this balance can result in the reconstruction of entire liquid crystal structures. For example, anchoring of common liquid crystals such as 5CB and 8CB on typical solid surfaces is strongly affected by the interaction between the first layers of liquid crystals with the surfaces. This interaction can be changed by macroscopic intrinsic properties such as hydrophobicity and hydrophilicity of the surfaces. It is also influenced by the molecular-level properties of the surface, for example, the nature of the terminal group of the self-assembled monolayer (SAM) (when it is used as substrate) and even the carbon number (and odd–even effect) of hydrocarbon chains within the SAM.

The changes of intermolecular interaction induced between this interface of the first layer of liquid crystals and the substrates can induce the structural change of entire liquid crystals (Abbott, 1997). The change that occurs on one edge of the liquid crystals obviously propagates to the whole system. This is also one characteristic phenomenon of typical self-assembled systems. There are other examples of this *cooperativity* of liquid crystal systems. Addition of small amounts of chiral "impurities" can sometimes change a whole achiral nematic state into a chiral state. It is also possible that an achiral state is formed by chiral molecules, and vice versa. These types of systems often provide ferroelectric and antiferroelectric properties, which have potential for many applications on nanoscale devices. This issue of "chirality in self-assembly" will be revisited in Chapter 6 with a molecular chirality induced by self-assembly at surfaces.

4.3.2. Lyotropic Liquid Crystals

The liquid crystalline state that is formed by the concentration change of its building unit is called *lyotropic* liquid crystal. The most popular ones are those found in surfactant solutions. Similar states are also found in solutions of amphiphilic polymers. Unlike in cases of thermotropic liquid crystals, the formation of lyotropic liquid crystals is, for the most part, the result of multistep self-assembly of primary building units. Some of the details have been discussed in previous chapters (Section 1.3, *Type II* and Section 3.4). This section will extend those discussions in the context of the general phase diagram of surfactant solutions. Surfactant liquid crystals have more variety of structural diversity than thermotropic liquid crystals and usually exist in equilibrium with monomers. They are often in equilibrium with other self-assembled aggregates, including normal micelles, reverse micelles, vesicles, and microemulsions. Structural transformations between surfactant liquid crystals and these other aggregates are not uncommon. Like other phase diagrams, the phase diagrams of surfactant solutions are also strongly system dependent (Tiddy, 1980; Fontell, 1990).

Their detailed pictures are delicately changed according to the individual system and conditions. However, the picture of the general phase diagram of surfactant solution that covers the entire range will certainly help provide a systematic understanding of surfactant self-assembled systems. Two general phase diagrams will be introduced here: the binary temperature-concentration phase diagram and the ternary one with oil (or co-surfactant). Also, this discussion will help in understanding the complicated phase issues raised during the application of surfactant self-assembled systems in various nanotechnology areas: synthesis of nanostructured materials and nanoparticles, drug-delivery systems, external field-induced changes in nanosystems, and so on.

4.3.2.1. Concentration-Temperature Phase Diagram. Figure 4.9 shows the general phase diagram of aqueous surfactant solutions. Surfactant concentration covers 0–100 wt. %, and temperature ranges –20–300 °C.

For most of the ionic surfactants, there exists a clear temperature boundary throughout the whole concentration range. Below this boundary, surfactant solution is in the form of hydrated crystal equilibrated with its monomer or in crystal

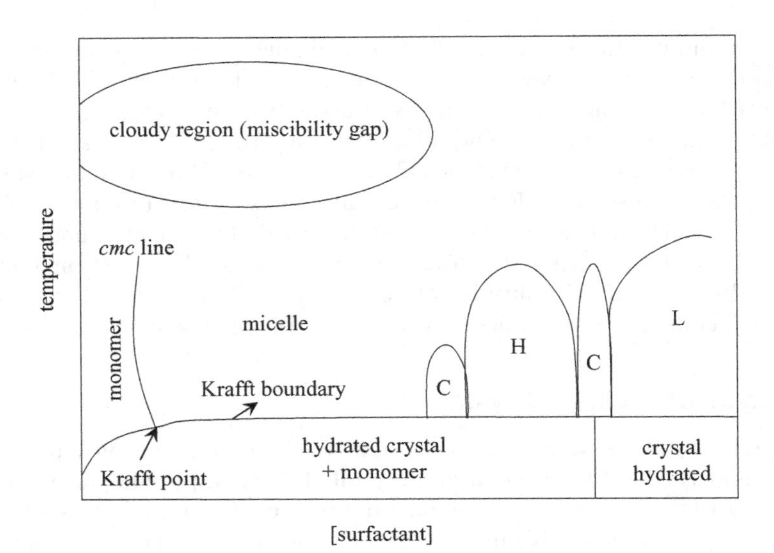

Figure 4.9. General concentration-temperature phase diagram of surfactant solutions.

hydrated form at high concentration. Above this boundary, surfactant solutions are clear and homogeneous. This distinctive temperature boundary of ionic surfactant solutions is called the *Krafft boundary*. The *Krafft point* is the point where the Krafft boundary intersects with the *cmc* line.

The upper limit of the *cmc* line is different, depending on the surfactants. It is also varied by the pressure of the system. Many ionic surfactant solutions show minima on this *cmc* line, usually at 20–40 °C. This is the result of the entropy–enthalpy compensation of micellization (Section 3.3.3). The micelle region includes spherical-, rod-, and wormlike micelles. The wormlike micelle region is often divided into dilute and semidilute regimes. This is based on its geometrical property. When the average dynamic distance between the micelles is longer than their average dynamic length, it is defined as a *dilute regime*. When the distance is shorter than the micelle length, the solution is defined as a *semidilute regime*. Entanglement, branching, and micelle chain exchange (through a mechanism such as scissor-recombination) are common in the semidilute regime. And this structural characteristic is one origin of the viscoelasticity of surfactant solutions in this regime (Cates and Candau, 1990).

For almost all types of surfactants, liquid crystal regions appear at their high concentrations after the semidilute regime. Three typical structures are hexagonal, cubic, and lamellar (see the schematic illustration in Figure 3.7). They are also called *middle phase, viscous isotropic phase*, and *neat phase*, respectively, based on their historical applications (middle and neat phases) and macroscopic properties (viscous isotropic phase). In view of ordering, these lyotropic liquid crystal structures can be understood as a smectic phase (ordering or alignment) of surfactant self-assembled aggregates that are anisotropic and finite sized. All known cases show that liquid crystals appear on the order of hexagonal, cubic, and lamellar with the increase of surfactant concentration. Hexagonal and lamellar always take major regions while the cubic region is relatively small. Depending on the surfactants, another cubic region can exist between semidilute and hexagonal regions. This early cubic region is a non-bicontinuous one. It is formed by arrangement of finite-sized micelles such as short rodlike micelles into cubic geometry. The cubic region between hexagonal and lamellar is always bicontinuous. Some systems also have intermediate phases between the three major phases. This includes tetragonal, orthorhombic, rhombohedral, and so on. But their regions are usually narrow in their phase diagrams, and not always easy to detect.

EXERCISE 4.1

From equation (1.4), show that the concept of surfactant packing parameter can be directly applied to the structural change of surfactant liquid crystals. Also, propose appropriate g-values for each liquid crystal geometry, including hexagonal, cubic, lamellar, and tetragonal.

Solution: Based on the basic micellization process, liquid crystals can be modeled to be formed by the intermicellar interaction between primary self-aggregates such as spherical, rod-shaped, and wormlike micelles. Also, thermodynamics predicts that most of those primary self-aggregates coexist with the liquid crystals formed. Thus, equation (1.4) should be valid to express this situation:

$$U_{total}(x) = f_P \cdot \lfloor U_{A,P}(x) + U_{R,P}(x) \rfloor + f_S \cdot \lfloor U_{A,S}(x) + U_{R,S}(x) \rfloor$$

Most of the liquid crystals of ionic surfactants are formed at the concentration range of ~20—~40 wt. % of surfactant, while the *cmc* usually is below ~1 wt. %. For nonionic surfactants, liquid crystals are formed at ~5—~20 wt. %, but the *cmc* is also much decreased below 0.1 wt. %. Thus, the contribution of the primary self-assembly process to $U_{total}(x)$ can be ignored and f_S can be set as unity. This yields

$$U_{total}(x) \cong U_{A,S}(x) + U_{R,S}(x)$$

As can be seen in Figure 3.7, the surface of all liquid crystals is contacted with water, while all hydrocarbon chains are embedded into the core region. As in the case of micelles, we can consider that the major attractive force arises from hydrophobic interaction, while the repulsive components can be electrostatic, hydration, or steric interaction. Thus, $U_{A,S}(x)$ can be replaced with $\gamma \cdot a$, where a is the area occupied by each monomer on the surface of liquid crystals and γ is the free energy at the oil–water interface. Similarly, $U_{R,S}(x)$ can be replaced with K/a, where K is a constant. This yields the exact same form of relation as for the micelles: equation (3.16). Estimation of a_o, and of v and l can follow the same track. Thus, the same concept of g-value can be used for the surfactant liquid crystals.

Since hexagonal and tetragonal consist of the same building unit of rod-shaped micelles with different two-dimensional lattices, the g-value for these geometries should be close to 1/2. Building units of cubics are curved planar sheets; thus g should be close to 2/3. Lamellar should have g of ~1.

For most of the aqueous nonionic surfactant solutions, there is a region in the temperature-concentration phase diagram where the clear homogeneous solutions become heterogeneously cloudy. This region usually appears at a high temperature of 50–300 °C with the concentrations usually below those of liquid crystal regions. This region is called the *cloudy region* or *miscibility gap*. The temperature at which this cloudy phase begins to appear at a defined concentration is the *cloudy point*.

Typical nonionic surfactants that show this cloudy phenomenon are those with an oxyethylene chain attached to long hydrocarbon chains. Oxyethylene chains are the head groups. When temperature is increased in their aqueous solutions, the oxyethylene group becomes less hydrated, so the degree of hydration of water begins to decrease. This means that the repulsive hydration force that balances the attractive hydrophobic force for the micelles (spherical, rodlike, or wormlike) decreases. Thus, the force balance inside the micelles shifts to less repulsive, and the interaction between the micelles becomes less repulsive, too. This provides enough driving force for the growth of individual micelles and at the same time to fuse them together. The solution begins to look cloudy when the size of this micelle reaches a certain point, with two phases of micelle-rich and micelle-deficient that are in equilibrium (emulsion of these two). Also, less water means the decrease of the a_o value, which means the increase of the g value, which is favorable for micelle growth. The system also gains entropy contribution during this process: (1) from water molecules released from oxyethylene chains and (2) from more flexible oxyethylene chains at high temperature.

The general trend of surfactant self-assembly can be reviewed in this phase diagram. Based on the concept of force balance, when this balance is shifted toward a more attractive force, such as in the case of double chains or longer hydrocarbon chains, the balance shift induces more favorability for the self-assembly. This results in a general shift of the liquid crystalline phases toward lower concentrations. When a repulsive force is dominant, the shift is toward higher concentrations. Nonionic or zwitterionic surfactants usually have their liquid crystalline phases at lower concentrations than ionic ones. This also can be understood as the result of the decreased electrostatic repulsive force between the head groups at the micelle surface.

4.3.2.2. Ternary Surfactant–Water–Oil (or Co-surfactant) Phase Diagram. When the third component is introduced into aqueous surfactant solutions, the self-assembled surfactant systems show much more rich (and more complicated) structural and phase behaviors. This third component can be either oil (simple hydrocarbon) or co-surfactant (another amphiphile). The concepts of micellar solubilization (Figure 3.10) and formation of mixed micelle (Figure 3.9) are critical to understanding those structural and phase changes. Figure 4.10 shows the general phase diagram of the water–surfactant–oil (or co-surfactant) ternary system.

The line of water–surfactant is the phase diagram of Figure 4.9 at a given temperature. When the amount of oil or co-surfactant that is a part of the surfactant micelles or surfactant liquid crystals is relatively small, the self-assembled systems can experience some degree of swelling, which can cause some shift of critical concentrations (*cmc*, the concentration where each liquid crystal first appears, etc.). But the general phase sequence in Figure 4.9 does not change significantly. Thus, normal micellar solutions are positioned in the left corner of the ternary phase diagram, while the right corner is taken by the normal liquid crystals. As the content of oil becomes high, water becomes the minor solvent, and oil becomes the dominant solvent. This condition does not always guarantee the formation of reverse-shaped self-assembled aggregates. But when the repulsive force on the micelle surface is suppressed enough (e.g., the decrease of hydration force or the decrease of electrostatic force by co-surfactant) and/or the attractive force is strong enough (e.g., by the action of solvent that is hydrophobic), the formation of reverse micelles and reverse liquid crystals becomes more favorable. This makes the top corner the region where many of those reverse-shaped self-assembled aggregates are found in many ternary phase diagrams. Most of the microemulsion is found near the line of water–oil with relatively small amounts of surfactants. More about microemulsion will be covered in the next section. The rest of the area, especially the middle area, is quite delicate and complicated. A variety of different structures and phases can appear in

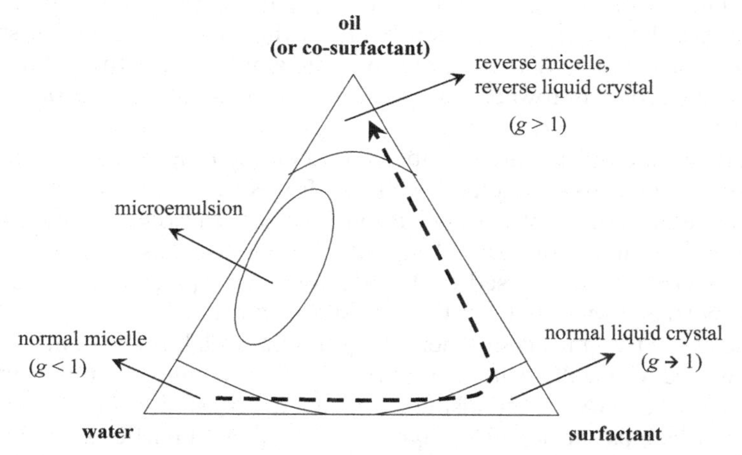

Figure 4.10. General ternary-phase diagram of surfactants. Co-surfactant here is not a co-emulsifier.

this region, which makes it almost impossible to draw the general trend of ternary phase behavior.

Of course, the individual phase diagram of each surfactant ternary system will show a wide range of differences. This is also strongly dependent on a variety of chemical and physical parameters and experimental conditions such as temperature, pH, salinity, molecular structure of each component, and so on. However, the schematics of this general ternary-phase diagram will help, with Figure 4.9, catch the general view of surfactant self-assembled systems. Also, it becomes useful to correlate with issues for application to nanotechnology. For example, the bold dashed line in Figure 4.10 represents the direction of the increase in *g*-value from spherical micelles to reverse micelles. By following this general guidance, the search for the specific micellar or liquid crystal conditions can be made easy. Instead of looking for the entire region, only a small portion of the region can be narrowed and experimented. Design of mesoporous materials using liquid crystals (Figure 9.9) and of macroporous materials (Figure 9.14) using microemulsion will be much more sophisticated by following this logic of surfactant self-assembly.

4.4. EMULSIONS

When two immiscible liquids are mixed together, the common sense of our experience is that they are not mixed up together, and soon after mechanical mixing is stopped, they will be separated into two clear phases. Mostly likely, the heavier water is on the bottom, and the lighter-density oil is on the top phase. This thermodynamic process can be dramatically altered when amphiphiles are introduced into this situation. The mixture of two immiscible liquids (usually oil and water) can be stabilized by this third component. The amphiphiles are self-assembled into protective layers on the interface of the two liquids, and dramatically decrease the interfacial tension between the two liquids, which stabilizes the water or oil droplets. The head group of the amphiphile is toward water side, while the tail group is toward the oil phase. This is called *emulsion* (Petsev, 2004).

Two types of emulsion are possible, based on the ratio of water to oil phase. When water is in a disperse phase while oil forms the droplets, it is defined as oil-in-water emulsion (O/W). Water-in-oil (W/O) is the reverse situation, with water droplets in the oil-disperse phase. Emulsion is also classified as microemulsion and macroemulsion, based on the size of the droplets and the thermodynamic properties. Both can form O/W or W/O emulsion.

To be clear, the third component, amphiphile, which is active at the liquid–liquid interface, is usually called *emulsifier*. For microemulsion, only surfactants are known to be active as emulsifier. For macroemulsion, a greater variety of components such as surfactants, polymers, and colloidal particles (or nanoparticle) can be active as emulsifiers. For many cases of macroemulsion, a fourth important component called *co-emulsifier* is needed to achieve satisfactory stability.

Supercritical fluid such as supercritical CO_2 has an ability to be in an oil phase with water to form an emulsion. Due to its easy-to-handle property and environmental benignity, it becomes a useful oil-phase substitute for many applications.

4.4.1. Microemulsions

Microemulsion is a thermodynamically stable, transparent, low-viscous, and isotropic emulsion. Its first appearance in the literature was almost seven decades ago (Hoar and Schulman, 1943). Since then, it has been widely used for a variety of applications such as consumer products, pharmaceuticals, agricultural products, and so on. And its use for the controllable confined media for the synthesis of nanostructure materials and particles has renewed its importance in the area of nanotechnology.

Figure 4.11 shows the schematic representation of the formation of emulsions. When a small amount of oil is introduced into normal micellar solution, most likely a swollen micelle is first formed as the oil (or hydrophobic molecule) is solubilized inside the normal micelle. This is a typical micellar solubilization,

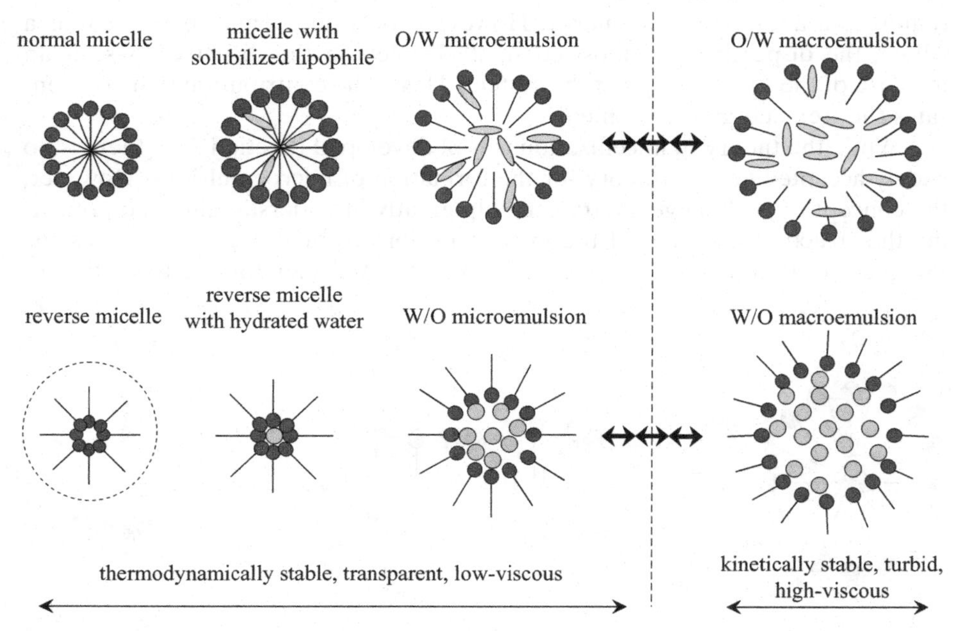

Figure 4.11. Schematic representation of the relation of micelles, microemulsions, and macroemulsions.

as discussed in Chapter 3. As more oil is solubilized, the micelle swells more, and oil droplets begin to be formed inside the micelle. This oil droplet stays in almost a pure oil phase, but not as the mixture of oil and hydrocarbon chain of surfactant. This state thus can be considered as the oil droplets emulsified (or stabilized) by surfactant layers in the water phase. This is the O/W type of microemulsion. The point of the amount of oil that can tell us where it is a swollen micelle or it is O/W microemulsion cannot be clear cut, and largely depends on each system. W/O-type microemulsion is its reverse version. This will be discussed in the next subsection on reverse micelles (Bourrel and Schechter, 1988).

For both types of microemulsions, the size of oil or water droplet usually ranges 5–25 nm. They show some degree of polydispersity that is comparable to that of typical normal micelles. They are also very dynamic objects. They continuously collide and coalesce with each other, and continuously break apart (Zana and Lang, 1987). For example, the interchange rate of water droplets in W/O microemulsion is very high with $k = 10^6$–$10^7\,M^{-1}s^{-1}$, where M is the concentration of water droplets in the system.

To be formed spontaneously, the free energy of formation of microemulsion should be negative. The oil–water interfacial tension is very low, that is, usually 10^{-2}–10^{-4} mN/m, but not zero or negative. Thus, there is a free energy requirement (energy needed) to create the high interfacial area. For example, 5–20 nm droplet diameter can have 1200–300 m^2/g of interfacial area (with the assumption of density of 1). Also, this is not a planar, but a highly curved surface, which requires an additional free energy. However, for a microemulsion system as a whole, the dispersing of nanosized small droplets in the disperse phase is an increase of the total entropy of the system. This is the contribution that compensates the free energy requirement.

While the theory of micellization is well developed (Section 3.3), there is no widely accepted general theory for the formation of microemulsions. However, the concept of packing geometry can help greatly in understanding and predicting the structure and type of the microemulsions formed. Figure 4.12 presents this picture. It shows the change of g-value of emulsifier for the formation of

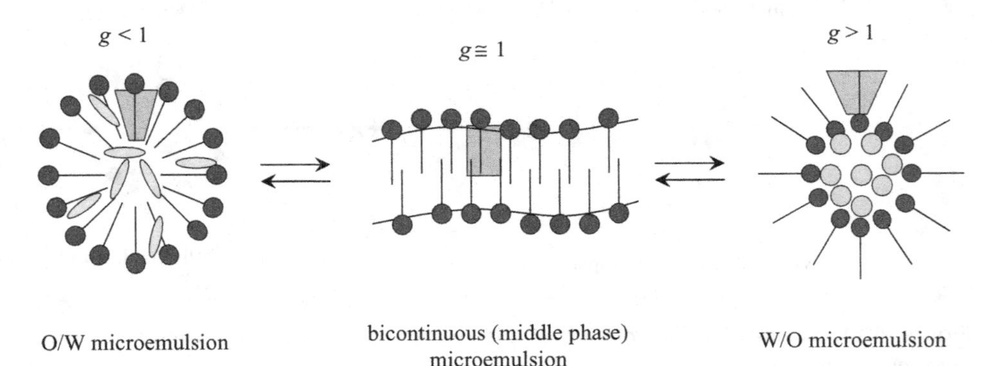

Figure 4.12. Consideration of packing geometry for the formation of microemulsions.

W/O, O/W, and bicontinuous (sometimes called middle-phase) structures. When the *g*-value is smaller than 1, the positive curvature of the interface is energetically favorable for the O/W type. When it is greater than 1, the W/O type can be the most favorable form. When it is close to 1, the low curvature of interface can make it stretch infinitely throughout the system. This is *bicontinuous microemulsion*. For this system, oil and water phases are interconnected with each other.

Both lamellar liquid crystal and this bicontinuous microemulsion have *g*-value close to 1. When there is a strong attractive or repulsive interlayer force between surfactant films (or layers), the layers can be arranged with a regular spacing with ordered geometry. Free energy of the system can be minimized this way. However, when the force is not strong enough, the system can achieve the minimum free energy by gaining entropy contribution by forming a disordered bicontinuous structure. This is the case for most of the nonionic surfactants or ionic surfactants with high salt concentration, and can be the bicontinuous type (of monolayer film), just like bicontinuous cubics of bilayers.

This simple exercise can provide, at least semiquantitatively, the facile design of microemulsion by the design of *g*-value of emulsifier. Just like the design of micellar structures (Section 3.4), all the basic concepts are also valid here. It can be done by the design of the molecular structure of the emulsifier itself, by the control of counterion binding (for ionic surfactant), by the change of temperature (control of hydration for nonionic surfactant), and by the design of micellar solubilization and cosurfactant (mixed micelle). This understanding of the structural control of microemulsion provides an important starting point for a variety of applications. For example, for the formulation of consumer products, the additional additive that is needed can be correlated with these factors to predict/tune the properties of the final products. Also, on the nanotechnology side, for the preparation of nanoparticles using microemulsion as a confined reaction medium, these factors play an important role in predicting the final outcome, which is strongly correlated with the interaction of reactant (or reaction precursor) with those components of microemulsion.

In many cases, microemulsions are in equilibrium with normal micelle solution or with reverse micelle solution in their phase diagrams (Figure 4.10). This leads typical ternary systems (water/oil/surfactant) that have microemulsion phases to show some characteristic features in their phase diagrams, such as the famous *fish-tail*. We will not explore the details of this phase behavior in this chapter. But understanding this is particularly important, not just for emulsion technology itself but for its applications as well (Solans and Kunieda, 1997).

4.4.2. Reverse Micelles

The distinction between a reverse micelle and a W/O microemulsion is not clear cut. There is no clear consensus among the scientific community to define the difference between the two. Also, there are no experimental techniques currently available that have enough resolution to solve this issue. This may be the reason why so many literature sources just do not distinguish these two or, in many cases, cause confusion by simply using two terms together. In view of the *symmetry* of self-assembled systems that we discussed in the previous section (Section 4.2.1), this issue can be reasonably approached by one-by-one comparison with the normal micelle and the O/W microemulsion (Figure 4.11). The ideal reverse micelle should be composed of the surfactant molecules that are positioned in reverse compared with those of normal micelles. This means that the surfactant molecules should be thermodynamically stable without the aid of water molecules around the head groups. This picture cannot be relevant since it will generate too strong repulsive forces between the head groups. Thus, to have thermodynamically stable reverse micelles that can be counterparts of normal micelles, there should be at least a minimum number of hydrated water molecules that can balance the repulsive force with the attractive hydrophobic force. This picture should be defined as reverse micelles. When the content of water molecules increases, the core region of reverse micelles can hold additional water molecules that are nonhydrated. This should be defined as W/O microemulsion, as long as it stays thermodynamically stable and transparent. This nonhydrated water droplet thus can be called a *confined* or *nanoscale* medium, as much of the literature and many applications refer to it.

Figure 4.13 shows the formation scheme of reverse micelles. It is a stepwise process and has no clear *cmc*. Like normal micelles, reverse micelles are also a dynamic object. There is a constant exchange of monomer with reverse micelles. Also, there is the constant breakup and reformation of entire structures. The timescales of these processes are generally similar to those of normal micelles.

Microemulsions are formed entirely by the interaction of weak intermolecular forces. Any shifts in their delicate force balance can dramatically change the entire structure and properties of the systems. Thus, their phase diagrams are very complicated and can contain regions that have different structures or phases, such as normal micelles, reverse micelles, liquid crystals (lamellar), and so on. This may be the reason why such a great variety of nanostructures are obtained when they are synthesized by using microemulsions. For example, only a small change of reaction conditions for nanoparticle synthesis can result in rod-shaped, needle-shaped, spherical-shaped, and many others. This in turn makes us produce a greater variety of structures, such as asymmetric nanoparticles and branched nanoparticles by control of these factors. Details will be found in Chapter 10.

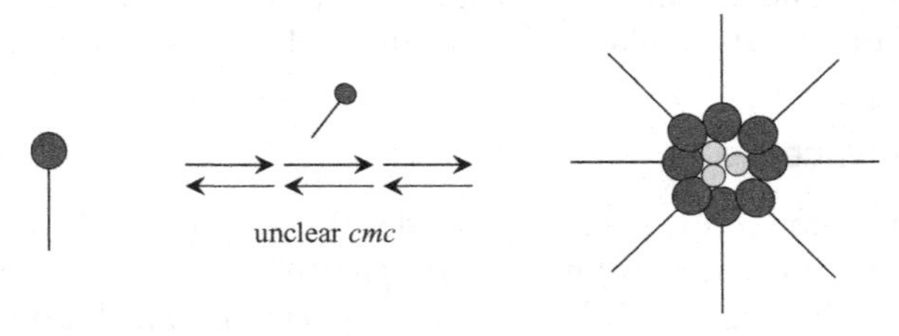

Figure 4.13. Formation of reverse micelles.

4.4.3. Macroemulsions

Macroemulsion is a thermodynamically unstable, turbid, and highly viscous system. Or, it can be stated as a kinetically stable or metastable dispersion of colloidal size of one liquid droplet in the second liquid stabilized by emulsifier. Figure 4.11 shows a schematic comparison with microemulsion. Like microemulsions, macroemulsions can have water-in-oil (W/O) and oil-in-water (O/W) types of structures. Supercritical fluid can be an oil phase, too. Depending on the systems and conditions, a double emulsion (water-in-oil-in-water—W/O/W) can be also structured. No stable (kinetically) bicontinuous-type of macroemulsion has been reported. This may be the result of subtle stability of macroemulsions that can be easily transformed into either O/W or W/O type from their possibly short-lived bicontinuous macroemulsions.

Stable macroemulsion can last for years, but eventually all macroemulsions will be separated into two (oil and water, with either or both phases with emulsifiers) or more phases (oil, water, and phases with mainly emulsifiers). Formation of macroemulsions is not a spontaneous process. It usually needs the input of mechanical or chemical energy. Thus, the determination of O/W or W/O type depends on both physicochemical and external parameters. The former can include temperature and composition, while the latter includes kinds of energy applied such as stirring, the way energy is applied, the amount of time applied, and even the mixing order of ingredients.

Formation kinetics and the stability of the self-assembled film of emulsifier (mixed monolayer usually) at the oil–water interface are the keys to the stability of macroemulsion. The packing geometry of surfactant or emulsifier does not play a significant role in this process. What determines O/W or W/O type is the competition between two opposing processes: stabilization of droplets (oil or water) by emulsifier and coalescence. Whatever the reason (internal or external) is, a phase that is more stabilized becomes a dispersed phase and the other becomes a continuous phase. The key phenomenon for this stabilization is the Marangoni effect (Gibbs-Marangoni effect for surfactants). Figure 4.14 shows the schematic representation. When a liquid droplet begins to be formed within another liquid medium (e.g., by stirring), the emulsifiers that may be at the state of micelles and/or monomers (from either or both phases) also begin to be adsorbed on the newly created oil–water interface. Since the diffusion of emulsifiers and the rate of droplet creation are comparable each other, a region where the density of emulsifier is high on the oil–water interface can be formed. As a result, a density gradient of emulsifier is created on the interface, which means the gradient of interfacial tension along the interface. Thus, the emulsifier molecules can move from the high-density region to the low-density region to compensate the interfacial tension difference. This process is faster than the adsorption of emulsifier from liquid phases, which makes it critical for the stabilization of macroemulsions.

Again, all macroemulsions are eventually separated into two or more phases: destabilized. This process is involved in a sequence of flocculation/creaming,

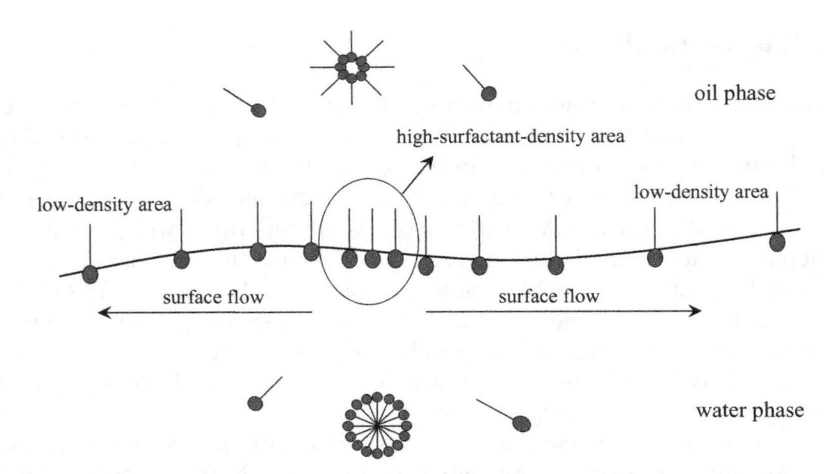

Figure 4.14. Formation of macroemulsion: Marangoni or Gibbs-Marangoni effect.

coagulation, coalescence, and phase separation. Let us understand these conventionally defined processes with the concept of force balance and the effect of external force. Droplets of macroemulsions are stabilized because there is a significant repulsive force (mainly electrostatic for O/W) between them. Also, the nature of the droplet surface bears attractive forces that are mainly van der Waals force. Thus, DLVO-type of force relation can be applied. First, the droplets can be drawn into the secondary minimum of the force curve by the attractive force, so the system can reduce the free energy. This is the same scenario as for the hard colloidal suspensions, and can be called *flocculation*. Since the size of a macroemulsion droplet is relatively big (thus, relatively heavy), external forces can affect the force balance of the systems (without any others, gravity becomes an important factor). It can bring the droplets to either the top or bottom of the container. This is *creaming*. Once the droplets are brought closely together either by flocculation or creaming (or both), they can be positioned even closer to the primary minimum of DLVO curve. This is *coagulation*.

As shown in Chapter 2, further process for shortening the distance is impossible for the hard colloidal suspensions due to the dramatic increase of the repulsive force. But, for the macroemulsion droplets, this process can combine/fuse the two approaching droplets. It is possible, in most cases, by molecular diffusion that is caused by the free energy difference between droplets with different sizes, which is caused by the difference of the surface curvature. This is *Ostwald ripening*. It is also possible by the direct fusion of the surfactant layers. This is the *coalescence* process. Thus, the system is finally separated into different phases.

This concept of force balance is also responsible for the long-term stability of W/O/W double emulsion. The stability of this emulsion is strongly affected by the interaction between the internal aqueous droplet and the external aqueous phase across the oil phase of the middle of the emulsion droplet. This is achieved by the force balance between the electric double-layer force and van der Waals force (Wen et al., 2004).

For microemulsion, surfactant monolayer is the layer protecting the water or oil droplets. But for macroemulsion, surfactant monolayer is not the only possible self-assembled layer. Other ones include surfactant multilayer such as lamellar-type, amphiphile monolayer and multilayer, layer of globular proteins, colloidal particles such as surface-modified spherical silica or fat crystals, polymer chains, and so on. As long as they have an ability to be located (adsorbed) at the oil–water interface and reduce the interfacial tension below the surface tensions of both oil and water phases, they can become effective emulsifiers (Aveyard et al., 2003).

4.4.4. Micellar Catalysis on Microemulsions

Micellar catalysis has been discussed in Chapter 3. Three main concepts for this process were micellar solubilization, counterion binding, and mixed self-assembly with co-surfactant. These concepts can be directly extended to the self-assembled monolayer for the formation of microemulsions. Figure 4.15 shows the schematic representation. All three can have significant influence on the packing geometry of the monolayer, and on the structures of microemulsions. Also, they can be applied to control the fine properties of microemulsions, such as viscosity, stability, turbidity, detergency, control of content of foreign component, and so on.

For emulsion systems, two types of micellar catalytic reactions are possible. With the W/O type, the small water droplet provides a well-confined nanosized "reactor." The controllability of the droplet sizes, broad spectrum of emulsion systems available, and their well-established mechanistic studies make this system extremely useful to prepare a variety of nanosized particles (Figure 4.16a). For example, the size of water droplet of an AOT-based W/O microemulsion system can be easily tuned by simply changing the mass ratio, [water]/[AOT].

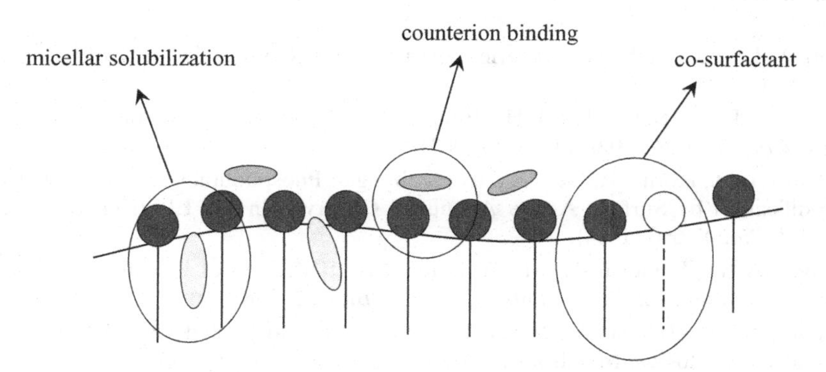

Figure 4.15. Concepts of counterion binding, micellar solubilization, and co-surfactant at the surfactant monolayer for the formation of bilayers and microemulsions.

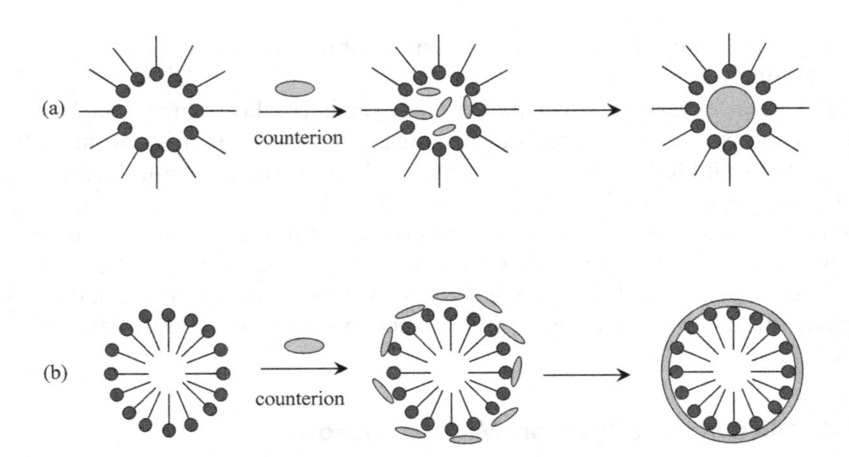

Figure 4.16. Concepts of micellar catalysis on emulsions.

For the O/W type (Figure 4.16b), most of the schemes in Figure 3.11 can be applied. Since the size range of emulsion systems is larger and broader than that of micellar systems, they can provide an ideal reaction "template" to extend the available length scale beyond the scale of normal micelles. Preparation of macroporous inorganic materials (Chapter 9) will be one good example. Also, the versatility of oil and emulsifier compounds available for emulsions provides an additional control factor, which can be directly applied for property control of the final materials. For example, using supercritical CO_2 as an oil phase can greatly simplify the template-removal process, which can improve the yield of the nanostructured or macrostructured materials. This method can minimize the structural damage, too.

REFERENCES

Abbott, N. L. "Surface Effects on Orientation of Liquid Crystals," *Curr. Opin. Coll. Inter. Sci.* **2**, 76 (1997).

Aveyard, R., Binks, B. P., Clint, J. H. "Emulsions Stabilised Solely by Colloidal Particles," *Adv. Coll. Inter. Sci.* **100–102**, 503 (2003).

Bangham, A. D., Horne, R. W. "Negative Staining of Phospholipids and Their Structured Modification by Surface Active Agents as Observed in the Electron Microscope," *J. Mol. Biol.* **8**, 660 (1964).

Bangham, A. D., Standish, M. M., Watkins, J. C. "Diffusion of Univalent Ions Across Lamellae of Swollen Phospholipids," *J. Mol. Biol.* **13**, 238 (1965).

Bangham, A. D., Hill, M. W., Miller, N. G. A. "Preparation and Use of Liposomes as Models of Biological Membranes," *Meth. Memb. Biol.* **1**, 1 (1973).

Bourrel, M., Schechter, R. S., eds. *Surfactant Science Series*, Vol. 30 (Microemulsions and Related Systems) (Marcel Dekker: 1988).

Cates, M. E., Candau, S. J. "Statics and Dynamics of Worm-like Surfactant Micelles," *J. Phys.: Condens. Matter* **2**, 6869 (1990).

Engberts, J. B. F. N., Kevelam, J. "Formation and Stability of Micelles and Vesicles," *Curr. Opin. Coll. Inter. Sci.* **1**, 779 (1996).

Fontell, K. "Cubic Phases in Surfactant and Surfactant-like Lipid Systems," *Coll. Polymer Sci.* **268**, 264 (1990).

Gennes, P. G. D., Prost, J. *The Physics of Liquid Crystals* (Oxford: 1993).

Hoar, T. P., Schulman, J. H. "Transparent Water-in-Oil Dispersions: The Oleopathic Hydromicelle," *Nature* **152**, 102 (1943).

Iampietro, D. J., Brasher, L. L., Kaler, E. W., Stradner, A., Glatter, O. "Direct Analysis of SANS and SAXS Measurements of Catanionic Surfactant Mixtures by Fourier Transformation," *J. Phys. Chem. B* **102**, 3105 (1998).

Katsaras, J., Gutberlet, T., eds., *Lipid Bilayers: Structure and Interactions* (Springer: 2001).

Kunieda, H., Nakamura, K., Evans, D. F. "Formation of Reversed Vesicles," *J. Am. Chem. Soc.* **113**, 1051 (1991).

Lasic, D. D. *Liposomes: From Physics to Applications* (Elsevier: 1993).

Petsev, D. N., ed. *Emulsions: Structure, Stability, and Interactions* (Elsevier: 2004).

Rosoff, M., ed. *Surfactant Science Series*, Vol. 62 (Vesicles) (Marcel Dekker: 1996).

Solans, C., Kunieda, H., eds. *Surfactant Science Series*, Vol. 66 (Industrial Applications of Microemulsions) (Marcel Dekker: 1997).

Tiddy, G. J. T. "Surfactant-Water Liquid Crystal Phases," *Phys. Rep.* **57**, 1 (1980).

Wen, L., Cheng, J., Zou, H., Zhang, L., Chen, J., Papadopoulos, K. D. "Van der Waals Interaction between Internal Aqueous Droplets and the External Aqueous Phase in Double Emulsion," *Langmuir* **20**, 8391 (2004).

Zana, R., Lang, J. "Dynamics of Microemulsions," *Microemulsions: Structure and Dynamics*, Friberg, S. E., Bothorel, P., eds., pp. 153–172 (CRC Press: 1987).

5

COLLOIDAL SELF-ASSEMBLY

When conditions are right, particles with a colloidal size range of nm–μm self-assemble into a defined morphology both in solutions and at surfaces (Collier et al., 1998; Dinsmore et al., 1998; Grier, 1998). Typical colloidal particles include silica sphere (the Stöber method is the first and the most-referenced synthetic procedure; Stöber et al., 1968), polystyrene latex sphere, and poly(methyl methacrylate) sphere. Most of the nanoparticles that will be discussed in Chapter 10 are also largely within the colloidal-particle category. Generally, colloidal self-assembly is a kinetic, random, and homogeneous process. This was briefly discussed in Chapter 1. Typical colloidal processes such as coagulation, flocculation, and precipitation can be viewed as one type of colloidal self-assembly, too. They yield a lot of useful information related to many conventional technologies such as flotation, oil recovery, and soil remediation.

What makes colloidal self-assembly critically different from molecular self-assembly is the strong involvement of other forces that are induced by various colloidal phenomena. These forces can be attractive driving forces and at the same time can act as repulsive opposition forces. They work with intermolecular/colloidal forces (see Chapter 2 of the details) cooperatively. Thus, the complete picture of the force balance of the colloidal self-assembly should be constructed with the incorporation of these forces. A significant portion of this chapter will be allocated to this issue.

Recently, new types of colloidal systems have been discovered, which shows that spontaneous thermodynamic self-assembly of colloidal particles is indeed possible in solutions. This process strongly resembles typical molecular self-assembly processes such as micellization. For example, the balance between attractive and repulsive forces among colloidal particles almost solely determines their critical formation concentration and morphology. One of the critical issues of nanotechnology is the assembly of colloidal particles in a controllable manner. Thus, thermodynamically driven self-assembly of colloidal particles is very important in nanotechnology.

This chapter is designed to address the general features of the colloidal self-assembly, its uniqueness related to colloidal phenomena, and the scheme of force balance in its process. It does not include the influence of external forces. The topic of colloidal self-assembly under external forces has certainly evolved from the issues of this chapter. But, since it has a more direct relation to a variety of applications, external force–induced colloidal self-assembly will be described in a chapter in Part II (Chapter 12).

Colloidal self-assembly also can be used as a practical model system for the study of nucleation, crystallization, and phase transition of atomic and molecular systems. The interaction processes and energetics are very similar. But colloidal systems can be visualized using typical techniques such as microscopy, so real-time observations can be made.

5.1. FORCES INDUCED BY COLLOIDAL PHENOMENA

Chapter 2 details the intermolecular forces and subsequent colloidal forces. They often have the same origin or a virial type of expansion based on a reasonable assumption. On the self-assembly side, both types of forces are, of course, critical.

Meanwhile, there are other types of forces that are equally critical. This is especially true for self-assembly in the colloidal-length scale. These forces do not necessarily have their origins in intermolecular interaction. Rather, they are *induced* as a result of a variety of phenomena that are typical and also unique in the colloidal regime. These phenomena include surface tension, capillary action, wetting, osmotic pressure, flow, and so forth. They have a long history of research, mainly because they have so much importance in a variety of industrial processes and in many critical biological processes as well. These *colloidal phenomena–induced forces* usually have a similar length scale of their action and a similar magnitude of strength compared with the colloidal forces. Thus, in view of force balance in self-assembly, they should be inevitably included within the scheme of its process. They are expressed in colloidal systems as if there were a direct interaction between colloidal objects such as particles and substrates. This interaction works in the system to increase or decrease the distance between them.

In this sense, they can be viewed as attractive and/or repulsive forces that act on colloidal objects.

The purpose of this section is to connect these colloidal phenomena–induced forces to colloidal self-assembly. Details and the full scope of definition, measurement, and applications of colloidal phenomena themselves are not the intention of this section. That has been supplied by excellent colloidal/surface chemistry textbooks (Israelachvili, 1992; Hiemenz and Rajagopalan, 1997).

5.1.1. Surface Tension and Capillarity

Since atoms or molecules at surfaces are always less stable than those in the bulk, surfaces of liquids (and solids) that are interfaced with gases (or vapors) have a natural tendency to get their surface area minimized by minimizing the number of atoms or molecules at the surfaces. *Surface tension* (γ) is a force that operates along the surface as a result of this tendency. This yields the common notion that there can be no such term as "surface tension of *what*." It should be always stated as "surface tension of what with *which interface*." Surface tension is always perpendicular to the surfaces. Thus, its unit is expressed as force/unit length. From the viewpoint of energy, this is work (or surface free energy) that is required to increase a given surface by unit area. Thus, it can be stated as energy/area as well. In the example of water interfaced with air, γ is 72.8 dyne/cm (or mN/m) or 72.8 mJ/m^2 (or erg/cm^2) at 20.0 °C.

Suppose that there is a liquid that is interfaced with its vapor (or gas phase) with surface tension γ. As shown in Figure 5.1, if any perturbation acts on this liquid surface to create either a concave (b) or convex (c) type of curved interface, there always exists a force that tries to restore the surface area to its minimum. The intensity of this force is determined by the absolute value of surface tension. This means that whenever there is a curved interface, a pressure difference is created (or begins to operate) from the convex side to the concave

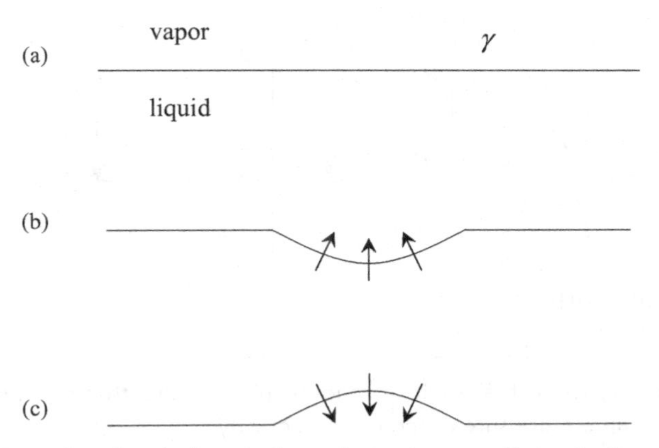

Figure 5.1. Curved surface induces a force that acts across it due to its surface tension.

side. This is true for all types of interfaces that are curved, and expressed quantitatively with the *Laplace* equation. It says

$$\Delta p = \gamma\left(\frac{1}{R_1}+\frac{1}{R_2}\right) \tag{5.1}$$

where Δp is the pressure difference across the interface (often called *Laplace pressure*) and γ is the surface tension. R_1 and R_2 here are the radii of the "two" curvatures, respectively. They are cut by two planes that are perpendicular to each other, so all types of curvatures can be expressed by modifying these two planes.

Now, let us focus on how this Laplace pressure works for capillary force and its direction. Figure 5.2 shows a typical capillary rise situation. For example, let us consider a system of a capillary with a liquid whose contact angle (θ; details in the next subsection) is less than 90°. This means that the liquid has a wetting capability on that surface of capillary. A typical example of this type of system is water with a glass capillary tube. The moment the capillary touches the surface of liquid (t = 0), the wetting of the liquid on the inner surface of the capillary tube creates a concave curvature (toward the gas phase). While its wall is wetted by the liquid, the center has an atmospheric height the same as the reference liquid level, which is the liquid level outside of the capillary tube. Thus, the hydrodynamic pressure at point 1 (P_1) is the same as the atmospheric pressure (P_{atm}). And the Laplace equation says that the pressure right above this point (P_2) should be lower than P_1. This pressure difference ($P_1 - P_2$) creates an instant "force" on the liquid mass, directing it upward. This force will be eventually

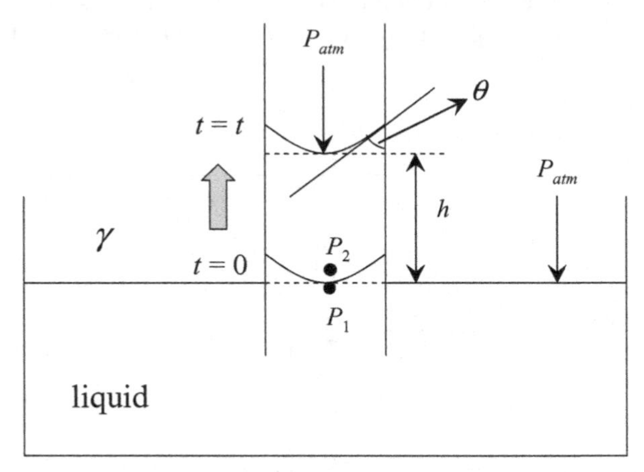

Figure 5.2. Capillary rise is the result of colloidal phenomena, that is, a process toward a balance between Laplace pressure difference and gravity.

compensated (or balanced) by the hydrodynamic pressure generated by the weight of the liquid (ρgh) (where ρ is the density of the liquid and h is the height of the risen liquid inside the capillary). Thus, the rise stops at this balance point ($t = t$). This can be rewritten as

$$\begin{aligned} &\text{at } t = 0;\ P_1 = P_{atm} > P_2 \text{ and} \\ &\text{at } t = t;\ P_1 - P_2 = \rho gh = P_{atm} \end{aligned} \tag{5.2}$$

This classical capillary rise can be viewed as a colloidal interaction that is driven by surface tension. The force that moved the liquid is induced by the surface tension of the liquid that has the direction toward concave side (from the liquid). The strength of this force is determined by the surface tension of the liquid, and the force works against the atmospheric pressure. Adaptation of these phenomena on the self-assembly of colloidal particles can be presented for a variety of situations, including immersion, flotation, and bridged capillary forces. These will be presented in Chapter 12 with details.

For real systems, liquid and vapor phases are in equilibrium. The *Kelvin* equation provides a quantitative solution for the pressure difference, that is, the pressure that works on the curved liquid surface. It can be simply derived from the Laplace equation as follows:

$$RT \cdot \ln\left(\frac{P_{V,C}}{P_{V,O}}\right) = \gamma V_m \cdot \left(\frac{1}{R_1} + \frac{1}{R_2}\right) \tag{5.3}$$

$P_{V,C}$ is the vapor pressure at the curved interface; $P_{V,O}$ is the normal vapor pressure, that is, the vapor pressure at the infinitely plat surface. V_m is the molar volume of that liquid. This equation says that the smaller the liquid particle is, the higher the vapor pressure on the curved surface. It thus explains *Ostwald ripening phenomena*, where smaller liquid drops that have higher vapor pressure constantly vanish while larger drops constantly grow. Also, it says that the pore or space that has negative curvature should have decreased vapor pressure. This explains the capillary condensation, for example, between colloidal particles or solid substrates, that can create a bridge between them.

On the nanotechnology side, these colloidal phenomena of surface tension and capillary condensation, and their quantitative relations, have important implications in various aspects. For example, they are critical to understanding the operating principles of nanodevices such as micro- or nanofluidic devices for the penetration and flow of liquid inside the nanoscale pore or channel. It includes determination of the proper pressure to be imposed on a given liquid in a given size of channel. The term *tortuosity* refers to the empirically determined radius of the pore with the assumption of cylindrical shape from these equations. Also, the well-known *Poiseuille* equation shows the relation of this value to the actual rate of the liquid inside the pore.

5.1.2. Contact Angle and Wetting

While surface tension is a quantitative value that is defined at the interfaces formed by two different phases, *contact angle* (θ) is an angle at the junction of three different phases. This is measured in the liquid, typically at the meeting point of gas–liquid–solid. Figure 5.3 shows the schematic representation. From this picture, horizontal force balance in equilibrium yields a simple relation called the *Young* equation, as follows:

$$\gamma_{LV}\cos\theta = \gamma_{SV} - \gamma_{SL} \tag{5.4}$$

γ_{LV}, γ_{SV}, and γ_{SL} are surface tension at the liquid–vapor, solid–vapor, and solid–liquid interface, respectively. This relation is based on the assumption of a perfectly smooth solid. Thus, structural factors such as surface roughness and surface heterogeneity have a significant impact on its validity.

Wetting is defined based on this picture of the contact angle. In the case of liquid on a solid surface, complete wetting means its contact angle on that surface is zero. Complete *nonwetting* requires its value to be 180°. Ideally, complete wetting means the formation of infinite liquid film on that surface, while complete nonwetting means the formation of liquid drops with perfect sphere shapes. In the case of water, when the surface shows a tendency toward complete wetting, it is generally called a *hydrophilic* surface. The surface with a tendency toward complete nonwetting is called *hydrophobic*. In the case where the water contact angle is above 150°, the name *superhydrophobic surface* is used.

The Young equation says that γ_{LV} is an intrinsic value. Thus, if there is a change in any of the values of γ_{SL} or γ_{SV}, a new force balance will take place. This could induce the movement of liquid drops on a solid surface. Figure 5.4 shows this situation. Suppose that a liquid drop sits on a surface, and this surface is patterned with a scale comparable to the size of that drop. Also, suppose that the tip of the drop touches the area, for example, with a smaller value of γ_{SV} (an

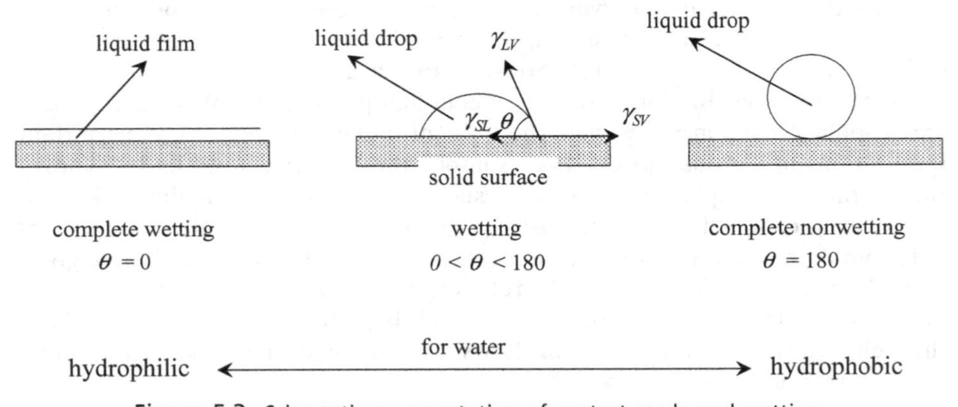

Figure 5.3. Schematic representation of contact angle and wetting.

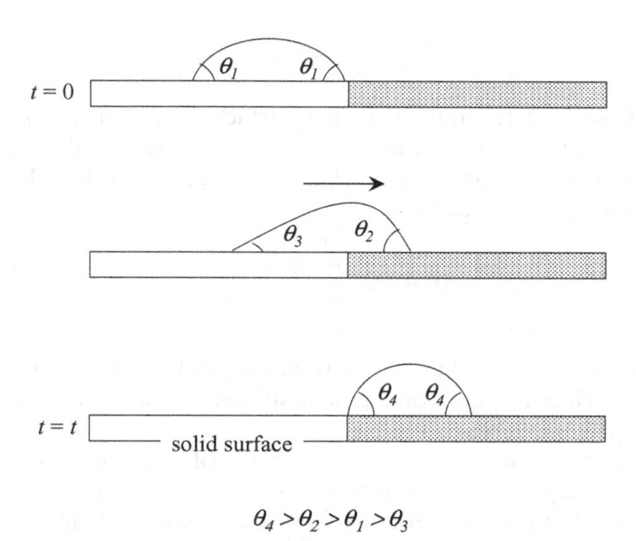

Figure 5.4. Contact angle difference on a well-designed surface can move a liquid drop.

area more hydrophilic in the case of water; $t = 0$). The difference on the right-hand side of the Young equation decreases its value; thus, the contact angle should be increased on this side to balance off that force change. This will result in a movement of the body of the liquid drop toward this side. It is another example of a movement of liquid mass by colloidal phenomena. With careful design of the surface, this colloidal interaction can induce a movement of liquid drops even against other forces, such as gravity (Chaudhury and Whitesides, 1992).

5.1.3. Adhesion

Adhesion is the work done to separate two different phases infinitely; it is a thermodynamic term and reversible. *Cohesion* is the case when the two phases are the same. Adhesion and cohesion are always referred to as the *attractive situation*, so the values are always positive. They measure the quantitative amount of the attractive force between the two phases with the colloidal scale. Atomic force microscopy (AFM) and its derivative versions, such as lateral force microscopy and friction force microscopy, now routinely measure them with the nano-Newton (nN) scale. More practically, this is done by attaching colloidal spheres on the cantilever of AFM, followed by approaching and retracting it on and from the substrates. This direct measurement greatly helps in understanding the role of these colloidal phenomena–induced interactions in the self-assembly of colloidal objects as well as further issues on materials and nanotechnology.

If two phases, A and B, are separated by a third phase C, the work of adhesion can be written as

$$W_{AB} = \gamma_{AC} + \gamma_{BC} - \gamma_{AB} \tag{5.5}$$

Here γ is the surface tension of each interface. W_{AB} (or W_{ACB}) also can be expressed as energy per unit area, and so can be replaced with $U(x)$ in equation (2.11) for Derjaguin approximation. Thus, the "adhesion force" between colloidal subjects can be expressed as

$$F = 2\pi\left(\frac{R_1 \cdot R_2}{R_1 + R_2}\right) \cdot W_{AB} \tag{5.6}$$

This is for the case of two spheres with radius R_1 and R_2, respectively. Expansion to other cases such as a sphere on a planar substrate is well developed elsewhere (Israelachvili, 1992).

In most cases, surface tension is in the range of a few tens of mN/m, and the size of colloidal subjects is in the range of sub- to a few micrometers. Thus, the adhesion forces that can be deduced from these numbers are in the range of nano-Newtons. For example, the friction of a typical long-chain (C_{10}–C_{20}) self-assembled monolayer (SAM) on typical substrates such as gold, silver, and platinum can range ~10—~100 nN. This value is comparable to the intermolecular and colloidal forces. It is also comparable to the forces that are induced by external forces (this will be covered in Chapter 12). These facts show that the interaction induced by surface tension–originated colloidal phenomena can take a strong role in colloidal self-assembly and also in external force–induced colloidal self-assembly.

5.1.4. Gravity and Diffusion

The size and density of ordinary colloidal particles provide enough weight that they can be affected by gravity. When a colloidal particle moves in the direction of gravity, it is called *sedimentation*. When the density difference between particle and medium pushes it in the opposite direction, it is called *creaming*. Gravity is always present on Earth, so its influence on the interaction between colloidal subjects is inevitable. The same effect is expected when a centrifugal force is imposed on colloidal systems, with arguably a faster and stronger response than with gravity.

With the assumption of the colloidal particle with neutral surface (no surface charge) and nonmodified surface (e.g., no surface-grafted polymer chains, which means no steric interaction; when there is a surface charge, the movement of the particles generates the *electrokinetic effect*; this will be presented in the next subsection), the effect of gravity on the dynamics of the colloidal particle is always counterbalanced (or compensated) by the action of diffusion. Figure 5.5 shows the schematic representation. At the moment when gravity is imposed on the colloidal particles, they begin to move in the direction of gravity. For spherical particles, this is well expressed with the *Stokes* equation. This particle move-

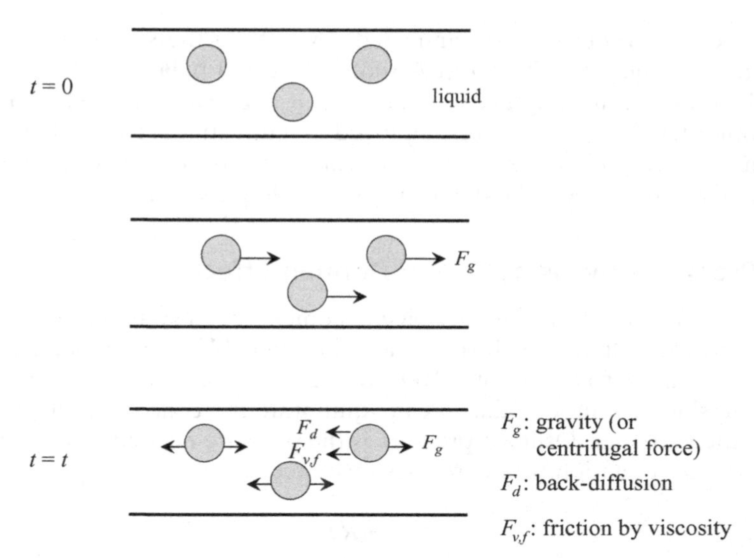

Figure 5.5. Force on a colloidal particle generated by gravity is balanced by back-diffusion and friction.

ment creates an instant concentration-deficiency behind it. And this concentration gradient generates a diffusive force in the direction opposite to gravity. This situation is well described by *Fick's first and second equations.* Diffusion is random, but under these circumstances, it can be considered as directional.

As the particle size increases, the effect of gravity begins to overtake the effect of diffusion, and as the particle size decreases, it will be the reverse. For gases or ions, the gravity effect will be almost negligible, but for large particles, the diffusion effect will be. The region where sedimentation and diffusion effects are comparable is the region of colloidal size, which is nanometer–micrometer. For most of the spherical colloidal particles with their radii in the submicrometer—a few tens-of-micrometers range, the sedimentation velocity under gravity can range ~1–~100 μm/s. Diffusion coefficients of the same particles under the same conditions can range a few tens of $\mu m^2/s$.

For real systems, the force balance between gravity and diffusion may not be pictured exactly as two opposing forces. Diffusion may show more random characteristic features, especially at elevated temperature. Also, the influence of the buoyancy and viscosity of the surrounding medium can act as a force on particles that should be opposite to gravity. But setting up a concept like this, which is based on the picture of opposing gravity and diffusion forces, can benefit many applications such as the control of materials processing and their properties by tuning their microstructures on the nanometer scale. One good example is the effect of gravity versus diffusion on sterically modified polymer colloidal particles, which was briefly discussed in Section 1.3. When an unnecessarily strong gravity force was removed (or minimized) by putting the system in space, the pure attractive and opposition forces took effect and balanced. This yielded a well-defined or -arranged lattice of colloidal self-aggregate as a colloidal crystal. On the other hand, gravity can be employed as a directional force when thermal diffusion and/or repulsive forces are too strong, and so can help overcome those force barriers to achieve colloidal aggregates with good quality.

5.1.5. Pressures by Osmotic and Donnan Effects

The classical definition of *osmotic pressure* comes from a system of solution that is separated from its pure solvent by a semipermeable membrane. This membrane allows the permeation of solvent but not of solute. Thus, solvent gets to flow to the side of solution, and so can equilibrate the concentration difference between the two sides. Osmotic pressure is the pressure needed to stop this flow. The *Van't Hoff* equation well describes this effect:

$$\pi V_s = nRT \tag{5.7}$$

where π is osmotic pressure, V_s is the volume of solvent, and n is the solute concentration. This equation works well within the limiting case of dilute solution. With simple logic, it says that more solute induces stronger pressure. When similar phenomena happen when the solution is separated by the membrane that allows the flow of both solvent and small ions but not of larger macroions, it is called the *Donnan effect*. Charged colloidal particles, proteins, and polyelectrolytes are typical examples of these macroions.

I will not repeat the detailed thermodynamic interpretation and related classical applications such as molecular weight (number average) determination using osmometry and biological colloidal purification such as dialysis and ultrafiltration (reverse osmosis) here. I will focus on the aspect of the forces generated by these phenomena, which have an important impact on the self-assembly of colloidal particles. For example, osmotic pressure always exists whenever there is a concentration perturbation in the solution (chemical potential difference) even without the need of membranes. A depletion attractive force between colloidal particles arises due to the difference of osmotic pressure that is caused by the existence of nonadsorbing polymers or small particles (Section 2.3). This is the case for colloidal particles with no surface charge; thus, the osmotic pressure–driven force will be coupled with the van der Waals force and possibly other forces such as hydrophobic, but not with electrostatic force. When the surface of the colloidal particle is charged, which is a realistic case for most particles, the Donnan effect occurs.

Let us assume that there is a homogeneous solution of polymers (nonadsorbing on the colloidal surface) or small particles. Figure 5.6 shows the schematic representation. Now, colloidal particles, which are much bigger than these, are introduced into this solution. At time zero, the space where the polymers or small

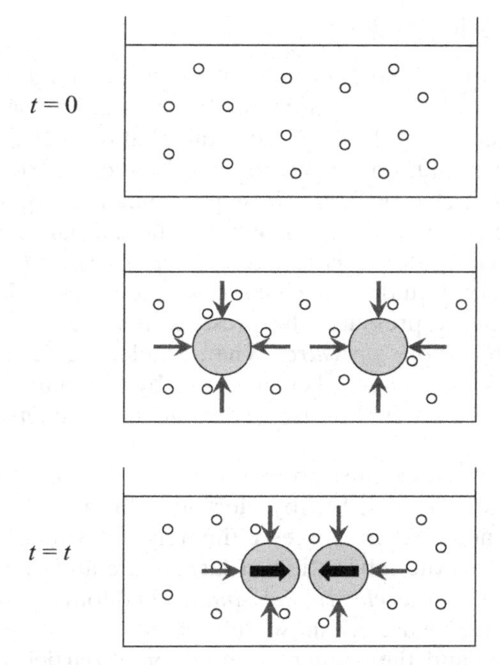

Figure 5.6. Small arrows indicate the direction of osmotic pressure or Donnan effect, while big arrows indicate the direction of movement of colloidal particles due to the action of this pressure.

particles were located is taken by the introduced colloidal particles, which makes the total available space smaller, and thus more concentrated. And the very near region of the surface of the colloidal particles will be instantly more concentrated, and so create the instant concentration fluctuation. Osmotic pressure is created by the action of equilibrium, the flow of solvent in the direction of solution. Thus, it is expected that the flow of solvent will be generated from the less concentrated area to the more concentrated area of the surface of colloidal particles. This induces a pressure around the surface of the colloidal particles. As long as the particles are distant enough, the pressure around the surface will be statistically even; thus, the particles will stand still. If, by any interactions, those particles get to be close, so they can squeeze out some of the small particles (or polymers) from the closing region (depletion region), then the pressure on this depletion region will be lost. Thus, the rest of the pressure that still acts will generate the force, which brings the colloidal particles closer. This is the *depletion force* (also see Figure 2.6). When the surface of the particles is charged, electric double-layer repulsion takes effect. And when the particles are close enough, this repulsion will be balanced with this depletion force.

5.1.6. Electrokinetic Force

It is natural that when colloidal particles are surface-charged, their movements will be affected by the applied electric field. In a similar analogy, when an electric field is applied on a colloidal-dimension channel that has a charged inside surface, the flow of the inside solution will be affected by the electric field. Or, the flow will create an induced electric field. These phenomena are generally referred to as *electrokinetics* (electro- + kinetics). When colloidal particles move under the influence of an electric field, it is called *electrophoresis. Electroosmosis* occurs when the solution or liquid with charged species moves inside the charged channel. As for osmotic pressure, the pressure needed to counterbalance this flow is called *electroosmotic pressure.* When an electric field is induced by the liquid that flows inside the charged channel and by the charged particles that are sedimented, it is called *streaming potential* and *sedimentation potential*, respectively.

As with the above cases, this subsection covers the forces that are generated on charged colloidal particles, but by electric field, and the concepts of their interplay. The balance of these forces ultimately determines their movements. Further issues, such as the fabrication of large-scale and/or patterned colloidal particles with the use of this *electric field-induced* colloidal self-assembly, will be covered in chapter 12. Figure 5.7 shows the schematic representation of electrophoresis. The forces and their balance on colloidal particles under an electric field are also shown. This is for a DC electric field. When charged particles begin to feel the electric field applied, for example, of positive charge, they will move

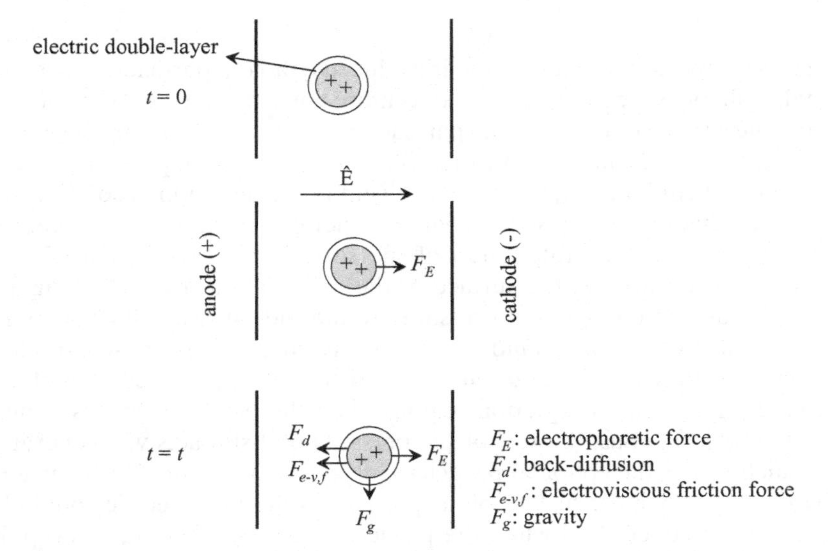

Figure 5.7. Force balance on a positively charged colloidal particle under external electric field.

in the direction of cathode. This will generate friction by the viscosity of the liquid and back-diffusion as well. *Electrophoretic mobility* refers to their movement as a result of this force interplay. Depending on the density and size of the particles, gravity (sedimentation force) can take a significant role in this force balance.

For this electric field–induced colloidal particle dynamics, the role of the electric double-layer that will be inevitably present on the charged particle is particularly important. This "soft" layer does not necessarily follow the same exact pathway as the "hard" particles do. There will be a distortion of this double-layer as the particles move. And there will be counterforces whose direction will be opposite to the direction of the particles. (The atmosphere of this layer is charged oppositely to the actual surface charge. See Figures 2.3 and 3.6.) This is called the *relaxation effect* and *retardation effect*, respectively. This shows that the force balance and particle movement will be affected by many factors that are controllable. It includes the type and density of the surface charge, and the solution condition that can affect this charge, such as dielectric constant, salt concentration, and pH. Density and magnitude of the electric field also affect this force. The term *zeta potential* is widely used of the quantitative description of the situation of ionic atmosphere around colloidal particles.

Figure 5.8 shows the schematic representation of electroosmosis (a) and streaming potential (b). Both are cases where the external force induces the movement of colloidal particles. When a DC electric field is applied across the channel, for example, on positively charged particles, they will move toward the cathode, and the solution flow that results from this particle movement will generate the electroosmotic pressure. When an AC electric field is applied, the electrohydrodynamic-origin flow is generated, and it, in many cases, acts on the particles as an attractive force between them. This can induce the assembly of

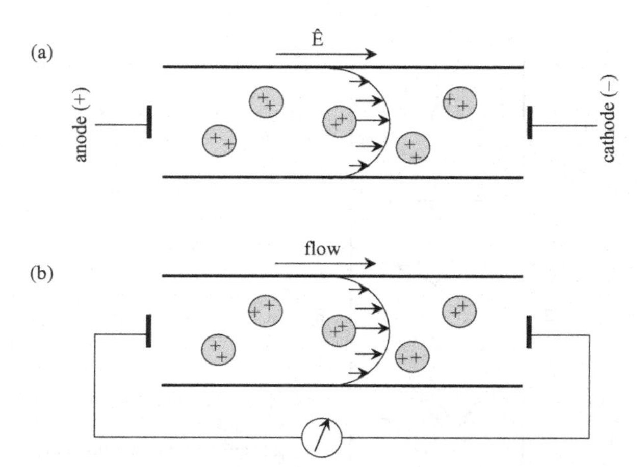

Figure 5.8. Electroosmosis (a) and streaming potential (b) for positively charged colloidal particles within a negatively charged inner surface of a meso-, nano-, or microchannel under DC electric current.

the particles on the surface of the channel. More details about this fact are found in Chapter 12. When flow induces the streaming potential, the generated potential can sometimes be high enough to cause an unwanted result such as a spark, with mainly nonpolar solvents, which can cause the disruption of the designed flow of the liquid inside the channel. Thus, both of these colloidal phenomena are particularly important when the operative principles are designed for nano- (or micro-) fluidic devices such as Lab-on-a-Chip. For example, surface tension often inhibits the entrance of a sample volume of colloidal suspension into the narrow inlet of a fluidic channel. But, when a proper threshold voltage is chosen and applied, the suspension can be forced to move into the channel. In this case, the generated electrocapillary force that can drag the suspension throughout the channel should be carefully considered as well.

5.1.7. Magnetophoretic Force

As with the electrophoretic movement of charged colloidal particles under an external electric field, it is a natural conception that magnetic particles will experience some degree of force when they are under an external magnetic field. The rheological behavior of suspensions of magnetic particles under an external magnetic field has been studied for decades mainly for practical application issues (Odenbach, 2004). When ferro- or ferrimagnetic colloidal particles of mainly nanometer size are dispersed in carrier liquids, they are commonly called *ferrofluid. Magnetorheological fluid* refers to the dispersions of magnetic particles of somewhat larger size, usually up to the range of micrometers.

Figure 5.9 shows the forces that are induced and their interaction when the magnetic particles are subjected to an external magnetic field. The general scheme in which each magnetic particle experiences under a magnetic field is

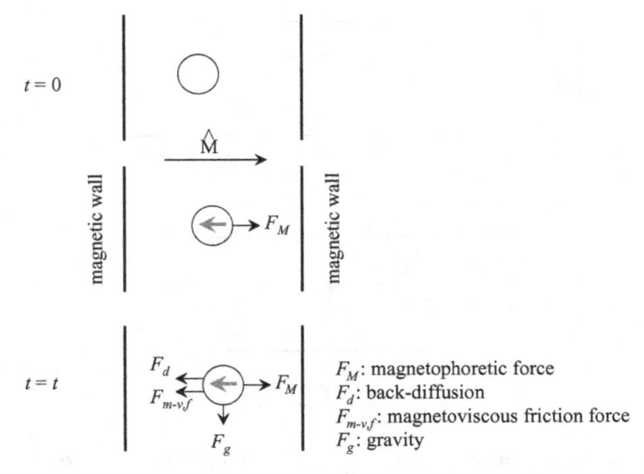

Figure 5.9. Force balance on a magnetic colloidal particle under an external magnetic field. The arrow inside the particle indicates the direction of magnetization.

quite similar to the above cases of gravity and electrophoretic forces. But the difference is that for gravity and sedimentation, the mass of particles (the density of the particles) was one critical factor on the side of the particles. For electrophoretic force, it was the surface charge and inherited electric double-layer. However, for magnetic particles, it is the magnetization of the particle, which is mainly determined by the type and composition of the particles. Once the magnetic particle begins to experience the magnetic field, it will start to move. Its direction will be dependent on the direction of the magnetism of the particle and magnetic field. This force then will be counterbalanced with the back-diffusion and viscous friction force, which is commonly called *magnetoviscous friction*. The role of gravity will be determined depending on the mass of the particles.

Self-assembly of these particles under the influence of a magnetic field is mainly the action of the induced magnetization, the permanent magnetism, and the relation of the direction between them. These factors also will be correlated with the hydrodynamic flow that is inevitably generated by the movement of the particles. This aspect of colloidal self-assembly under a magnetic field is one important difference compared with self-assembly under an electric field.

5.1.8. Force by Flow

There are two typical types of flow that can be applied to the general rheology of pure liquid, solution, and colloidal suspensions: shear and elongational flows. Figure 5.10a shows the schematic representation. Both types have a velocity gradient. Shear flow has its maximum velocity at one side of the surface while the other side has its minimum. But elongational flow has its maximum velocity at the middle of its flow profile, and it gradually decreases toward both sides of

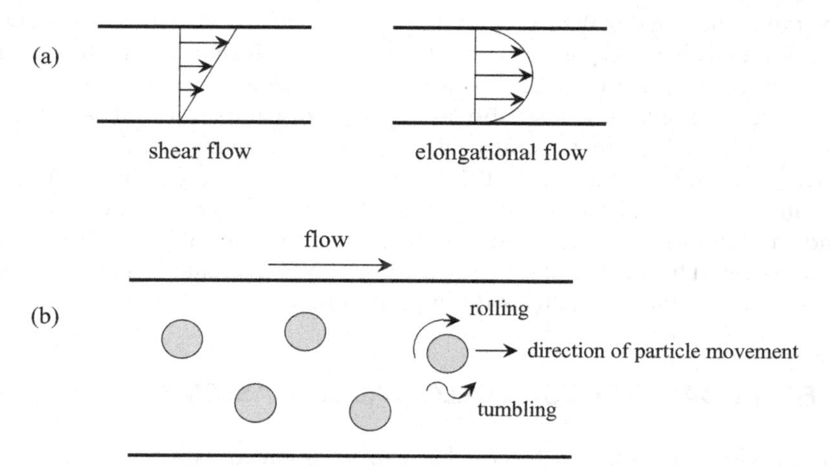

Figure 5.10. (a) Two typical types of flow and (b) the movement of colloidal particles under external flow.

the surface. For solution rheology, these two types of flow can sometimes generate a significant difference in the rheological properties of a given system. But, as long as interplay between the forces is concerned in the colloidal self-assembly, they do not cause that much difference. An exact description of the rheology of colloidal suspension is not within the scope of this section, nor are the basic concepts of rheology, such as Newtonian fluid, non-Newtonian fluid, shear-thinning, shear-thickening, and the viscoelastic property. Readers will find details of these concepts in the excellent Macosko textbook (1994).

Figure 5.10b shows the schematic picture of the interaction that colloidal particles experience under the flow force. The hydrodynamic force originated by flow is usually directional with the direction of flow. But other motions, such as rolling and tumbling, that are also originated by flow around the particles can act as random or less directional force for the interaction between them. For nonsymmetric particles, such as rods or tubes, these modes are more complicated. These hydrodynamic-origin forces compete with colloidal forces such as van der Waals, electrostatic repulsion, and depletion force. One difference from the above cases of gravity, osmotic pressure, electric field, and magnetic field is that those are the forces that primarily act on the particles, so it is the particle that has the initial movement. The balancing forces, such as back-diffusion and viscous friction, are generated as a result of this initial motion, with direction opposite to the particle movement. But, for the flow, the entire suspension (both of the particle and of the liquid) is subject to the flow force. This means that these balancing forces are much smaller than in the other cases, and somewhat random. In the case of charged particles, the out-of-sync situation of the particle within its surrounding double-layer causes additional friction, known as the *electroviscous effect*.

Flow force (that is generated by hydrodynamics) participates in the picture of force balance in a cooperative way. It can be a part of the attractive force with, for example, the van der Waals force; thus, it can reinforce the attraction between particles. It can also be a part of the repulsive force with electrostatic force, which can disrupt the self-assembly of the particles. This leads to the general picture of flow-induced colloidal self-assembly as depicted in Chapter 12. Flow can act to construct colloidal aggregates; also, it can disrupt a given colloidal aggregate. This leads to the important aspect that it can also be employed to minimize the defect in a given or about-to-given colloidal aggregate (colloidal crystal) or to extend the limitation of the scale length that can be assembled with just interparticle forces. This means that the flow force can help create larger-area and/or bigger-size colloidal aggregates with minimum defects.

5.2. FORCE BALANCE FOR COLLOIDAL SELF-ASSEMBLY

In the previous section, the forces that can be induced as a result of colloidal phenomena have been described. These forces act on the colloidal objects at the colloidal-length scale, and their magnitude of strength is comparable to that of

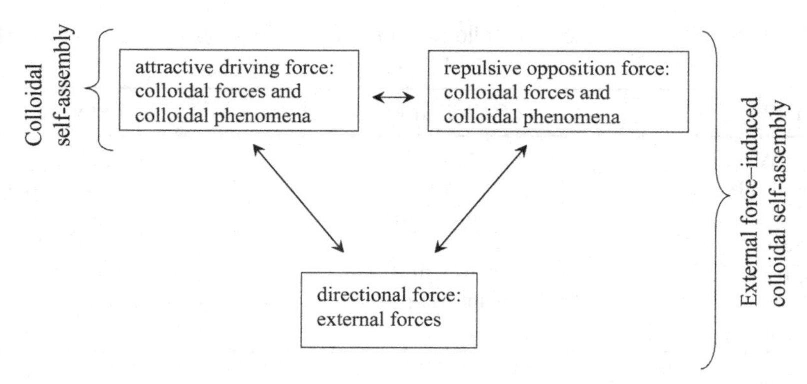

Figure 5.11. Force balances for colloidal self-assembly and external force–induced colloidal self-assembly.

colloidal forces. Thus, these forces can contribute to the process of the self-assembly of colloidal particles with the degree of strength and length comparable to those of colloidal forces. This leads to the general picture of the force balance for colloidal self-assembly as depicted in Figure 5.11. Both the colloidal phenomena–induced forces and the colloidal forces can operate as attractive driving forces and as repulsive opposition forces as well. Self-assembly of colloidal particles is a process toward balance between these two classes of forces in a given system and condition. And the self-assembled aggregates of the colloidal particles are formed at the balance point (more likely a narrow region). Their morphology and further structural transformation are also determined by this force balance and its evolution toward rebalance (if there is any type of disruption in the existing balance).

It is rare that colloidal self-assembly occurs in a highly directional pattern. This is especially true when there are no external forces involved. Some of the colloidal phenomena–induced forces are directional, but not enough to be a directional force for colloidal self-assembly. This is the reason that the colloidal self-assembly is primarily a random process. The type and effect of those external forces will be introduced in Chapter 12. These external forces also act on colloidal objects at the colloidal-length scale, and their magnitude of strength is comparable to that of colloidal forces and colloidal phenomena–induced forces. Thus, they are incorporated into the picture of colloidal self-assembly. The most distinctive feature of those external forces is that they are, for the most part, highly directional. They are primarily directional forces in this force balance picture of colloidal self-assembly. Table 5.1 lists typical examples of these three different groups of forces.

Molecular self-assembly is induced by intermolecular forces (Chapters 3 and 4). Origins of intermolecular and colloidal forces are usually the same, but their affecting length scale and area are sometimes orders of magnitude different (Chapter 2). This fact helps colloidal particles to be self-assembled within a wide

TABLE 5.1. Typical examples of colloidal forces, colloidal phenomena, and external forces.

Colloidal Force	Colloidal Phenomenon	External Force
Van der Waals	Adhesion	Flow
Electric double-layer	Wetting	Electric field
Steric	Contact angle	Magnetic field
Depletion	Surface tension	Capillary force
Hydrophobic	Sedimentation	Mechanical force
Solvation	Osmotic pressure	etc.
Hydration	etc.	

range of particle sizes. Molecular self-assembly is limited to within a few nm to ~20 nm of molecular size (the size of its primary building unit). But colloidal self-assembly is possible within a particle-size range of nm–μm. However, this fact also makes it difficult to achieve a well-defined self-assembled system of colloidal particles. Quite often the intended system falls into the sedimented, coagulated, or precipitated colloidal aggregates.

5.3. GENERAL SCHEME FOR COLLOIDAL SELF-ASSEMBLY

Figure 5.12 shows the general scheme for self-assembly of colloidal particles in solution. Usually from a few to ~50 colloidal particles participated in the primary self-assembly process. Unlike the case of surfactant micelles, the aggregation number is not clear cut and is quite polydisperse. The resultant primary self-aggregate is usually a compact aggregate of various sizes and shapes, depending on the particle and solution conditions. This aggregate can be understood as the state wherein the sum (vectorial) of the attractive forces is delicately compensated for by the sum of the repulsive forces within the aggregate. All the forces are random here, while they are somewhat localized in the case of micellization (hydrophobic part and hydrophilic part). A typical example can be found in the Bernal spiral formed by spherical particles of submicron size (Bernal, 1964; Campbell et al., 2005; Sciortino et al., 2005). Short-range depletion is the attractive driving force for this self-assembly, and long-range electrostatic repulsion serves as the repulsive opposition force.

In many cases, the secondary self-aggregate is developed into fractal-like structures whose detailed dimensions are changed by many factors that affect colloidal forces. It can be assembled from the primary self-aggregates with significant involvement of monomers (individual particles). Again, less-localized forces compared with micellization may be responsible for this. Further increase of the particle concentration usually induces the higher-order self-assembly that often produces the higher-order self-assembled aggregates such as networked gel-like and close-packed structures. For example, for hard-sphere colloidal par-

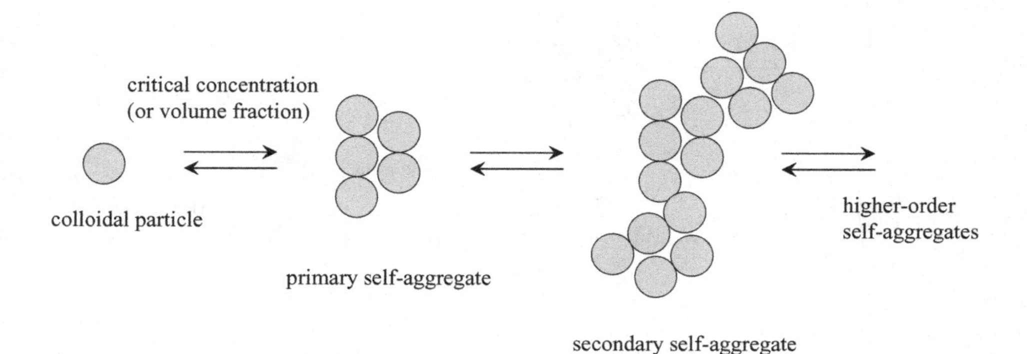

Figure 5.12. Schematic representation of colloidal self-assembly in bulk solution.

ticle self-assembly with no surface charge involved, as the volume fraction of the particles reaches 0.58, 0.637, and 0.74, the self-assembled structures have glassy, random close packing, and face-centered cubic structures, respectively. For surfactant self-assembly, liquid crystals are formed as a result of this higher-order self-assembly.

5.4. MICELLE-LIKE COLLOIDAL SELF-ASSEMBLY: PACKING GEOMETRY

With careful synthesis, well-designed colloidal particles can be obtained that can have both hydrophilic- and hydrophobic-like characters in a single particle. This may be called *amphiphilic colloidal particles* for surfactant or amphiphilic polymers. By proper selection of solvent and environmental conditions, these can self-assemble with thermodynamically driven processes in solution, just like surfactants or amphiphilic polymers do. Their hydrophobic parts can provide enough driving force for this self-assembly, and the repulsive force on their hydrophilic parts can counterbalance it. As shown in micellization of surfactant molecules, the structure of self-assembled aggregates of these types of clear bicharacter (hydrophilic and hydrophobic) building units is strongly affected by their shapes. This is well modeled into packing geometry (Chapter 3). The structures of self-assembled aggregates of amphiphilic colloidal particles are also strongly affected by their physical geometry. Figure 5.13 shows the schematic representation. As well described in Figures 3.8 and 9.8, packing geometry, g-value, provides a semiquantitative prediction for the structures of self-assembled aggregates. The definition of the three subparameters is the same as for the surfactants or polymers. a_o is the top area of the hydrophilic part (head group), l is the length of the particle, and v is the volume of the particle. As for micelles, for g-value close to 1/2, the resultant self-assembled aggregates show the spherical type of structure. For g-value close to 2/3, their structure has great possibility for having a

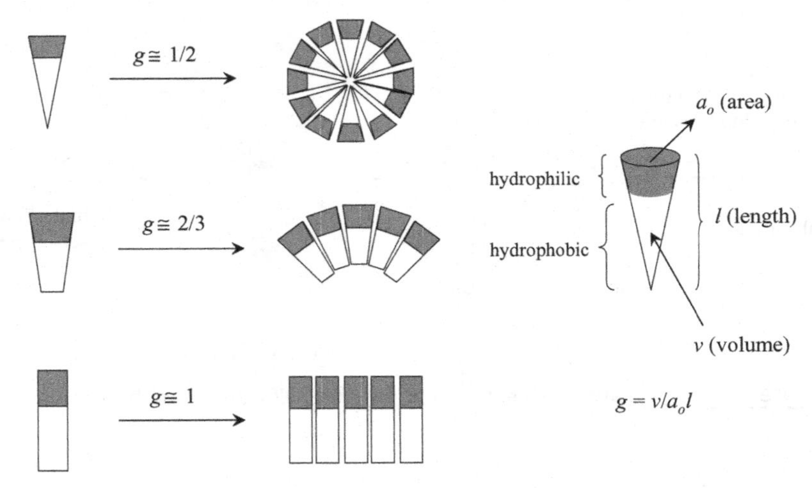

Figure 5.13. Schematic representation of the packing geometry of colloidal self-assembly.

curved geometry, which may result in a cubic or vesicle type of overall structure (Liu et al., 2003). As g-value approaches to 1, self-assembled aggregates will have more planar structures. These are all primary self-aggregates.

If the force balance condition is set adequately, secondary and even higher-order self-assemblies should be possible for these micelle-like structures, like the analogy of micellization toward the higher order. But, since this is for particles the size of tens of nanometers–micrometers, there is a high possibility that the colloidal phenomena–induced forces will get strongly involved with a complex pattern. This can make a reasonable prediction for the secondary and higher-order self-assemblies very difficult at least for now.

5.5. SUMMARY

This chapter has described the role of colloidal phenomena–induced forces in the self-assembly of colloidal particles. Thus, the scheme of the force balance in colloidal self-assembly incorporates these forces along with colloidal/intermolecular forces. Understanding colloidal self-assembly in light of this force balance provides simple but clear routes toward its further development and applications. This is true not only for conventional applications of colloidal self-assembly, but more importantly for nanotechnology as well. Colloidal self-assembly that is driven thermodynamically and the subsequent validity of packing geometry for certain types of amphiphilic colloidal particles have an important impact on the development of a variety of nanotechnology-related issues.

This chapter mainly has considered spherical particles. However, a variety of different shapes of colloidal particles are readily available. They will create more complex force interactions and directionality in their outcome of force balance, which will result in more complex structures and self-assembly patterns. For example, for rodlike colloidal particles such as the nanotube, longitudinally directional capillary force can often dominate the rest of the force, and so can prompt highly directional self-assembly. Also, this chapter examines mainly *homogeneous* self-assembly; but the self-assembly that is involved with more than just one type of particle, a *heterogeneous* colloidal self-assembly, will have even more meaning for functional self-assembled systems in nanotechnology applications.

REFERENCES

Bernal, J. D. "The Bakerian Lecture, 1962: The Structure of Liquids," *Proc. Roy. Soc. Ser. A* (1964).

Campbell, A. I., Anderson, V. J., van Duijneveldt, J. S., Bartlett, P. "Dynamical Arrest in Attractive Colloids: The Effect of Long-Range Repulsion," *Phys. Rev. Lett.* **94**, 208301 (2005).

Chaudhury, M. K., Whitesides, G. M. "How to Make Water Run Uphill," *Science* **256**, 1539 (1992).

Collier, C. P., Vossmeyer, T., Heath, J. R. "Nanocrystal Superlattices," *Annu. Rev. Phys. Chem.* **49**, 371 (1998).

Dinsmore, A. D., Crocker, J. C., Yodh, A. G. "Self-Assembly of Colloidal Crystals," *Curr. Opin. Coll. Inter. Sci.* **3**, 5 (1998).

Grier, D. G. "From Dynamics to Devices: Directed Self-Assembly of Colloidal Materials," *MRS Bull.* **23**, 21 (Oct. 1998).

Hiemenz, P. C., Rajagopalan, R. *Principles of Colloid and Surface Chemistry*, 3rd ed. (Marcel Dekker: 1997).

Israelachvili, J. N. *Intermolecular and Surface Forces*, 2nd ed. (Academic Press: 1992).

Liu, T., Diemann, E., Li, H., Dress, A. W. M., Mūller, A. "Self-assembly in Aqueous Solution of Wheel-Shaped Mo_{154} Oxide Clusters into Vesicles," *Nature* **426**, 59 (2003).

Macosko, C. W. *Rheology: Principles, Measurements, and Applications* (Wiley: 1994).

Odenbach, S. "Recent Progress in Magnetic Fluid Research," *J. Phys.: Condens. Matter* **16**, R1135 (2004).

Sciortino, F., Tartaglia, P., Zaccarelli, E. "One-Dimensional Cluster Growth and Branching Gels in Colloidal Systems with Short-Range Depletion Attraction and Screened Electrostatic Repulsion," *J. Phys. Chem. B* **109**, 21942 (2005).

Stöber, W., Fink, A., Bohn, E. "Controlled Growth of Monodisperse Silica Spheres in the Micron Size Range," *J. Colloid Interface Sci.* **26**, 62 (1968).

6

SELF-ASSEMBLY AT INTERFACES

Self-assembly can occur at any type of interface with any type of building unit. Interfaces provide a geometrical, physical, and sometimes chemical space that can direct self-assembly along their directions. It is usually referred to as *interfacial* self-assembly or *surface* self-assembly. It is also commonly called *two-dimensional* self-assembly, since at least one of the self-assembly dimensions is intrinsically restricted within the length scale of the building units. For well-defined surface micelles, there can be a surface-critical micellar concentration (surface *cmc*), but mostly *cmc* is not clearly defined. Depending on the type of building units, interfacial self-assembly can be either a thermodynamic or kinetic process. Some self-assembly in bulk also occurs in a two-dimensional way and produces self-assembled aggregates with seemingly two-dimensional geometry. A good example is the bilayers described in Chapter 4. But, since they eventually form three-dimensional structures as a result of enclosing or local fluctuation, they are not classified as two-dimensional self-assembly. Self-assembly within confined spaces (confined self-assembly) will be described in Chapter 12. However, this is about the effect of spatial confinement on three-dimensional self-assembly; thus, it will be classified as *three-dimensional self-assembly*.

Self-Assembly and Nanotechnology: A Force Balance Approach, by Yoon S. Lee

What makes interfacial self-assembly so unique compared with bulk self-assembly (three-dimensional self-assembly) is the inevitable interactions between the building units and the interfaces where the self-assembly takes place. The strength of this interaction can vary but is usually comparable to the intermolecular and colloidal forces among the building units. This intrinsic factor yields three typical characteristics of interfacial self-assembly that are significantly differentiated from bulk self-assembly:

First, for bulk self-assembly, there always have to be amphiphilic characteristics encoded into each building unit to ensure the proper force balance. However, for interfacial self-assembly, this is not necessarily prerequisite. Whether the building units are amphiphilic or not, once they are attracted onto the given interfaces, the force interplay among themselves in many cases is enough to induce self-assembly. This fact substantially helps us expand the scope of the building units for interfacial self-assembly.

Second, the interactions of the building units with the interfaces can alter the intermolecular and colloidal forces among the building units themselves. For example, even nonpolar molecules can have dipole moments when they are adsorbed onto the interfaces—through their symmetry breakups. This fact may make the force balance scheme a little bit more complicated, but it can provide additional means to control the self-assembly process at interfaces.

Third, the self-assembled aggregates from the interfacial self-assembly always exist in a state of confinement (or attachment) on the interfaces. Thus, for many cases, they have significantly different physical and chemical properties from those analog from bulk self-assembly. The tensile strength of the Langmuir monolayer is different from what of bilayers formed in bulk solution from the same building units. And the magnetism of metal nanoparticles assembled at solid surfaces can be greatly changed from that of those formed in bulk solution.

This chapter mainly presents the interfacial self-assembly of atomic, molecular, and bio-mimetic building units. Colloidal building units will be described, too, in selected sections. Formation of the self-assembled monolayer (SAM) is essentially involved with chemisorption or other strong bonding of the building units with the interfaces (mostly solid surface). Since a similar type of bonding (not intermolecular interaction) often dominates the assembly of many nanostructured films, SAM will be discussed in Chapter 11 as one of their classes.

6.1. GENERAL SCHEME FOR INTERFACIAL SELF-ASSEMBLY

6.1.1. Surfaces and Interfaces

The definition of surfaces and interfaces can be straightforward. The term *surface* is to be used for any "interface" that is spontaneously formed between any two phases. It will cause no scientific trouble when the term *interface* is viewed as expressing details about the two phases that form the surface.

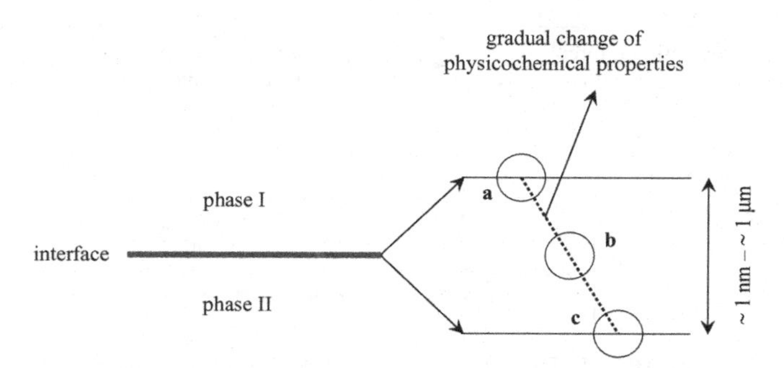

Figure 6.1. Macroscopically, surfaces are two-dimensional; but microscopically, they are three-dimensional with nanoscale thickness.

Figure 6.1 shows the schematic illustration of interfaces. Geometrically and macroscopically, they are two-dimensional with no thickness. However, physicochemically and microscopically, they are three-dimensional. The interface is the region where the physicochemical properties of one phase vary from those of the adjacent phase. This change is gradual. And this region usually has the thickness range of ~1 nm–~1 μm (Myers, 1999).

6.1.2. Force Balance with Interfaces

The thickness range of interfaces happens to be the dimension of the colloidal domain, in other words, the nanometer scale. This coincidental fact provides a very important aspect of interfacial self-assembly. As shown in Chapter 1, almost all of the self-assembly building units have their size ranges within the nanometer scale, except for some of the atomic building units. Thus, once those self-assembly building units are adsorbed within the interfaces (as the way their energetics become the most favorable), the interfaces inheritably become a nanometer-scale *surface well*, where the interplay of intermolecular forces is confined. The interaction among building units becomes more favorable along the direction of the interfaces, but less favorable through the direction of either phase. This is based on the fact that, since the change of physicochemical properties such as density within the interfacial region is gradual (The pattern of this change should be variable, depending on the system and given conditions; it can be close to linear, exponential, etc.), whether the actual adsorption of building units is close to either phase (a or c) or close to the middle of the interfacial region (b), the actual adsorption site of building units inside the interfacial region should be determined within the size range of the building units. This can explain why the monolayer (or monolayer-based finite structure) is always the only primary self-assembled aggregate that can be obtained from most of the interfacial self-assemblies. A higher level of self-assembled aggregates such as bilayer, multilayer, and others are assembled when there is a significant manipulation of intermolecular

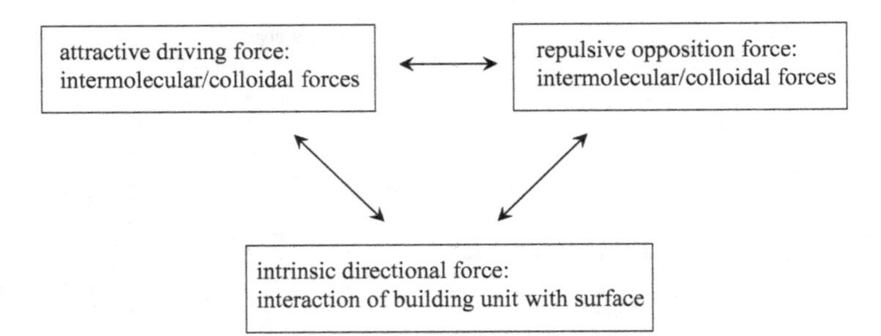

Figure 6.2. Force balance of interfacial self-assembly.

interaction, for example, by the external forces or a significantly increased concentration of the building units.

Figure 6.2 shows the general scheme of the force balance for interfacial self-assembly. As for the typical molecular and polymeric self-assembly in bulk (Chapters 3 and 4) and colloidal self-assembly (Chapter 5), the intermolecular and colloidal forces among the building units can act as both an attractive driving force and a repulsive opposition force. However, unlike in cases of bio-mimetic self-assembly (Chapter 7), it is the interaction of the building units with the interfaces that determines the direction of this self-assembly process. Thus, the term *intrinsic directional force* introduced in the figure can be justified.

As with micelle formation (Figure 3.4), the self-assembled aggregates formed at any interfaces are also dynamic structures. There exists a constant exchange of building units between the self-assembled aggregates and the monomeric phases. Thus, the surface diffusion of both entities (aggregate and monomer), which is restricted two-dimensionally and is usually slower than the bulk diffusion, can have a better chance to affect the kinetics of the interfacial self-assembly process and the properties of the self-assembled aggregates (Suo and Lu, 2000; Barth et al., 2005). If the surface diffusion of the building units is much faster than the thermodynamic equilibrium with the interface, the building unit–interface interaction might have a limited influence on the force balance between the building units. On the other hand, if the thermodynamics takes over the kinetic diffusion process, the building unit–interface interaction will have a much stronger impact on the force balance between them.

Figure 6.3 presents the general scheme for interfacial self-assembly. Regardless of the types of interfaces where the interfacial self-assembly takes place, the building units first have to be located at the interfaces. This prerequisite condition is exclusively acquired by adsorption. The building units at the interfaces then self-assemble into the self-assembled aggregates as the force balance between them is fulfilled. This force balance also determines the packing mode of the building units, the detailed structure of the self-assembled aggregates formed, and the possibility of subsequent higher-order self-assembly.

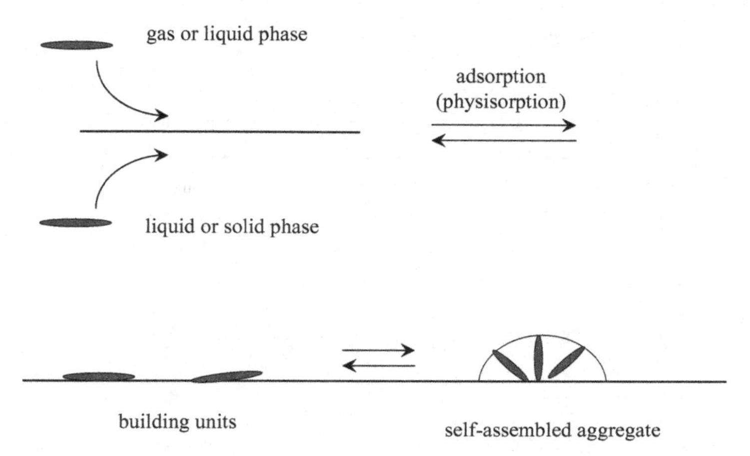

Figure 6.3. General scheme of self-assembly at interfaces.

6.2. CONTROL OF INTERMOLECULAR FORCES AT INTERFACES

The building units for interfacial self-assembly can be categorized in three different groups: amphiphilic building units, building units with functional group(s), and nonamphiphilic ones. Each group shows a distinctive packing pattern. This section describes its details and the fine-tuning possibility of the intermolecular forces among the building units in each group, which ultimately could provide us a better opportunity to control the structures and properties of the self-assembled aggregates.

6.2.1. Packing Geometry: Balance with Attractive and Repulsive Forces

Figure 6.4 shows two different types of packing geometry that can be anticipated from the interfacial self-assembly of amphiphilic building units. Each mode is determined based on the types of interfaces, the external forces that are applied during self-assembly, and the degree of the surface coverage (surface density). It is also affected by the molecular or experimental factors such as molecular structure and properties, deposition rate and time, pH, types and concentration of salts, and so on. The upright mode (a) is typically expected when there is an applied external pressure on the lateral direction (such as the Langmuir monolayer) or when the surface coverage is very high. Depending on the specific interaction of the head groups or the tails with the interfaces, two submodes are possible: head-down and head-up. For both submodes, the amphiphiles are usually self-assembled with some degree of tilted geometry. The flat mode is expected when the surface coverage of the building units is relatively low.

For both upright and flat packings, the concept and application of the packing geometry (g) stand exactly the same, as shown in Figure 3.8. It should be possible

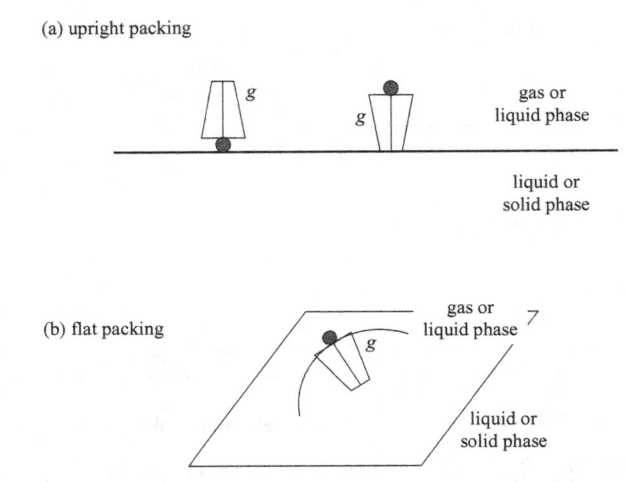

Figure 6.4. Two different types of amphiphile packing at interfaces.

to predict the structure of the self-assembled aggregates from the change of this value. It should also work for the colloidal building units, once the force balance is well designed and played as described in Figure 5.13.

6.2.2. Packing with Functional Groups: Balance with Directional Force

This subsection examines interfacial self-assembly of the building units with functional groups, mainly those with hydrogen-bonding capability. Attractive hydrogen bonding during interfacial self-assembly can be much stronger than the other intermolecular forces (Yaminsky et al., 2006), which provides the building units the ability to overcome possible repulsive forces such as electrostatic repulsion and the steric effect. This is the main reason that interfacial self-assembly of this type of building unit is highly directional and selective, and generates self-assembled aggregates with greater structural richness.

6.2.2.1. Building Units with Multifunctional Sites. Figure 6.5a shows the schematic representation of porphyrin-based derivatives. These are maybe the most characteristic examples that can show how the simple but rational structural design of the building unit can provide the fine-tuning of intermolecular forces between them, and thus can control the interfacial self-assembly process. Other examples include phthalocyanine-, triphenylene-, hexa-peri-benzocoronene-based building units. Porphyrin-based building units can provide at least three different strategies for the fine-tuning of intermolecular forces. First, site III is for the position of ligands. Metal ligands are the most common ones. They provide the complexation between the building units and, by the choice of proper metal ligands, its strength and direction can be tuned as well. This complexation is

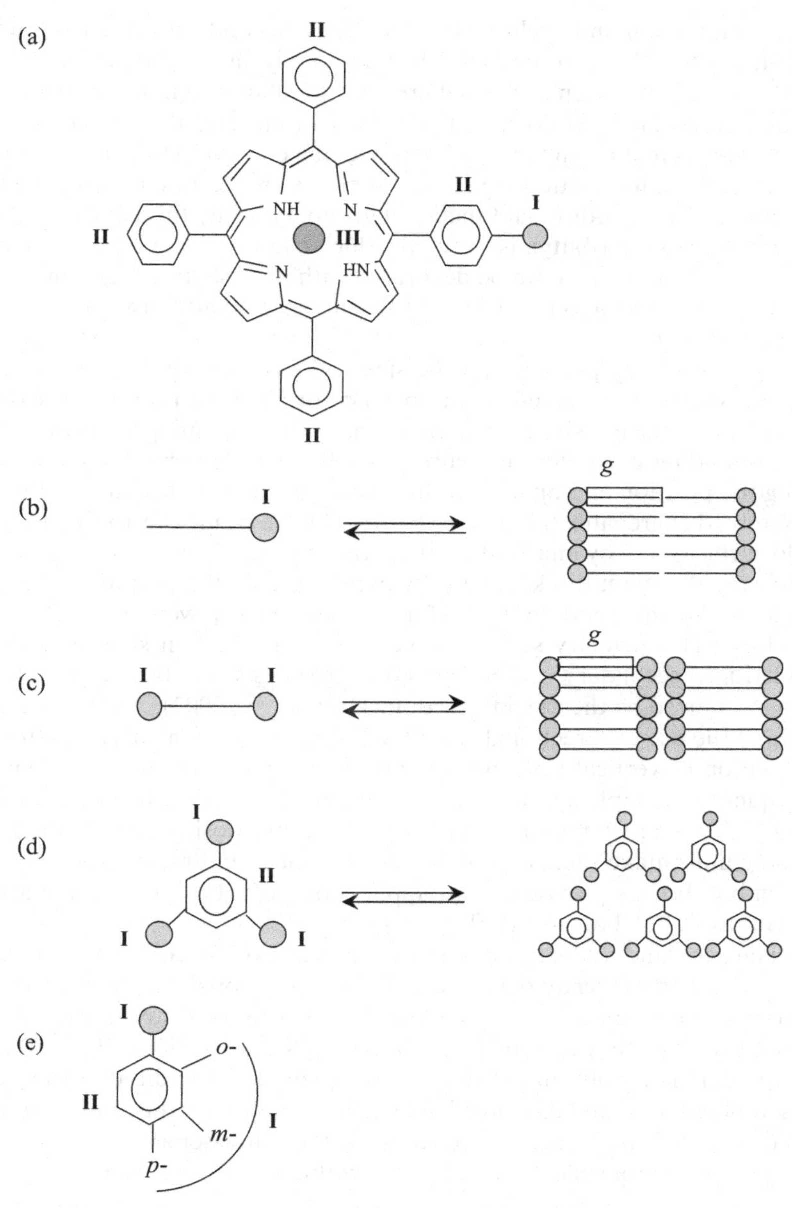

Figure 6.5. Schematic representation of the interfacial self-assembly of molecular building units with the site of directional force capability: (a) porphyrin-based derivatives, (b) monofunctional hydrocarbons, (c) bifunctional hydrocarbons, (d) trifunctional rings, and (e) bifunctional rings.

attractive interaction and exclusively directional. Second, site II is for four possible positions for bulky groups. The phenyl group is the most typical for this position. But other rings such as pyridyl are also popular. The number of the bulky group definitely can be controlled from one to four, and this global structural design provides great room for molecular packing control. Also, each group provides π–π stacking interaction, which can serve as an additional attractive force. Third, site I is for an additional functional group; typically this can be one of the bulky ones such as tert-butyl, iso-propyl, which can provide a delicate steric repulsive force. This site also can be decorated with the selective functional groups such as hydroxyl or cyano that can provide site-specific attractive force through hydrogen bonding.

The self-assembly process and the size and structure of the self-assembled aggregates will be determined by the balance of all these forces. For example, by selectively replacing site I with the cyano group, meaning by providing the hydrogen-bonding capability and subsequently controlling its directionality by choosing the position and number of the cyano group, the structure of the given self-assembled aggregates was able to be tuned from triangular to ring to wire at the gold surface (Yokoyama et al., 2001). Also, by the incorporation of the complexation capability on this site with the pyridyl group, the porphyrin derivative was able to be designed to be self-assembled into a well-defined nonamer morphology. This primary self-assembled aggregate is then subsequently self-assembled into secondary self-assembled aggregates with stacked morphology, which was induced by the π–π interaction (Drain et al., 2002). Another excellent example of the use of directional force for self-assembly control can be found in the formation of vertical stacking of a porphyrin derivative that is designed to have sequential complexation capability on site III (Langford et al., 2002). A variety of choices and types of forces and their flexible position within these types of building units make them one of the most unique building units for interfacial self-assembly. It thus provides great application potential, such as for artificial photosynthesis (Balaban and Buth, 2005).

Calixarene- and fullerene-based derivatives usually are not amphiphiles (unless they are specifically designed so), but with proper but relatively simple structural modification, they, too, can be designed to have strong capability for self-assembly, once they are adsorbed onto solid surfaces (Wan, 2006). Depending on the detailed molecular structures, the picture of the directionality of the self-assembly process and the force balance that determines the final morphology will be different. But the general concept for the self-assembly process of these highly applicable molecules can be viewed within the same scheme.

6.2.2.2. Building Units with Single Functional Sites. Another typical example of interfacial self-assembly with directional force is that with an alkyl chain or aryl group that has hydrogen-bonding capability functional groups. By tuning the position, number, and types of functional groups such as hydroxyl, cyano, carboxylic, or dicarboxylic groups, meaning by tuning the directionality or circularity of the most possible hydrogen-bonding networks, the interfacial self-assembly can be tuned to be directional, and so can obtain the self-assembled aggregates with a planned geometry (De Feyter and De Schryver, 2005).

Figure 6.5 also shows the representative structures of long alkyl chain–based (b and c) and phenyl ring–based (d and e) building units. When the long alkyl chain with one functional group on the one end (monofunctional building unit) is adsorbed (confined) onto liquid or solid surfaces, its hydrogen-bonding tendency with the functional group from the adjacent building unit makes primary self-assembly occur along the direction of its long axis (the axis that runs though the alkyl chain). Then, the balancing process of the hydrophobic attractive force with the hydration force and/or the possible electrostatic repulsive force from the functional group arranges them into most likely the bilayer type of self-assembled aggregates. Typical examples of this type of interfacial self-assembly can be found in palmitic acid (hexadecanoic acid), stearic acid (octadecanoic acid), arachidic acid (eicosanoic acid), behenic acid (docosanoic acid), and so forth. It is highly possible that the primary and secondary self-assemblies cannot clearly be defined. It may be rather a concurrent process. But, since the strong hydrogen bonding acts not only as a directional but as an attractive force as well, this pseudo–second step self-assembly (after the adsorption) should be the most likely scenario. The alkyl chain, of course, of this type of building unit can be further modified for an additional hydrophobic and steric effect—by tuning the alkyl chain length, by introducing the bulky group, or by modifying its molecular geometry. For example, if oleic acid (9-*cis*-octadecenoic acid), elaidic acid (9-*trans*-octadecenoic acid), and stearic acid are systematically self-assembled at the interfaces, the different molecular packing (obviously two-dimensional) as encoded by their molecular geometry significantly rules the secondary process, which thus makes the self-assembled aggregates have different spacing between the dimeric primary self-assembled aggregates but with the same form of bilayer-type structure.

For bifunctional building units that have two functional groups on both ends of the alkyl chain (homo- and hetero-) (c), the linear chainlike structures formed by the one-dimensional hydrogen-bonding network are most likely primary self-assembled aggregates. The subsequent hydrophobic force from the alkyl chain and possible steric and electrostatic forces then determine the final geometry of their self-assembled aggregates. This also should be close enough to the monolayer type of morphology.

When the phenyl ring–based building units (d and e) are adsorbed onto the interfaces, the linear or circular characteristics of the possible functional groups on the ring provide a strong possibility for tuning the self-assembled aggregates with unique linearity or circularity. For example, building units with two functional groups that are linearly aligned (at 1- and 4(*para*)-positions) (e) will be most likely self-assembled into the monolayer type of structures, as in the case of the bifunctional alkyl chain. The building units with their functional groups at 1, 2 or 1, 3 positions will have a possible tendency to be assembled as a circular type with defined aggregation numbers. Those with three functional groups (d), for example at the 1, 3, 5 position (trimesic acid), have enough packing and molecular geometrical requirements and directional force for the circular type of structure that their network could run through the entire structure (Lackinger et al., 2005).

Again, the interaction of the above building units with the interfaces can be considered as the primary reason that interfacial self-assembly of these building units almost exclusively occurs along the direction that is predetermined by the given interfaces. For the long alkyl chain–based building units, the flexibility of the chain will certainly contribute to this characteristic, and prevent self-assembly vertical to the interfaces. But for the phenyl ring–based building units, this can be solely explained with the geometrical confinement that is provided by the interaction with the interfaces. Also, their relative weakness compared with the chemisorption or complexation gives the self-assembly units enough room to be accommodated into the minimum energy geometry. For most cases, the building units are entirely physisorbed at the interfaces with a "flat" mode. But, even though some specific functional group may have more of a tendency to be physisorbed on the specific site of the interface, especially on the solid surfaces with physical or chemical inhomogeneity, the nearest physisorption of the adjacent building unit is likely to happen in a way that provides those building units enough room to be positioned along the surface by the interaction of the rest of the building units. It may also be possible that the adsorption site is favorably rearranged by the adsorption of the next building unit.

For cases where specific functional groups have the capability to be chemisorbed or complexed with the specific site on the solid surfaces, and where the surface sites are distributed closely enough and thus are less than the dynamic molecular length of the building units, there can be a strong possibility that the units are self-assembled with the geometry of being upright to the surfaces. The formation of the self-assembled monolayer (SAM) is one good example. This will be described in Chapter 11.

6.2.3. Packing of Nonamphiphilic Building Units

For nonamphiphilic building units, the concept of *g*-value can be useful, as long as the structures of those building units are well defined. There is no functional group that can be chemisorbed or complexed on the surfaces. Thus, there can be no upright mode of packing for this type of interfacial self-assembly, unless there is a strong external force that can overcome the adsorption force and intermolecular/colloidal forces. For example, the strong mechanical pressure in the formation of the Langmuir monolayer can sometimes induce the upright mode of packing. Figure 6.6 shows the typical schematic representation. For the spherical building units, the hexatic-type of self-assembled aggregates should be the typical structure, regardless of the types of interfaces and building units (atoms, molecules, colloids; Gyarfas et al., 2005). For the nonspherical building units, there can be relatively weak directional interactions between them. This structural factor can be understood as the two-dimesnional projection of the formation of liquid crystals (Chapter 4). Thus, the self-assembled aggregates usually have

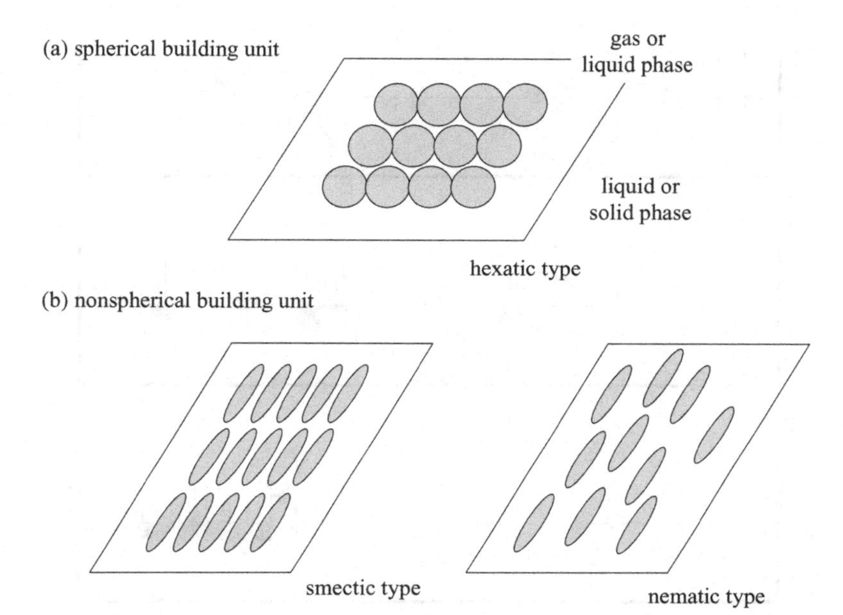

Figure 6.6. Typical packing of (a) spherical and (b) nonspherical nonamphiphiles at interfaces.

structures that are analog with those of liquid crystals. (For details on smectic and nematic types of arrangement, see Figure 4.8.) They also often appear with surface defects such as spiral.

6.3. SELF-ASSEMBLY AT THE GAS–LIQUID INTERFACE

Two typical cases of interfacial self-assembly at the gas–liquid interface will be discussed in this section: the Langmuir monolayer and the surface micelle. The Langmuir monolayer can be formed at the liquid–liquid interface, too, usually at the aqueous solution–oil interface. And the surface micelle can be also assembled at the liquid–liquid interface if the conditions are right.

6.3.1. Langmuir Monolayer

Figure 6.7 shows the schematic representation of the formation of a Langmuir monolayer through the external force (mechanical force)–induced self-assembly of amphiphiles at the gas–liquid interface. For certain amphiphiles that have extremely low solubility with a given liquid, such as bilayer-forming amphiphiles in water, depending on the types and interplay of intermolecular and colloidal forces, they can form the surface micelle through flat packing, too. This will be described in the next subsection. It is especially true when the hydrophobic

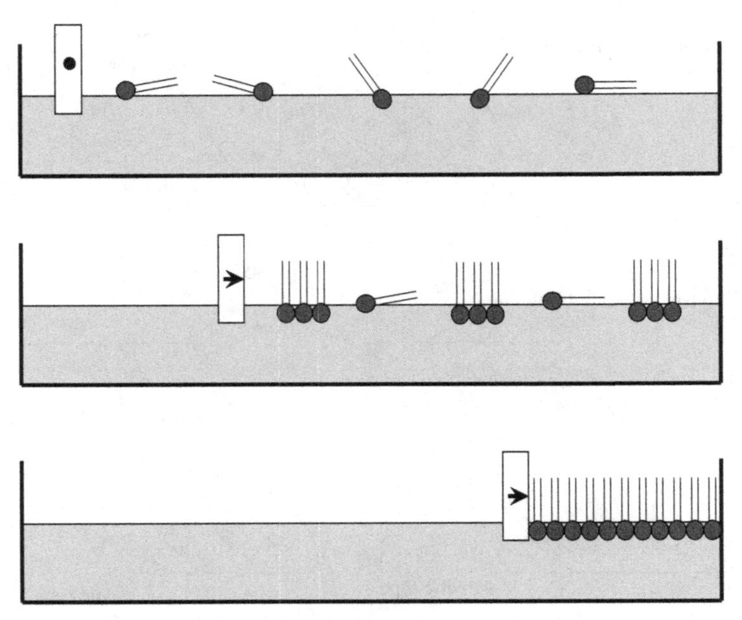

Figure 6.7. External force–induced self-assembly of amphiphiles at the gas–liquid interface: Langmuir monolayer.

interaction that is the primary attractive force for this self-assembly is strong enough to attract the building units closer. Electrostatic repulsion (in cases of ionic units) and hydration (in cases of nonionic units) forces are played as main repulsive balancing forces in the formation of surface micelles. But, when the repulsive forces are dominant over the attractive force at the flat mode, the building units are simply dispersed on the liquid surface in mainly monomer forms. This can happen by the molecular origin, or simply when the surface concentration is far below the *cmc* of surface micellization.

However, when the external force is applied so as to systematically decrease the total area of the liquid surface that is occupied by the monomeric forms of the building units, or as the area occupied by each building unit is decreased, the dominant repulsive forces between them leave as the only choice the upright packing of those building units. They simply cannot escape to the gas phase, and their extremely low solubility prevents them from being extracted out into the liquid phase (often called *subphase*) either. Once those building units are aligned in upright mode, side-by-side, they can *now* induce the dipole–dipole interaction between them, which originates from the aligned dipole moment of the individual building units. This acts as an additional attractive force, which, with the already-presented hydrophobic force, is being balanced with the repulsive forces. Also, once self-assembly occurs, the line tension around the self-assembled aggregates acts as another additional attractive force. Thus, the critical surface concentration(s) (critical surface area per molecule) of the self-assembly, and the size and shape of the self-assembled aggregates, are determined by the balancing of all these forces involved. Simply changing any of these forces will trigger the whole change toward a new force balance, which will result in new self-assembled aggregates. More about external force–induced self-assembly and the role of colloidal phenomena–induced forces such as the line tension will be presented in Chapter 12. The approach toward the formation of the Langmuir monolayer with this force balance scheme is useful as it can help quantify the individual factors for the entire process; thus, it can provide better control of this popular self-assembly process.

Figure 6.8 shows the experimental results of the formation of the Langmuir monolayer of dipalmitoyl phosphatidylcholine (DPPC), which is one of the typical lipids. The subphase is water, and the results in aqueous silicate solution are also given. The photographs in the figure are Brewster Angle Microscopy (BAM) pictures. This diagram of the change of surface pressure as a function of the surface area occupied by individual molecules is often called a *surface pressure–area isotherm*. For both subphases, DPPC shows a typical change of surface phases from liquid expanded (LE), to its mixed region with liquid condensed (LC), to pure liquid condensed, to solid (S) phase. Details for this type of phase behavior and the general concept of the Langmuir monolayer can be found in excellent resources (Kaganer et al., 1999; Vollhardt, 1999). Typical Langmuir monolayer–forming amphiphiles include long-chain alcohols or carboxylic acids

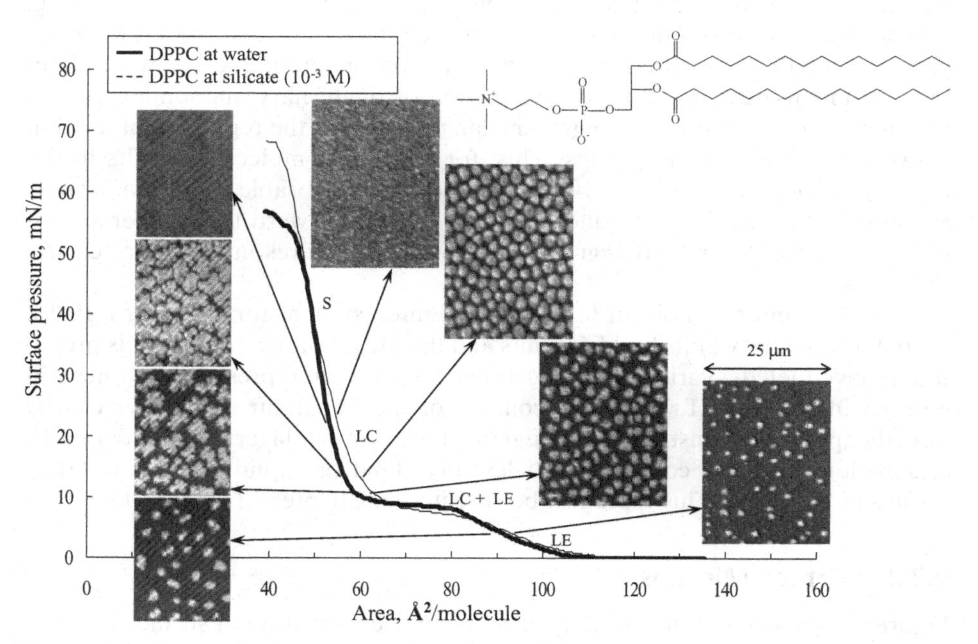

Figure 6.8. Formation of a Langmuir monolayer of dipalmitoyl phosphatidylcholine (DPPC) at the air–water interface.

with chain length usually more than 18 carbons, surfactants with double or triple long alkyl chains, and most of the biological lipid molecules.

What is obvious from these results is that they clearly show that the general self-assembly scheme in Figure 1.3 can be directly applied to this interfacial self-assembly. For both subphases, the clear surface micelles are observed at the low surface pressure region of the LE phase. As the surface pressure is increased, in other words, as the surface concentration is increased, the sizes of these primary self-assembled aggregates grow substantially (LE/LC mixed regions), and of course the concentrations of the self-assembled aggregates are increased, too. This can be understood as their growth due to the increase in the external pressure, somewhat like the transition of spherical micelles to rod-shaped and/or wormlike micelles in bulk solution (Chapter 3). As the surface pressure is further increased to the region where the LC/LE mixed phase is transformed into the LC-only phase, the picture clearly shows that those primary self-assembled aggregates begin to be assembled together. This can be viewed as secondary self-assembly. And it finally forms secondary self-assembled aggregates at the high surface pressure region of the S phase, which is a monolayer in this case.

Figure 6.8 also shows the effect of counterion binding on the Langmuir monolayer. DPPC is a zwitterionic amphiphile. Its head group has both cationic quaternary ammonium and anionic phosphate groups. When the DPPC is placed at the solution with silicate ions that are anionic, the quaternary ammonium parts attract the negatively charged silicate ions. This is clearly a counterion binding, only at the surface (Figure 3.6). Thus, the repulsion between the DPPC head groups should be decreased, but the negative phosphate parts that were effectively compensated with the positive charge of quaternary ammonium groups through the zigzag type of arrangement can now govern the repulsive interaction between the DPPC head groups. This, for the whole molecule, results in the increase of the repulsive force. It is, of course, a less favorable condition for self-assembly. Thus, the force balance should be shifted toward the smaller size of the primary self-assembled aggregates. The BAM pictures in the figure clearly show this.

The Langmuir monolayer has a rich and interesting history (Tanford, 1989). Also, the versatility of its building units and the facile characteristics of its preparation have fueled a variety of applications. Some of the representative ones will be seen in Chapter 11. Also, this concept of the Langmuir monolayer can be directly applied to construct the Langmuir type of monolayer of colloidal particles, as long as those colloidal particles stay afloat on liquid surfaces (Gattás-Asfura et al., 2005). This topic will be revisited in Chapter 12 (Figure 12.5).

6.3.2. Surface Micelles

Figure 6.9 shows the self-assembly process for the formation of surface micelles of amphiphiles at the gas–liquid interface. This can include surfactants and amphiphilic polymers. The term *surface micelle* can be used to encompass a wide

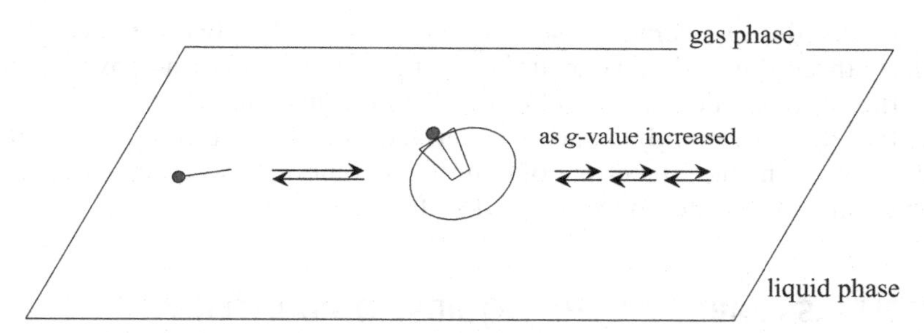

Figure 6.9. Self-assembly of amphiphiles at the gas–liquid interface for the formation of surface micelles.

range of self-assembled aggregates at various surfaces. But it is generally reserved for the amphiphile micelles at the gas–liquid interface (Fontaine, 2005; Kato and Iimura, 2005). It has been predicted and confirmed that the three-dimensional concept of molecular packing for self-assembly of amphiphiles in bulk can be directly applied to the two-dimensional analogs (Israelachvili, 1994). For typical amphiphiles, their monomers should be in equilibrium with self-assembled micelles at a defined surface concentration (surface *cmc*) at a given condition. As explained in Figure 3.8 for the case of bulk micellization, structural evolution by control of the *g*-value should be possible, too. As for the case of the liquid expanded phase of the Langmuir monolayer, the line tension around the surface micelles should be included as an additional attractive force.

6.4. SELF-ASSEMBLY AT THE LIQUID–SOLID INTERFACE

Two typical types of self-assembled aggregates of amphiphiles formed at liquid–solid interfaces through upright packing are *hemimicelles* and *semimicelles* (Figure 3.7) (Rabe, 1998; Manne and Warr, 1999; Miller and Paruchuri, 2004; Paria and Khilar, 2004). Hemimicelles are formed when the amphiphiles in liquid are adsorbed onto hydrophobic solid surfaces. The strong hydrophobic interaction of their alkyl chain with the solid surface forces the initially adsorbed monomers into the lying-down mode. Subsequent monomers are now packed within the scheme of their force balance. This should be true, as long as the interaction between their alkyl chains is dominant over their interaction with the solid surface. The molecular packing by following their molecular geometry should bring this type of hemimicelle when the *g*-value is close to 1/2. Again, as the *g*-value is gradually increased, this hemisphere structure should evolve into higher-order structures, such as rod-shaped or wormlike ones. The same is true for the formation of semimicelles, except that they are formed when the ionic amphiphiles are adsorbed onto hydrophilic solid surfaces. When amphiphiles have high *g*-value, self-assembled structures of a well-defined monolayer or even bilayer can be imagined as forming at liquid–solid surfaces.

Hemimicelles and semimicelles can be considered as two-dimensional projections of three-dimensional micelles in solution. They are formed by an entropy-driven thermodynamic process, like micellization in bulk (Zhu and Gu, 1991). Thus, all of those thermodynamic facts derived from it stand true just as for bulk micelles. These include the effect of counterion binding, the effect of concentration, synergism for mixed micelles, and so forth.

6.5. SELF-ASSEMBLY AT THE LIQUID–LIQUID INTERFACE

With a careful selection of building units, interfacial self-assembly for the formation of the Langmuir monolayer at given liquid–liquid interfaces can be induced. It is possible with both molecular and colloidal building units, and the interfaces are usually aqueous phase contacted with nonpolar fluids. External force (surface pressure) should be able to be applied onto the building units that are dispersed on the interfaces with proper experimental setup, which in many cases induces the sequence of phase transition (or multistep interfacial self-assembly) that is quite similar to that of Figure 6.8.

The building units should, of course, be afloat at liquid–liquid interfaces and fulfill the requirement of the force balance. This means that, as long as they are confined at the interface by possibly physisorption, a variety of building units should be able to be self-assembled. This includes amphiphiles, amphiphiles with functional groups, nonamphiphilic molecules, and colloidal particles. For the amphiphilic building units, since they can sit at aqueous solution–oil interfaces with a thermodynamic favorability that originates in their amphiphilic nature, they can be self-assembled into the surface micelle or further assembled into the Langmuir monolayer. Once formed, the concept of molecular packing (g-value) should be directly applicable for the amphiphile surface micelles as for the cases at the gas–liquid interface. Amphiphilic building units with functional group(s) will be self-assembled by the dominance of the functional forces and their directionality. Nonamphiphilic building units such as nanoparticles also will be able to be self-assembled into a variety of morphologies but with a long-range structural order based on the shift-direction of their force balance (Binder, 2005; Dryfe, 2006), as shown in Figure 6.6.

6.6. SELF-ASSEMBLY AT THE GAS–SOLID INTERFACE

For self-assembly at gas–solid interfaces, since the building units are deposited onto the solid surfaces as gas phase or in high vacuum, it can hardly be spontaneous. Self-assembly at the gas–solid interface is, in almost all cases, a kinetically driven process. Furthermore, competition with thermodynamics greatly affects the whole process and the structures and properties of the self-assembled aggregates formed. This fact often leads to *self-organization*. Its process shows quite different characteristics depending on the relation between the two groups of interactions: intermolecular interaction among the building units and surface-building unit interaction. When the intermolecular forces among the building units that are adsorbed onto the solid surface are greater than the interaction with a solid surface, the adsorption force is only able to play as the *template force* that provides the temporary confinement of those building units at the surface. The building units are almost free to be rearranged on the surface toward their force balance. Thus, the self-assembly process itself, and the size and shape of the self-assembled aggregates, are mainly determined by this interplay of intermolecular foreas (or colloidal forces in cases of colloidal building units) between them. Previous cases of self-assemblies on gas–liquid and liquid–solid interfaces also belong to this category. A variety of atomic and molecular building units such as porphyrin derivatives, and alkyl chain–based functional units from Figure 6.5 can be introduced for this type of interfacial self-assembly, as long as they can be adsorbed from their vapor phase or in high vacuum. The formation of a *molecular corral* by the self-assembly of haloalkane on a semiconductor surface provides an excellent example (Dobrin et al., 2006). It proves that self-assembly indeed can create a nanostructure with delicate functional structure without time-consuming one-by-one manipulation of its building unit. This is one of the key points I will make in Chapter 8: why self-assembly is such an important topic for the development of nanotechnology.

When the strength of the intermolecular forces is similar to that of solid surface–building unit interactions, this interaction, which is mainly physisorption, becomes an important part of the force balance scheme as an attractive force. When force interplay becomes the case for the idealistic unpatterned smooth solid surface, it is very possible for it to be self-assembled as an evenly developed monolayer. However, this cannot always be the case for realistic systems. When the solid surface is prepatterned with a higher physisorption capability component, or if there is a preexisting nanoscale pattern such as a stepped surface or a well-structured characteristic pattern such as on a single crystal surface, which usually has higher affinity toward adsorption, this higher adsorption site (or area) becomes a *surface well* for interfacial self-assembly. The building units can have the tendency to be adsorbed more exclusively onto this site at the initial stage of the self-assembly; then the force boundary provided by the sites confines those building units that are adsorbed onto the site, and forces them to progress toward the force balance within this space, which is usually nanoscale. Most of the molecular and atomic building units can be introduced for this type of interfacial self-assembly. One of the special but well-developed cases can be found in the epitaxial growth of crystals of one material on the crystal face of another (heteroepitaxy) or the same (homoepitaxy) material (*epi* and *taxis* mean *equal* and *in ordered manner*, respectively, in Greek). Along with the comparative strength of the intermolecular forces with solid surface–building unit interaction, which leads to the condition of an intermediate contribution ratio of kinetics to thermodynamics, the geometrical matching between the interaction points on both sides of the building units and the solid surface is key to realizing this highly tricky but sophisticated self-assembly process. This geometrical matching consideration will be revisited for the sake of understanding nanostructured materials in Figure 9.10.

This epitaxial deposition technique has a huge impact on the development of semiconductor film–based devices such as silicon germanide, gallium nitride, gallium arsenide, indium phosphide, and so on. At some sense, it can be viewed as interfacial self-assembly with a directional force. But, in this case, this directional force (the highly selective building unit–surface interaction) can help not only self-assemble patterned surfaces but create the pattern-on-pattern type of self-assembled aggregates as well, if it is well designed along with functional groups that have directional force capability. Also, in this case, it is possible that the electron transfer would occur between the building units and the solid surfaces, especially for metals and semiconductors. This obviously reveals the importance of electric repulsion as a major repulsive force, and even if there is no electron transfer, the induced dipole moment can have a significant impact on its repulsive force component.

For the next situation, where the intermolecular forces are smaller than the solid surface–building unit interaction, the building units that are adsorbed onto the solid surface have little tendency to be reorganized following their force balance. In this case, instead of assembly of building units, *surface reconstruction* by possible surface instability and a coarsening process often results in the formation of a periodic nanoscale surface pattern (Soukiassian and Enriquez, 2004). This can be understood as the self-assembled aggregate that is driven by the force balance among the forces originated on/from the solid surfaces. This process is often triggered by an external perturbation such as sputtering or annealing (Deak et al., 2006). *Surface relaxation* is the case in the rearrangement of the uppermost atoms in a direction perpendicular to the surface. Since it is also a process toward the force balance between the atoms that are rearranged and those beneath the surface, it can also be categorized within this situation. But this process does not induce any type of distinctive surface structure.

6.7. INTERFACE-INDUCED CHIRAL SELF-ASSEMBLY

Molecules, self-assembled aggregates, or any other hard objects are called *chiral* when their structures cannot be superimposed on their mirror symmetry. The importance of this chirality in nature and in many modern technological applications has long been recognized (Lough and Wainer, 2002; Pérez-García and Amabilino, 2002). Chirality of self-assembled aggregates can be expressed both in bulk and at surfaces, and there are very interesting examples of how the delicate force balance between self-assembly building units and surfaces can create a variety of structural and functional diversities.

Chiral self-assembly can be defined as any self-assembly (bulk or interfacial), whether its building unit is chiral or achiral, that forms self-assembled aggregates that are chiral. Bulk chiral self-assembly can be understood as part of directional self-assembly, and many bio-mimetic self-assemblies show chiral nature, too.

This will be discussed in the next chapter. Chirality of individual molecules and hard objects does not belong to the scope of this book. This section mainly addresses interfacial chiral self-assembly—especially, how the view of force balance can be adopted to explain the formation of chiral self-assembled aggregates from achiral building units at interfaces. This process can be called *pseudo-bio-mimetic self-assembly*, since the key is not the chirality of the building units. Rather, it is the chiral symmetry breaking by the "asymmetric packing" of those building units (Spector et al., 2003). This term also embraces the formation of chiral self-assembled aggregates from chiral building units, and of achiral ones from chiral or achiral building units. For bulk chiral self-assembly, there should be three components expressed within the building units. These are hydrophilic moiety, hydrophobic moiety, and functional moiety. For induced chiral self-assembly at interfaces, the building units need to have a functional moiety, such as for hydrogen bonding, in many cases. But they do not need to have hydrophilic and hydrophobic moieties at all times to be self-assembled. For some cases of induced chiral self-assembly, none of these structural prerequisites is required.

First, let me briefly mention the chirality expression of the individual self-assembly building unit within self-assembled aggregates at surfaces. Most of the chiral building units retain their chirality within self-assembled aggregates. This is the case for achiral building units, too. Most of the achiral building units directly transfer their nonchirality into their self-assembled aggregates. However, when the packing mode of the building units within self-assembled aggregates changes their molecular symmetry, the chirality expression can be changed dramatically, too. When the chiral building units are assembled too tightly through strong intermolecular forces such as hydrogen bonding, or in such a way as to be a specific geometry, the transfer of their chiral information is prevented, which can induce the loss of their molecular chirality (Zhang et al., 2005). On the other hand, when the achiral building units are confined at interfaces (or adsorbed onto surfaces) in a way so as to break (reduce) their molecular symmetry, or in such a way as to be geometrically mismatched with the lattice structure of the substrates (especially solid substrate), they can be expressed as chiral entities.

This notion of the molecular chirality of individual molecules within self-assembled aggregates can be directly expanded to understand the formation of the chirality expression of self-assembled aggregates from achiral building units. The key point is how the building units that are achiral in three dimensions break (or reduce) their symmetry at interfaces, and how this symmetry breaking propagates throughout the all self-assembled aggregates within the scheme of the force balance for the formation of all self-assembled aggregates. Two main mechanisms can be envisioned. The first is the case where the building units are assembled through "distorted" or "tilted" packing. In most cases, this mechanism is strongly related to the hydrogen-bonding capability of the building units. A hydrogen bond is a typical directional force in self-assembly. It forces self-assembled aggregates to have a directional structural characteristic. But, since this force is much stronger than the other intermolecular forces that are involved in the self-assembly process, it, in many cases, can bring those building units too close. If, in this situation, there happens to be another intermolecular force(s) within this shortened space in that self-assembly system, there will be a great chance that this intermolecular force(s) does not play to keep the directionality of the hydrogen bonding intact. The tendency should be the same whether that force is attractive, such in as π–π stacking (Huang et al., 2005), or repulsive, such in as the steric effect. Thus, the only way to compensate (or to reach new force balance) is by asymmetric packing of the building units, so this extra force can be relaxed evenly. Self-assembly will be propagated throughout the self-assembled system with this way of packing until the force balance for all aggregates is reached. Thus, the result is self-assembled aggregates that have chirality.

The second mechanism is strongly involved in shape of the building units. Figure 6.10 shows the schematic representation. When the building units have a "bent" or "kinked" shape, which can be generic, such as in the case of many liquid crystals (Koshima, 2000) or dicarboxylic acid (Stepanow et al., 2005), or induced by the adsorption onto surfaces such as long alkyl chain–based building units (Ca and Bernasek, 2004), they can be adsorbed onto the surfaces in such a way that the two geometries cannot be superimposable on each other, even though they are achiral in three dimensions. Each of the geometries can be developed as its own self-assembled aggregate, thus, forming self-assembled aggregates that are chiral. Many cases of this process are also involved in hydrogen bonding. But other forces, such as van der Waals, can sometimes induce this type of chiral interfacial self-assembly, too, especially where the van der Waals force is stronger than usual with, for example, much longer alkyl chains or multiple chains (Tao and Bernasek, 2005).

The already-given chirality of the building units can be expressed as achiral self-assembled aggregates. This can be understood as that the chiral information of the building units has been prevented from being transferred into the self-assembled aggregates. This can happen when the intermolecular interaction between the building units is smaller than the building unit–surface interaction. And it occurs in such a way that the self-assembled aggregate has a morphology whose structural conformation that is needed for chiral expression is limited or restricted. Simply speaking, too much packing either by hydrogen bonding or by adsorption can induce the symmetric assembly of building units.

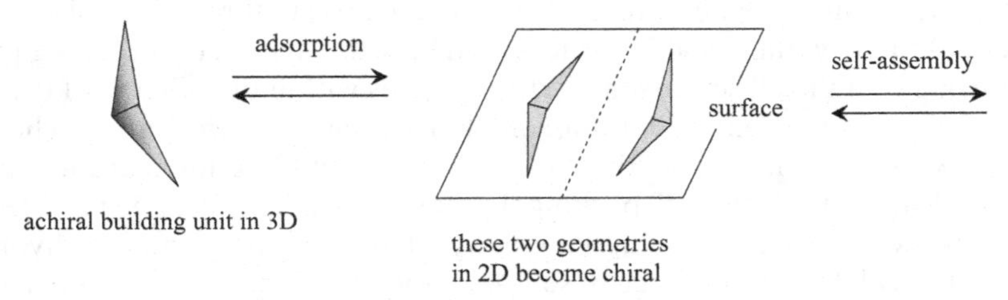

Figure 6.10. Chiral self-assembly induced by adsorption.

Chiral self-assembly can be precisely controlled once the interplay between involved forces is fully identified and understood. Asymmetric packing of building units can also be manipulated by the application of external forces such as magnetic field (Berg and Patrick, 2005). This fact can help vastly expand the diversity of self-assembled aggregates (which happens to be a nanometer scale) and the availability of the building units for chirality expression in self-assembly, not only for molecular but for polymeric, biological, and colloidal building units as well, which in turn can provide a variety of potential applications in the nanotechnology field.

REFERENCES

Balaban, T. S., Buth, G. "Biomimetic light harvesting," *Nachrichten—Forschungszentrum Karlsruhe* **37**, 204 (2005).

Barth, J. V., Costantini, G., Kern, K. "Engineering Atomic and Molecular Nanostructures at Surfaces," *Nature* **437**, 671 (2005).

Berg, A. M., Patrick, D. L. "Preparation of Chiral Surfaces from Achiral Molecules by Controlled Symmetry Breaking," *Angew. Chem. Int. Ed.* **44**, 1821 (2005).

Binder, W. H. "Supramolecular Assembly of Nanoparticles at Liquid–Liquid Interfaces," *Angew. Chem. Int. Ed.* **44**, 5172 (2005).

Ca, Y., Bernasek, S. L. "Adsorption-Induced Asymmetric Assembly from an Achiral Adsorbate," *J. Am. Chem. Soc.* **126**, 14234 (2004).

Deak, D. S., Silly, F., Newell, D. T., Castell, M. R. "Ordering of TiO_2-Based Nanostructures on SrTiO3(001) Surfaces," *J. Phys. Chem. B* **110**, 9246 (2006).

De Feyter, S., De Schryver, F. C. "Self-Assembly at the Liquid/Solid Interface: STM Reveals," *J. Phys. Chem. B* **109**, 4290 (2005).

Dobrin, S., Harikumar, K. R., Jones, R. V., Li, N., McNab, I. R., Polanyi, J. C., Sloan, P. A., Waqar, Z., Yang, J. (S. Y.), Ayissi, S., Hofer, W. A. "Self-Assembled Molecular Corrals on a Semiconductor Surface," *Surf. Sci.* **600**, L43 (2006).

Drain, C. M., Batteas, J. D., Flynn, G. W., Milic, T., Chi, N., Yablon, D. G., Sommers, H. "Designing Supramolecular Porphyrin Arrays that Self-Organize into Nanoscale Optical and Magnetic Materials," *Proc. Natl. Acad. Sci. USA* **99**, 6498 (2002).

Dryfe, R. A. W. "Modifying the Liquid/Liquid Interface: Pores, Particles and Deposition," *Phys. Chem. Chem. Phys.* **8**, 1869 (2006).

Fontaine, P., Goldmann, M., Muller, P., Fauré, M.-C., Konovalov, O., Krafft, M. P. "Direct Evidence for Highly Organized Networks of Circular Surface Micelles of Surfactant at the Air–Water Interface," *J. Am. Chem. Soc.* **127**, 512 (2005).

Gattás-Asfura, K. M., Constantine, C. A., Lynn, M. J., Thimann, D. A., Ji, X., Leblanc, R. M. "Characterization and 2D Self-Assembly of CdSe Quantum Dots at the Air–Water Interface," *J. Am. Chem. Soc.* **127**, 14640 (2005).

Gyarfas, B. J., Wiggins, B., Zosel, M., Hipps, K. W. "Supramolecular Structures of Coronene and Alkane Acids at the Au(111)-Solution Interface: A Scanning Tunneling Microscopy Study," *Langmuir* **21**, 919 (2005).

Huang, X., Li, C., Jiang, S., Wang, X., Zhang, B., Liu, M. "Supramolecular Chirality of the Hydrogen-Bonded Complex Langmuir-Blodgett Film of Achiral Barbituric Acid and Melamine," *J. Coll. Inter. Sci.* **285**, 680 (2005).

Israelachvili, J. "Self-Assembly in Two Dimensions: Surface Micelles and Domain Formation in Monolayers," *Langmuir* **10**, 3774 (1994).

Kaganer, V. M., Möhwald, H., Dutta, P. "Structure and Phase Transition in Langmuir Monolayers," *Rev. Mod. Phys.* **71**, 779 (1999).

Kato, T., Iimura, K. "Micro-Phase Separation in Two-Dimensional Amphiphile Systems," *Surfactant Science Series*, Vol. 124 (Mixed Surfactant Systems), 2nd ed., pp. 59–91 (Marcel Dekker: 2005).

Koshima, H. "Generation of Chirality in Two-Component Molecular Crystals from Achiral Molecules," *J. Mol. Struct.* **552**, 111 (2000).

Lackinger, M., Griessl, S., Heckl, W. M., Hietschold, M., Flynn, G. W. "Self-Assembly of Trimesic Acid at the Liquid–Solid Interface: A Study of Solvent-Induced Polymorphism," *Langmuir* **21**, 4984 (2005).

Langford, S. J., Lau, V.-L., Lee, M. A. P., Lygris, E. "Porphyrin-Based Supramolecules and Supramolecular Arrays," *J. Porphyrins Phthalocyanines* **6**, 748 (2002).

Lough, W. J., Wainer, I. W., eds., *Chirality in Natural and Applied Science* (CRC Press: 2002).

Manne, S., Warr, G. G. "Supramolecular Structure of Surfactants Confined to Interfaces," *ACS Symposium Series*, 736 (Supramolecular Structure in Confined Geometries), pp. 2–23 (American Chemical Society: 1999).

Miller, J. D., Paruchuri, V. K. "Surface Micelles as Revealed by Soft Contact Atomic Force Microscopy Imaging," *Recent Res. Devel. Surf. Coll.* **1**, 205 (2004).

Myers, D. *Surfaces, Interfaces, and Colloids: Principles and Applications*, 2nd ed. (Wiley: 1999).

Paria, S., Khilar, K. C. "A Review on Experimental Studies of Surfactant Adsorption at the Hydrophilic Solid–Water Interface," *Adv. Coll. Inter. Sci.* **110**, 75 (2004).

Pérez-García, L., Amabilino, D. B. "Spontaneous Resolution under Supramolecular Control," *Chem. Soc. Rev.* **31**, 342 (2002).

Rabe, J. P. "Self-Assembly of Single Macromolecules at Surfaces," *Curr. Opin. Coll. Inter. Sci.* **3**, 27 (1998).

Soukiassian, P. G., Enriquez, H. B. "Atomic Scale Control and Understanding of Cubic Silicon Carbide Surface Reconstructions, Nanostructures and Nanochemistry," *J. Phys.: Condens. Matter* **16**, S1611 (2004).

Spector, M. S., Selinger, J. V., Schnur, J. M. "Chiral Molecular Self-Assembly," *Materials-Chirality: Volume 24 of Topics in Stereochemistry*, Green, M. M., Nolte, R. J. M., Meijer, E. W. eds., pp. 281–372 (Wiley: 2003).

Stepanow, S., Lin, N., Vidal, F., Landa, A., Ruben, M., Barth, J. V., Kern, K. "Programming Supramolecular Assembly and Chirality in Two-Dimensional Dicarboxylate Networks on a Cu(100) Surface," *Nano Lett.* **5**, 901 (2005).

Suo, Z., Lu, W. "Forces that Drive Nanoscale Self-Assembly on Solid Surfaces," *J. Nanoparticle Res.* **2**, 333 (2000).

Tanford, C. *Ben Franklin Stilled the Waves: An Informal History of Pouring Oil on Water with Reflections on the Ups and Downs of Scientific Life in General* (Duke University Press: 1989).

Tao, F., Bernasek, S. L. "Chirality in Supramolecular Self-Assembled Monolayers of Achiral Molecules on Graphite: Formation of Enantiomorphous Domains from Arachidic Anhydride," *J. Phys. Chem. B* **109**, 6233 (2005).

Vollhardt, D. "Phase Transition in Adsorption Layers at the Air–Water Interface," *Adv. Coll. Inter. Sci.* **79**, 19 (1999).

Wan, L.-J. "Fabricating and Controlling Molecular Self-Organization at Solid Surfaces: Studies by Scanning Tunneling Microscopy," *Acc. Chem. Res.* **39**, 334 (2006).

Yaminsky, I., Gorelkin, P., Kiselev, G. "Concurrence of Intermolecular Forces in Monolayers," *Japn. J. App. Phys.* **45**, 2316 (2006).

Yokoyama, T., Yokoyama, S., Kamikado, T., Okuno, Y., Mashiko, S. "Selective Assembly on a Surface of Supramolecular Aggregates with Controlled Size and Shape," *Nature* **413**, 619 (2001).

Zhang, J., Gesquière, A., Sieffert, M., Klapper, M., Müllen, K., De Schryver, F. C., De Feyter, S. "Losing the Expression of Molecular Chirality in Self-Assembled Physisorbed Monolayers," *Nano Lett.* **5**, 1395 (2005).

Zhu, B.-Y., Gu, T. "Surfactant Adsorption at Solid–Liquid Interfaces," *Adv. Coll. Inter. Sci.* **37**, 1 (1991).

7

BIO-MIMETIC SELF-ASSEMBLY

Bio-mimetic self-assembly is the area where we can find the most diverse self-assembly processes and the richest spectrum of self-assembled aggregate structures.

Biological living systems are full of self-assembled aggregates with sophisticated levels of design and a wide range of structural and functional diversities. Many of their living mechanisms are also based on self-assembly processes at molecular and nanometer scales. All of these related processes where the operation originates in the balancing of intermolecular forces can be defined as "self-assembly from biological systems" or simply *biological self-assembly*.

There have long been efforts to express those *in vivo* features of biological self-assembly in terms of *in vitro* features (Zemb and Blume, 2003). This is where the new term(s) come into play. Whenever those entities in biological systems or those entities that are derived from the structural features of those biological entities are subject to the force balancing of intermolecular forces, that is, self-assembly, these processes can be defined as *bio-mimetic self-assembly* or *bio-inspired self-assembly*, or even *chiral self-assembly*, obviously to emphasize the chiral aspect. Throughout this chapter and this book, the term *bio-mimetic self-assembly* will be used for clarity.

Bio-mimetic self-assembly is a stepwise, hierarchical, chiral, and directional process. It forms highly functional self-assembled aggregates such as fibril, tube, column, helix, fiber, sheet, and so on, with well-controlled, predetermined sizes and shapes. These are usually much bigger than conventional molecular self-assembled aggregates such as surfactant micelles. They have ~1—~1000 μm range of length along their long axis and ~10—~100 nm range of size along their small axis. They are also more stable than others at a wider range of pHs and temperatures. Largely, bio-mimetic self-assembly is a molecular self-assembly. But the scope of its potential primary building units is much larger than that of typical molecular self-assemblies. They can range from the pure biological molecules such as amino acids, peptides, lipids, and sugars to those "bio-mimetic" molecules that are derived from these biological molecules. Some typical examples of these biologically derived molecules are peptidic amphiphiles, that is, amphiphiles with their head groups modified with peptides, sugar-group–based amphiphiles, and molecules with multiple hydrogen-bonding sites.

The critical difference between bio-mimetic self-assembly and the "conventional" self-assemblies described in previous chapters is that here the process occurs through asymmetric packing of its primary building units. This is the key to understand bio-mimetic self-assembly. And all of the fascinating features of bio-mimetic self-assembled aggregates are primarily derived from this aspect. Three major factors are strongly tied in with this: chirality of building units, asymmetric structure of building units, and asymmetric directional interaction between building units. A significant portion of this chapter will be allocated to this issue, along with the critical role of the directional forces of hydrogen and coordination bonds.

This chapter is designed to cover mainly bio-mimetic self-assembly, and its implication for nanotechnology. However, bio-mimetic self-assembly is inevitably integrated with biological self-assembly. Fundamental aspects and typical examples of biological self-assembly will be described throughout this chapter, too, whenever it is necessary to clarify bio-mimetic self-assembly.

7.1. GENERAL PICTURE OF BIO-MIMETIC SELF-ASSEMBLY

Figure 7.1 shows the schematic representation of the general process of bio-mimetic self-assembly. The primary building units, that is, the monomers, of bio-mimetic self-assembly come with a variety of structural and compositional diversities. Details on this will be found in the third section of this chapter. In most cases, hydrogen bonding is largely responsible for the primary self-assembly step, which makes the aggregates have a unique morphology and a fixed aggregation number. Rod-shaped dimers and disk-shaped oligomers are among the abundant forms. However, depending on the structure and functionality of the monomers and environmental conditions around them, that is, the conditions that can induce too loose or too tight packing at this step, conventional forms of self-assembled aggregates such as normal micelles, vesicles, and others are often

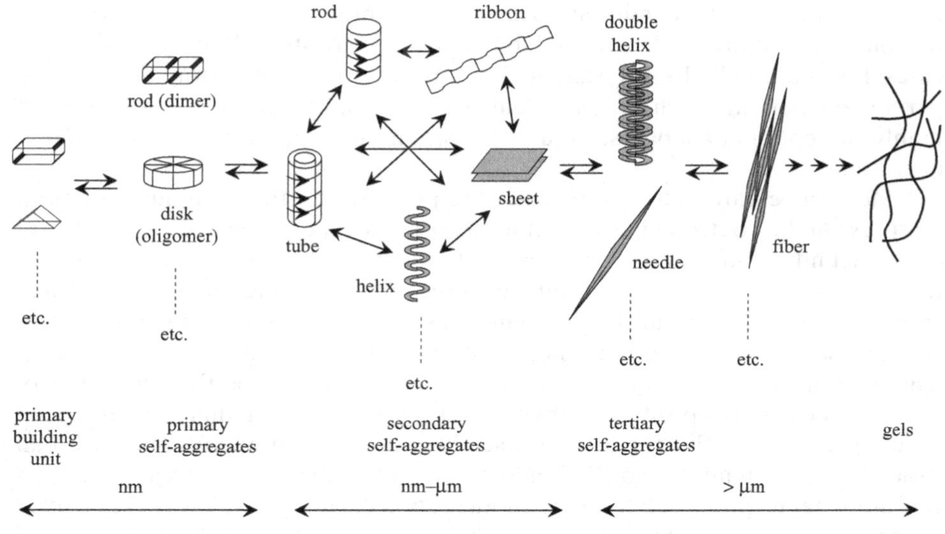

Figure 7.1. General scheme for bio-mimetic self-assembly.

assembled, too, even though hydrogen bonding is still involved. For a certain class of bio-mimetic self-assembly, the process occurs with only this first step. A typical example is avidin-biotin binding, which is a highly specific but strong one-way self-assembly with a strong action of hydrogen bonding with structural matching. Binding based on the cyclodextrin host can also be classified within this type of self-assembly. "Lock-and-key" or "nut-and-bolt" self-assembly might be a good name for this.

For most of the building units of bio-mimetic self-assembly, these primary self-aggregates possess enough intermolecular bonding capability to go through the secondary self-assembly step. The secondary self-aggregates formed at this step show a wide variety of structural diversities and a wide range of length scales at the same time. These include rod, hollow tube, ribbon, twisted tape, sheet, helix, and more, with their length ranging from nanometer to micrometer. Many of the unique functionalities of the bio-mimetic self-assembled aggregates are believed to have originated from the morphological uniqueness acquired at this step via this structural diversity. Tubular aggregates that beautifully serve as transmembrane channels (or pores) are one good example that can be found in living systems. Additional hydrogen bonding and π–π interaction are the forces that are responsible for the directionality of the secondary self-aggregates, their orientationality, and their highly asymmetric morphology.

Also, depending on the individual systems, the weak intermolecular forces such as van der Waals attraction take an important part in the force balance process. The morphology of each secondary self-aggregate is a unique result of force balancing between each primary building unit. It is also highly sensitive in each environmental condition, such as pH, counterion, ionic strength, concentration, temperature, and so forth. However, these are still self-assembled aggregates formed solely by intermolecular forces without the formation of any permanent bonding such as a covalent bond. Thus, for the most part, they are highly susceptible to subtle structural transition due to a change in any of these conditions.

Modern research has not reached the point where these unique structural features can be controlled at will, at least, at the level of structure/phase control of surfactant self-assembled aggregates. It will take the total understanding of the complete balance between all the forces involved. Geometrical matching should be included in the force balance, as well. Sometimes, this cooperative primary self-assembly and secondary self-assembly are involved with a fixed aggregation number. A typical example in living systems is the formation of microtubules. α- and β-tubulin subunits first form α,β-tubulin dimer by hydrogen bonding, and these dimers are self-assembled into helical microtubules with an exact aggregation number of 13 dimers in each 13 columns. This is a spontaneous thermodynamic process. Ferritin is another good example that always has a fixed aggregation number. Also, there can be a process without the formation of dimers or oligomers. One example is the formation of actin filaments. This linear helix is formed by chiral self-assembly of actin subunits, which is also a spontaneous thermodynamic process. As the concentration of filaments is increased, they are assembled into fiberlike or networked gel-like structures with the help of other proteins, fodrin and filamin.

Secondary self-aggregates can undergo a further self-assembly step, in many cases, by change of temperature and by the increase of the concentration of the building units. This forms tertiary self-aggregates. As is the case for the secondary ones, these tertiary structures also show a variety of diversities including double helix, needle, fibril, and superstructures such as vesicle-in-tube, and have much larger sizes that can easily pass beyond the micrometer scale. Directional hydrogen bonding and π–π interaction are not significantly involved in this step of self-assembly. The role of weak intermolecular forces becomes more important. A further step of assembly, whenever it is induced by another action of weak intermolecular forces, induces an even higher degree and larger size of self-assembled aggregates. Fiberlike structures whose size often reaches the millimeter range is a typical example. For example, in living organisms, the self-assembly of collagen all the way to the functional tissues fits with this scheme. Beyond this point, gelation is often the case for the next step of bio-mimetic self-assembly. Formation of gels depends, of course, on the intermolecular bonding capability between building units, but also on the type of solvents, that is, their intermolecular interaction with solvent molecules as well. This will be further described in the fifth section of this chapter.

Throughout the overall self-assembly processes of bio-mimetic building units, there almost always is chirality, whether it is molecular or morphological. It can be generic, and it can be induced, as well. Also, when it comes to the higher

TABLE 7.1. Bio-mimetic self-assembly versus conventional (non-bio-mimetic) self-assembly.

	Bio-mimetic Self-Assembly	Non-bio-mimetic Self-Assembly
Building unit	Chiral, asymmetric	Achiral, symmetric
Packing mode	Asymmetric	Symmetric
Directional force	Strong	Weak or none
Intermolecular force	Anisotropic	Isotropic
Assembly rate	Hr–week	< second
Aggregate structure	Chiral	Non-chiral
Aggregate size	nm–mm	nm
Stability	High: isolation possible	Low: isolation not possible
Monomer exchange	Slow	Fast
Disassembly	Not likely, slow	Constantly

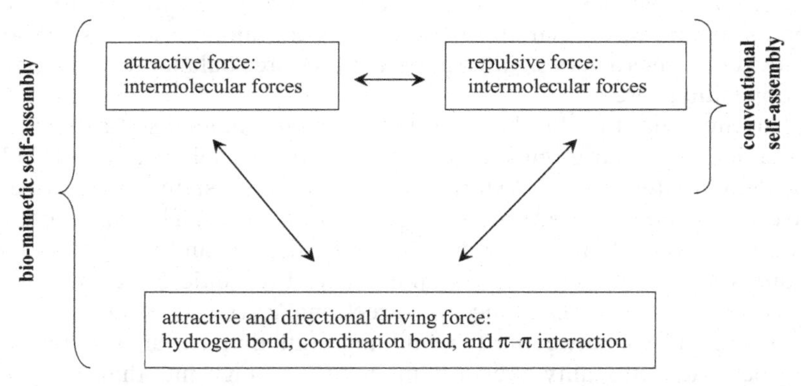

Figure 7.2. Force balances for bio-mimetic self-assembly and conventional self-assembly (non-bio-mimetic).

steps of self-assembly, this chirality characteristic is often combined with a *networkness* of the self-assembled aggregates. Table 7.1 summarizes the variety of unique characteristics of bio-mimetic self-assembly as compared with conventional (non-bio-mimetic) self-assembly.

7.2. FORCE BALANCE SCHEME FOR BIO-MIMETIC SELF-ASSEMBLY

Figure 7.2 shows the force balance scheme for bio-mimetic self-assembly. The scheme for conventional (non-bio-mimetic) self-assembly is also presented for comparison. For conventional self-assemblies, described in chapters 3 through 5, the self-assembled aggregates are formed as a result of the subtle balance of attractive driving forces with repulsive opposition forces. Both classes of forces are almost solely intermolecular forces (colloidal forces/phenomena in cases of colloidal self-assembly), and the directional forces are rarely involved during the balancing processes.

For bio-mimetic self-assembly, the role of relatively strong intermolecular forces becomes critical. These forces include hydrogen bond, coordination bond, and π–π interaction. These are attractive forces, but much stronger than the driving forces (such as van der Waals) for conventional self-assemblies. Thus, these forces become the major driving force for bio-mimetic self-assembly. Van der Waals force, in many cases, is present and becomes a part of the driving forces, but its contribution to the driving force factor is relatively less significant. The strong intermolecular forces also have a directional nature, which makes them act as directional (and functional) forces for bio-mimetic self-assembly as well. As in cases of conventional self-assemblies, the repulsive forces act as the opposition forces for bio-mimetic self-assembly, too. But here the repulsive forces are to be balanced with much stronger, directional driving forces. This may be the main reason that, in many cases of bio-mimetic self-assemblies, the steric effect becomes a very significant part of the force balance, as a major player in the opposition force.

It is quite usual that the building units for bio-mimetic self-assembly have much bulkier molecular groups than those of conventional ones. This is believed to provide a greater degree of steric repulsion to the system, so it can be more effective to counterbalance those stronger driving forces. They also often have those bulkier groups that seem to be "logically shaped" and "logically located." For many systems, this seems to be quite critical to guide the direction of the overall self-assembly process. Since the directional forces are local, meaning they are operating mostly between a few building units that are directly interacting, the degree of directionality over a number of building units that is formed by successive directional bondings can have a likelihood of geometrical distortion. This might even create self-assembled aggregates that have "chaotic" geometry. Cooperation with the *logical* steric effect seems to be another key to the formation of bio-mimetic self-assembled aggregates whose overall structures are also highly directional.

The force balance scheme for bio-mimetic self-assembly can provide a meaningful approach toward the phenomenological understanding of self-assembly processes in living organisms (biological self-assembly) as well. It can be a useful scheme to follow many biological activities. For example, biological cell membranes can be considered self-assembled aggregates of multiple biological building units. Cooperative balancing of the intermolecular forces that are involved is mainly responsible for their mechanical integrity, and for their flexibility at the same time. Many of the physiological functions of these membranes such as ion transport and responses toward incoming foreign molecules can be reasonably understood by identifying the intermolecular forces involved and following their shift toward the new balance.

7.3. ORIGIN OF MORPHOLOGICAL CHIRALITY AND DIVERSITY

The most characteristic feature of bio-mimetic self-assembly is the great structural diversity of its self-assembled aggregates. This diversity originates almost solely from the asymmetric packing of its building units. Asymmetric packing can be induced by various different factors. The two most essential ones are stereoselective interaction and multiple hydrogen bonding between the building units. This section will be devoted to this issue with details.

One of the critical differences between bio-mimetic self-assembly and conventional self-assembly can be found in the directionality and orientationality, not by the composition of building units themselves. As described in the previous section, bio-building units can embrace not only those from biological systems but those derived based on the composition and segments of biological molecules. They usually self-assemble within the scheme of the general bio-mimetic self-assembly process with directionality and/or orientationality. However, if these bio-building units self-assemble in a random manner (which can be real, depending on conditions), it had better be classified as conventional self-assembly. On the contrary, if conventional building units, even though they are colloidal particles, self-assemble in a directional and/or orientational manner through an asymmetric packing, it should be understood as bio-mimetic self-assembly. As will be shown next, this approach toward different self-assembly processes can greatly help clarify the origin of the different morphologies of self-assembled aggregates and possibly provide a means to control them.

7.3.1. Chirality of Building Units

Most of the self-assembled entities in nature have their unique handedness. And this morphological chirality originates from the asymmetric packing characteristic of their primary building units. The primary source is the existence of chirality. Amino acids are the basic building units of proteins, peptide hormones, and more. And most of them have a unique handedness. Nineteen out of 20 amino acids are chiral except for glycine. Two five-carbon sugars, which are the basic building units of nucleic acids (α-D-ribose and 2-deoxy-α-D-ribose), are also chiral compounds. The same is true for the parent sugar, α-D-glucose. Most of the synthetic building units for bio-mimetic self-assembly are designed to retain these basic functional segments (such as amino acids and sugars) within their molecular architecture, with the intention of having maximum reproduction of their unique functionalities (Shimizu and Hato, 1993). This makes them building units with intrinsic chirality. The scope of chiral building units, of course, is not limited by this. There are wide varieties of molecules that possess chiral centers, and so have the capability of being packed asymmetrically (Brizard et al., 2005).

Figure 7.3 presents the schematic illustration of symmetric and asymmetric packing of different building units. The first three cases (a) are for the packing of building units having at least one symmetry: three, two, and one symmetry,

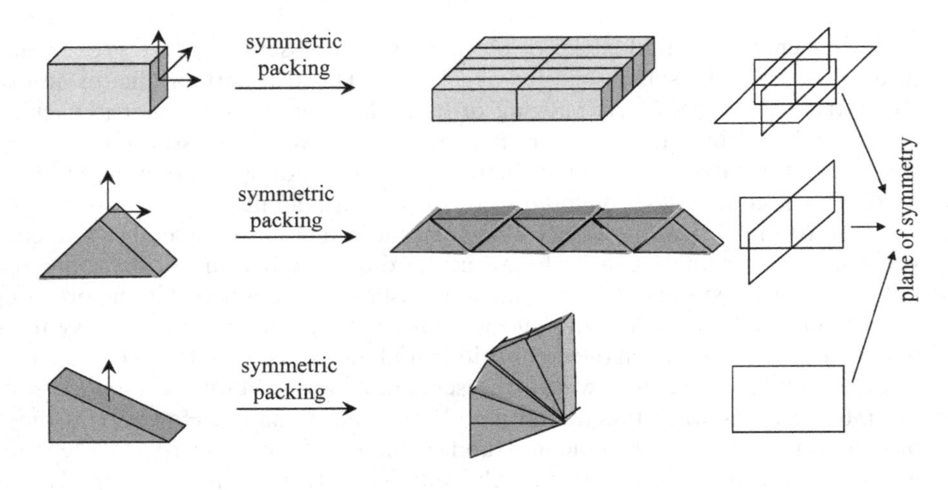

(b) Chiral or asymmetric building unit

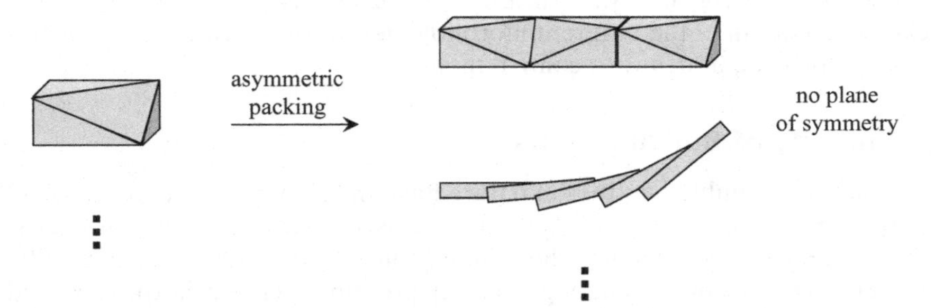

Figure 7.3. (a) Symmetric packing of symmetric building units; (b) asymmetric packing of building units by chirality or asymmetric molecular shape.

respectively. When the building units have symmetry, the energetically favorable way to be packed is to retain their symmetry within the self-assembled structures unless there are other significantly strong forces that can break that symmetry. Thus, the self-assembly occurs with symmetric packing that results in self-assembled aggregates with at least one plane of symmetry. Most of the conventional self-assembled aggregates that are described in previous chapters, including micelles, liquid crystals, vesicles, and colloidal crystals, are the result of this symmetric packing.

Figure 7.3b shows the cases for asymmetric packing. When the building units have chiral center(s) and/or asymmetric molecular morphology, they do not show any symmetry. All possible combinations of packing with these types of building units yield self-assembled aggregates without any plane of symmetry. One critical feature of this asymmetric packing is its diverse ways of being assembled. Since there can be no symmetry preference, a variety of different of ways of packing are possible depending on the conditions of the balance between the intermolecular forces. This might be the direct reason for the wide variety of structural diversity that is generated from asymmetric packing. It also might have something to do with their larger structures when compared with those from symmetric packing, because, that way, the self-assembled aggregates can be stabilized more favorably by the participation of other interactions and/or by self-stabilization through structural adaptation such as helicity, twisting, and circularity.

When asymmetric packing is primarily induced by the chirality of building units, it is often called *chiral packing* or *chiral self-assembly*. (+)-building units usually give a right-handedness to the morphology of their self-assembled aggregates. (−)-building units induce a left-handedness. Racemic mixtures are somewhat tricky. Depending on the actual system and its environmental conditions, that is, depending on the individual packing condition, they either can show two separate entities of right- and left-handedness, or can show a completely vanished handedness. As in the case of chiral interfacial self-assembly in Section 6.7, for bulk bio-mimetic self-assembly here, both chiral and achiral building units can be assembled into chiral self-assembled aggregates and into achiral self-assembled aggregates, too. Formation of chiral nematic (or cholesteric) liquid crystals from chiral building units (Section 4.3.1) is also a good example of asymmetric packing induced by its chirality.

7.3.2. Asymmetric Structure of Building Units

Asymmetric packing of building units is also induced by the molecular (or structural) chirality that originates in the asymmetric geometry of the building units, and not from the presence of chiral carbon or centers. The packing mode is the same as for the chiral building units in Figure 7.3b.

A typical class of this type of molecule is *helicene* (Balaban, 2003). These are mainly orthocondensed, nonplanar polycyclic aromatic molecules. The steric repulsion between the nearest bulky aromatic groups induces the helically shaped asymmetric molecules, which makes them available for asymmetric packing. Another type of molecule, closed rings of benzenes (circulene), also belongs to this class of compound.

Cis–trans isomers are another class of molecules that can be assembled through asymmetric packing. They are *stereoisomers* without any symmetry, but not enantiomers. Because of the steric repulsion between the nearest neighboring groups attached to the double-bond carbon, they are always at the geometrical condition with some degree of distortion along the double-bond line. This makes the entire molecule, whether in *cis* or *trans* form, asymmetric. Oleic acid–based lipid and lipidic amphiphile molecules are typical classes of this configurational asymmetric building unit. Besides *cis–trans* isomers, there are also molecular building units that have asymmetric molecular structure, and so can have a tendency to be assembled in asymmetric packing mode. This is especially true when the multiple bulky groups are connected through a flexible but rigid spacer, so there cannot be a free rotation along that line. The steric repulsion between the bulky groups wholly distorts the geometry of the molecules, which makes them asymmetric. The same effect can often be expected in non-pure-carbon-based multiple-bond molecules, too.

7.3.3. Multiple Hydrogen Bonds

Figure 7.4 shows the schematic relation of packing symmetry with the number of hydrogen bonds. A single hydrogen bond always induces symmetric packing between its building units, unless there are significant perturbation factors such as steric repulsion. Multiple hydrogen bonds from a single building unit can also induce symmetric packing, as long as all the hydrogen bondings are entirely even, as shown in the second case. However, since a hydrogen bond is a strong inter-molecular bond, and easily affected by environmental conditions, multiple hydrogen bonding easily can become uneven. The strength of a hydrogen bond also varies by its geometrical factor. It shows the greatest strength when the three atoms involved lie along one straight line. These are the reasons that many of the self-assembly systems involved with multiple hydrogen bonds occur through asymmetric packing. Examples include aromatic derivatives (Koshima, 2000; Zimmerman and Corbin, 2000; Bushey et al., 2004), calix[4]arene-based building units (Timmerman and Prins, 2001), and peptide tubes (Balbo Block and Hecht, 2005).

Again, nature provides us a wealth of examples of this. Five nitrogenous bases that are the basic building units of nucleic acids (uracil, thymine, cytosine, adenine, and guanine) are not chiral compounds. But they have the capability of multiple hydrogen bonding. DNAs that are formed by the assembly of these bases are always helical. Single-strand DNA is made of four building units bonded covalently: adenine (A), guanine (G), cytosine (C), and thymine (T).

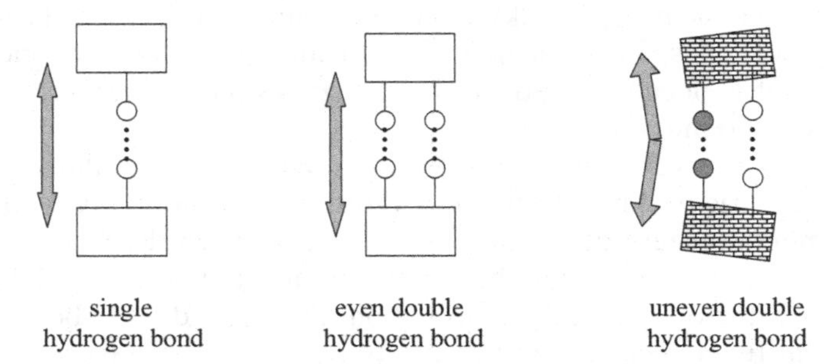

Figure 7.4. Asymmetric molecular packing of building units by asymmetric hydrogen bonding.

Double-strand DNA is formed by the complementary bonding between these four units along the strand. Bondings only between C–G (three hydrogen bonds) and T–A (two hydrogen bonds) are allowed because of the geometrical constraint imposed by these multiple hydrogen bonds. The chemical nature of each bonding site on each unit and geometrical packing compensation by this bonding (bases are bonded covalently, which is much stronger than hydrogen bonding, and each base has a different geometrical structure and size; thus, the geometrical constraint that should be inevitably imposed during the formation of a double strand can be relaxed only through a twisting action of the strands) cause asymmetric packing that happens to be completely relaxed by the twisting of a linear chain of 10.5 base pairs. The unique handedness of DNA can be understood as the result of the asymmetric packing of the building units induced by the multiple hydrogen bonds. Other intermolecular forces including hydrophobic, van der Waals, and dipole–dipole interaction also take part in this assembly process. But they are more likely related to the stability of the DNA. The helical conformation of RNAs can be understood as following the same principles with the additional role of uracil.

7.3.4. Cooperative Balance of Geometry and Bonding

For the majority of biological self-assembly, the chirality and asymmetric geometry of the building units work complementarily, which seems to be critical to achieving the structural diversities and functionalities. The unique handedness of a polypeptide (or protein) α-helix can be understood as the result of asymmetric packing induced by the chirality of its primary building units (amino acids) and uneven double hydrogen bonding between them. All 20 amino acids have at least two hydrogen bonding sites. For example, when thymine and adenine are complementarily bonded through two hydrogen bondings, the different chemical environments of (N–H . . . N) and (O . . . H–N) cause different bonding strengths. Many of the other building units that have chirality also possess this multiple-hydrogen-bonding capability. They include α-D-ribose, 2-deoxy-α-D-ribose, α-D-glucose, and most of the lipid molecules. Multiple hydrogen-bonding is the most abundant case that works with chirality. But the cooperative balance is often induced by an asymmetric coordination bonding as well.

Again, three major intrinsic molecular factors for asymmetric packing are chirality, configurationality, and multiple-hydrogen-bonding capability. For many cases of bio-mimetic self-assembly, any one of these three factors is enough to induce asymmetric packing. However, for the majority of self-assembly systems, asymmetric packing is induced by building units with not just one factor, but multiple factors together (Feiters and Nolte, 2000; Shimizu et al., 2005; Jonkheijm et al., 2006). For example, many classes of biological lipid molecules have all three of these factors in one molecule. Thus, many of the biologically derived amphiphiles that are designed with the intention to transcribe the structural characteristics of the biological molecules also have multiple asymmetric packing factors in one molecule (Bong et al., 2001; Fuhrhop and Wang, 2004; Keizer and

Sijbesma, 2005). This multiple-factor-bearing characteristic might be a primary reason that these building units are self-assembled with much a greater variety of structural diversities than are those building units with just one factor. Formation of helical graphitic nanostructures from hexabenzocoronene derivatives serves as another excellent example (Fukushima, 2006). The same arguments can be claimed for surface self-assembly of biologically derived building units with multiple asymmetric factors (Nandi and Vollhardt, 2003).

Another example can be found in saccharides. All naturally existing monosaccharides, whether they are in linear or cyclic forms, have multiple-hydrogen-bonding capability, and all of them but dihydroxyacetone have from one to four chiral carbons. Thus, the same factors are retained for all of the disaccharides and oligosaccharides, which makes them form a rich range of diverse stereoisomers. *Sugar-based* amphiphiles are those with the sugar groups on their head groups. They are also self-assembled with a rich variety of structural diversity and almost always with a unique handedness. Also, the six-carbon cyclic rings have boat and chair forms of conformational diversity, which adds even more structural diversity to their self-assembled aggregates. Nature provides good examples, too. The unique helical structure of starch and glycogen, which are the helical chains of long D-glucose-based polysaccharides, can be understood as self-assembled aggregates induced by the co-action of multiple hydrogen bonding ($\alpha 1 \rightarrow 4$ linkage) and chiral packing. When the multiple hydrogen bonding occurs between the D-glucose chains of polysaccharides (interchain hydrogen bonding; $\beta 1 \rightarrow 4$ linkage), the spatial constraint from asymmetric packing often is compensated inside the self-assembled aggregates. This yields the straight fiber of cellulose. The sheet type of β-conformation of some polypeptides can be understood using the same principles.

Polymers that can form self-assembled aggregates with a morphological handedness have great potential for diverse application fields. The same principles can be applied to designs such as *chiral polymers*. The key is to design the polymer backbone with either one of or all of molecular chirality, configurational isomeric backbone or substituent, or proper multiple-hydrogen-bonding sites. This task can be done also with the complexation or binding of achiral polymers with chiral dopant or chiral solvents (Cornelissen et al., 2001).

There are many molecules that are achiral but have fascinating physical/chemical properties that can bring tremendous benefits to areas of nanotechnology. One example is *thiophenes*. These have outstanding organic-electric properties. But, once they are able to be self-assembled into superstructures with a chirality such as a helix, they will be able to show even more variety in those functional properties. This could be done through understanding the above factors for asymmetric packing.

7.3.5. Induced Asymmetric Packing

The primary building units described so far in this section have at least one of the intrinsic asymmetric packing factors within their molecular structural features. Building units without any of the intrinsic asymmetric packing factors are usually self-assembled with the symmetric packing mode. However, when these building units interact with secondary units that have any of the intrinsic asymmetric packing factors, they can be self-assembled with the asymmetric packing mode, too. These secondary units can include counterions, catalysts, initiators, and solvents that are chiral and/or asymmetric. One example can be found in cationic gemini surfactants. These are typical achiral/symmetric molecules without any hydrogen-bonding capability. Thus, they usually form nonchiral self-assembled aggregates. However, when they are self-assembled with chiral counterions such as tartrate, the self-assembled aggregates formed show clear morphological chirality (Oda et al., 1999). The structural features of the self-assembled chiral ribbon of this system could actually be controlled by the chiral counterions (Pfeiffer effect). Also, polymers with helical morphology can be synthesized from achiral monomers when chiral catalysts or chiral initiators are properly employed.

Asymmetric packing of building units without any of the intrinsic asymmetric packing factors can also be induced by their mixed self-assembly with building units with any of those intrinsic factors. A good example can be found in the effect of chiral impurities whose addition can change the achiral nematic state formed from achiral building units into a chiral cholesteric state (Dörfler, 2002).

7.4. SYMMETRIC BIO-MIMETIC SELF-ASSEMBLED AGGREGATES

The bio-mimetic self-assemblies described so far were the asymmetric directional packing processes of their primary building units. However, there are some exceptional cases where the self-assembly occurs still through a directional but symmetric packing mode (or with a least degree of asymmetric packing). Even though the cooperative interaction with the structural constraint of the molecular building units is critical to guide the direction of the self-assembly, the major attractive forces are the directional ones such as π–π interaction, hydrogen bonding, or coordination bonding, which can justify their classification as another type of bio-mimetic self-assembly. Two representative cases will be described: H- and J-aggregates, and molecular capsules.

7.4.1. H- and J-Aggregates

Figure 7.5 shows the schematic representation of H- and J-aggregations. When the building units are self-assembled in a face-to-face (or parallel) type of arrangement, the self-assembled aggregate is called an *H-aggregate*. When they are arranged in an end-to-end (or head-to-tail) mode, the resultant self-assembled aggregate is called a *J-aggregate*. The common building units for H- and J-aggregations are dye molecules such as cyanine, merocyanine, or porphyrin derivatives. Usually, H-aggregates are induced by π–π interaction between them, while electrostatic interaction is the usual driving force for J-aggregates. But it

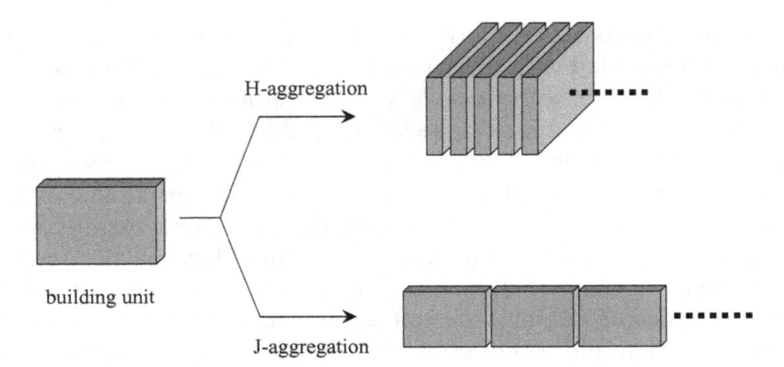

Figure 7.5. Schematic representation of H- and J-aggregations.

is not uncommon that their formation is induced by hydrogen or coordination bonding that happens to be in a very specific position between the building units.

Within the general scheme of bio-mimetic self-assembly (Figure 7.1), H- and J-aggregations can be understood as the process that occurs only with the primary self-assembly step. No further processes toward higher orders of self-assembly occur unless the environmental condition greatly favors a semidilute-like state, for example, a high concentration of the building units. The size, overall shape, and aggregation number of H- and J-aggregates are usually not clearly defined. Depending on the force balancing scheme of each system, they can be widely changed. Since there are always other intermolecular forces involved within the force balancing process, and they can provide the likely possibility for asymmetric packing, both H- and J-aggregates sometimes show morphological chirality. But its degree is much smaller than for the self-assembled aggregates formed by clear asymmetric packing.

H- and J-aggregates are often formed with the aid of external forces. For example, Langmuir monolayers or Langmuir-Blodgett films (details in Chapters 6 and 11) with their building units well assembled in either H- or J-aggregate mode can be obtained through the careful manipulation of surface pressure or capillary force (Kuroda, 2004). The self-assembly within the confined space (details in Chapter 12) can induce them, too. Whether they are self-assembled purely by the balancing of intermolecular forces or by the employment of external forces, the dynamic geometry of the building units is, for most cases, critically important along with the main driving forces. The dye molecules usually have planar geometry. Thus, the π–π interaction between them can be maximized when they are arranged in H-aggregation mode.

The interaction between the building units within H- and J-aggregates is highly directional and geometrically specific. Thus, they can have unique physical and/or chemical properties that can be quite different from the isotropic solution of the building units or their self-assembled aggregates packed through another mode (Knoester, 1995). In the example of spectroscopic properties, the H-aggregates of chromophores show blue shift, while their J-aggregates show red shift (J-band). This is the result of the directional ordering of the transition dipole moment. This fact can provide a useful tool for the development of nanostructured materials or nanodevices with controllable photoelectronic or optoelectronic functionalities. Also, their similarity to natural photosynthetic mechanisms can make them ideal for the study of artificial photosynthetic systems (Yonezawa et al., 2003).

7.4.2. Molecular Capsules

When the directional bonding between the molecular building units is perfectly coordinated with the geometrical matching between them, self-assembled aggregates with precisely defined structures can be formed. They mostly have a spherical or cylindrical shape of morphology with a hollow space inside. The size of the inner space is in the range of few molecules and their physical properties are usually tunable. This provides them with an ideal condition for encapsulation of guest molecule(s) in a highly size- and/or shape-selective manner. This is called a *molecular capsule* (Rebek, 2005).

As in cases of many other bio-mimetic self-assemblies, this process occurs with building units that have multiple numbers of hydrogen (Rebek, 2005) or coordination bondings (Sato et al., 2006). But, unlike the other bio-mimetic self-assembly processes, these multiple directional bondings are highly symmetric, which is believed to be enforced largely by their spatial geometrical features. It can be understood as the multiple hydrogen or coordination bondings between the building units being balanced with their well-designed structural constraint. It is a thermodynamically reversible process, and the self-assembled aggregates have fixed aggregation numbers that can range from a few to a few tens of molecules.

7.5. GELS: NETWORKED BIO-MIMETIC SELF-ASSEMBLED AGGREGATES

Excellent reviews are available for comprehensive details about gels (Terech and Weiss, 1997; Estroff and Hamilton, 2004; De Loos et al., 2005). In this section, however, as a continuation of the previous sections, the focus will be on the gelation process, which can be understood as the asymmetric packing of primary building units. Gels are self-assembled aggregates formed at the high end-step of the general scheme of bio-mimetic self-assembly (Figure 7.1). The *low molecular mass gelator* fits well within this picture of gelation, and will be the main subject of this section. Polymer gels can be considered as self-assembled aggregates that went through much simpler steps of assembly. One significant difference will be that the linearity of the building units for the polymer gels is, for most cases, intrinsic.

Gels can be defined as the state between solid and liquid whose inner structures are formed by the networked (or entangled) components (generally, the concentration of this component ranges ~0.1–~10 wt. %) that have immobilized (or trapped) a large volume of solvent molecules. This immobilization of solvent molecules is the action of surface tension; thus, weak intermolecular forces are the main interactions involved here. When the networked components trap water, they are called *hydrogelators* (the gels formed are called *hydrogels*). When they trap nonaqueous solvents, they are *organogelators* (the gels are *organogels*). The gelling process is commonly called *gelation*. For organogelators, hydrogen bonding (often coordination bonding) is almost the sole driving force for gelation. However, for hydrogelators, the role of other weak intermolecular forces becomes significant. *Xerogel* is the solid state that can be obtained from the gels by the evaporation of the liquid components.

Whether they are hydrogels or organogels, the key to their formation is the ability of their primary building units to be self-assembled into networked structures. This may be the primary reason that most of the primary building units of the low molecular mass gelators belong to the category of those for bio-mimetic self-assembly. As for the cooperative balance of geometry and directional bonding for many real systems (Section 7.3.4.), the primary building units for most of the low molecular mass gelators pose multiple factors for asymmetric packing; chiral centers, asymmetric structures, or multiple hydrogen (or coordination) bonding sites. However, not all of these building units, even with all three factors, always are gelled. Also, building units with single directional bonding usually do not form gels. For most cases, they must have at least two of the directional bonding sites per building unit to be able to form gels. It seems that this is a minimum requirement to form fibrous self-assembled aggregates that eventually can be networked as gels.

Through network formation, gels can provide a large surface area to self-assembled aggregates where the solvent molecules can be immobilized. The networked structure is also responsible for the characteristic properties of most of the gels, such as viscoelasticity, swelling, shrinking, and a gellylike appearance. This structure can be more easily formed when the self-assembly units have fibrous structures, as compared with spherical, rod, or disklike structures. (This self-assembly unit refers to the unit for "the" self-assembly step where the gels are formed. Thus, it can be either a primary building unit itself, or primary, secondary, tertiary, and even a higher degree of self-assembled aggregates.) And fibrous structures are most likely generated by building units with asymmetric packing.

Many of the organometallic-based gelators show cases where uneven multiple coordination bonding (like uneven multiple hydrogen bonding) can cause asymmetric packing of the building units, and thus form gels with morphological chirality. Organogel formation of binuclear copper tetracarboxylate provides an excellent example (Terech, 1994).

Gels also can be formed by induced asymmetric packing. When building units without any intrinsic gelling capability (no intrinsic asymmetric packing factors) are co-self-assembled with the second or third building units that can induce uneven directional bonding, most likely through hydrogen or coordination bonding, they can form fibrous self-assembled aggregates, which can be further networked and so form gels (Hirst and Smith, 2005). One example can be found in the formation of fibrous structures by complementary directional bonding, as in the formation of DNA, which will be eventually networked into gels when the condition is right. Similarly, building units without any gelling capability can often form gels when they interact with counterions that can induce asymmetric packing of the building units (Berthier et al., 2002).

The gels described so far are kinetically stable, bio-mimetic self-assembled aggregates. They are formed through asymmetric packing of their building units, so they have morphological chirality. However, gels are often formed through symmetric packing, too. Semidilute regimes of surfactant micellar solutions are formed as a result of the entanglement of flexible, long, wormlike micelles.

It is a thermodynamically stable, non-bio-mimetic self-assembled aggregate. But it usually shows a viscoelastic property. Most of the surfactants do not have intrinsic factors for asymmetric packing; neither do their common counterions.

Details are in Chapter 3. *Semidilute regime* is commonly defined as the region where the average contour length of the micelles is larger than the average distance between them. When the contour length is smaller than the average distance, it is defined as a *dilute regime*.

The difference between the gels and the liquid crystalline phases (Chapter 4) is often unclear. Many of the liquid crystals obviously have birefringence and a clear x-ray diffraction pattern since they have well-ordered inner structures. Gels have a randomly networked fibrous structure. Thus, they have circular dichroism when the morphological chirality is clearly expressed. But they show no birefringence, and their x-ray diffractions show no clear pattern; they are able only to provide a rough internal structure such as the average spacing between the dispersed fibers.

7.6. PROPERTIES OF BIO-MIMETIC SELF-ASSEMBLED AGGREGATES

The asymmetric packing nature of bio-mimetic self-assembly provides a variety of unique physicochemical properties to its self-assembled aggregates. These sometimes are very comparable to those of symmetric self-assembled aggregates, and sometimes somewhat parallel. This section provides a brief description of this fact. Its implications for nanotechnology will be presented throughout Part II.

7.6.1. Directionality, Site-Specificity, and Chirality

As emphasized throughout this chapter, the foremost characteristic of bio-mimetic self-assembly is its directional nature. Also, since it is largely those functional groups in the building units that bear the source of this directionality, the bio-mimetic self-assembled aggregates, in the majority of cases, have functionalities that are specific to these sites of functional groups. These unique properties, as opposed to symmetric self-assembled aggregates, can provide very significant implications for a variety of applications. Cooperative manipulation of directionality and site specificity can be useful to *guide* (or *fabricate*) the secondary (or tertiary and more) entities that are otherwise practically difficult to arrange against their own interaction. Production of a variety of minerals with directed (or oriented) morphology by mimicking the processes in living systems (biomineralization) will be one prime example. Directional progress of intermolecular forces with unique functionalities also might yield valuable operative principles for nanoscale devices. For example, a ligand or a host unit could be built up along the oriented surface of bio-mimetic self-assembled aggregates. And the interaction with an acceptor or a guest could be represented reversibly by controlling the conditions of the force balance change.

Directionality of bio-mimetic self-assembled aggregates always comes with their morphological chirality, such as twisted or helical structures. Why they and also biological self-assembled aggregates have to be this way is not clear. But, on the nanotechnology side, this feature can open up new technical possibilities for nanoscale materials and devices that can be stereoselective.

7.6.2. Hierarchicality

Another typical characteristic of bio-mimetic self-assembly is that, for most cases, it occurs with a clearly defined hierarchical mode. The most important difference that is provided by this feature, as opposed to symmetric self-assemblies, is the rich morphological diversities of bio-mimetic self-assembled aggregates. It is also this hierarchicality that makes their formation possible over such a wide range of length scale, from nanometer to beyond centimeter.

Usually, bio-mimetic self-assembled aggregates show higher stability against changes in environmental conditions such as pH, temperature, and pressure, compared with the nonhierarchical self-assembled aggregates. They also show much greater strength against external stresses such as mechanical, electric, or magnetic force. Formation of strong intermolecular interactions between the building units is certainly one reason for this. But the structural integrity that is generated by this hierarchicality is a major source of this stability and strength. For example, the majority of bio-mimetic self-assembled aggregates have enough stability to be easily separated from their solution, and have enough strength to be subjected to subsequent operations with external stresses. Nature also provides plenty of examples. Collagen molecules generate a variety of strength-requiring tissues within living organisms through their hierarchical self-assemblies; and a spider web acquires its striking strength also through the hierarchical self-assembly of collagen molecules.

The structural diversity of bio-mimetic self-assembled aggregates definitely will be able to expand the scope of the potential nanostructures, also over a wide range of length scale. As to the issue of effective connection between nanoscopic objects and macroscopic ones, which will be critical for many real-life applications, this feature of bio-mimetic self-assembly might provide many insightful implications. Also, their high stability and strength could substantially increase their practicality in many applications, especially when such a dynamic aspect of non-bio-mimetic self-assembled aggregates becomes a major hurdle. Templating using self-assembled aggregates is one of the major routes for the practical production of nanostructured materials (details in Chapter 9). Minimizing unnecessary phase changes during the synthesis, possibly by employing these aspects of bio-mimetic self-assembled aggregates, could ensure a very efficient outcome for the generation of templated materials with delicate details such as helix, twist, or molecular imprint. This would make it useful for the generation of functional nanostructured materials.

7.6.3. Complementarity

Double-helix strands of DNA self-assemble with almost perfect complementarity. A certain building unit self-assembles only with its complementary partner. This has been understood as the outcome of the "well-programmed" force balance that can originate from its elegant structural and functional designs. Most bio-mimetic self-assembly also occurs with a certain degree of complementarity. This complementarity of biological and bio-mimetic self-assembly processes strongly implies that the biological and bio-mimetic self-assembled aggregates can be designed to be formed with almost perfect selectivity among their building units. It also could mean that the intermolecular interactions for their assembly and disassembly can be designed to carry logic. The *selectivity* provides an effective route to fabricate the nanoscale building units for 1-, 2-, and 3-dimensionality (nanofabrication; details in Chapter 13) with great precision. The "logic" could provide a novel way to devise nanoscale devices that might be able to perform information transcription for communication between them. It might make even a logical function possible (some details in Chapter 14).

7.6.4. Chiroptical Properties

Optically active molecules have their characteristic optical property. It is usually called the *chiroptical property*, and circular dichroism and optical rotatory dispersion are the most common forms. The characteristic change near their absorption bands is called the *Cotton effect*.

The majority of the building units for bio-mimetic self-assembly have chiral center(s); thus, they are optically active. In most cases, their molecular chirality is transferred into their self-assembled aggregates, which have morphological chiralities. Also, the self-assembled aggregates that are formed from the achiral building units (by asymmetric packing) can show chiroptical properties through their morphological chirality, as long as the environmental conditions do not work against them.

The technical potential of chiroptical properties is important in many areas of application (Smith, 1998; Fujiki, 2001). For bio-mimetic self-assembled aggregates, the chiroptical properties are largely determined by their morphological chirality, which can be controlled (to a reasonable degree) through the proper design and manipulation of intermolecular forces involved in the self-assembly process. One implication for nanotechnology is the development of nanoscale devices that can take advantage of the controllable chiroptical properties. Examples can include optical information storage, wavelength guide, optical switch, and stereospecific sensor.

7.7. FUTURE ISSUES

Ironically, one of the key challenges for the future not only in the area of bio-mimetic self-assembly itself but also in its application comes from its greatest advantages; the richness of the building units available and the structural diversity. For conventional self-assemblies, the structures of the self-assembled aggregates and their transitions can be predicted from the structures of the primary building units with reasonable accuracy. However, for bio-mimetic self-assembly, the prediction of the self-assembled aggregate structures is not as practical as for conventional ones. For example, whether or not there will be a morphological chirality could be answered just by analyzing the structural features of the primary building units. But, since the asymmetric packing of building units could go in any unexpected directions even under the same conditions, the reasonable prediction of their detailed structures often becomes unlikely. The same is true of following their structural transition as the force balance condition is shifted. Better understanding regarding this challenge will be critical for the further development of this field.

REFERENCES

Balaban, A. T. "Theoretical Examination of New Forms of Carbon Formed by Intra- or Intermolecular Dehydrogenation of Polycyclic Aromatic Hydrocarbons, Particularly Helicenes," *Polycyclic Aromatic Compounds* **23**, 277 (2003).

Balbo Block, M. A., Hecht, S. "Wrapping Peptide Tubes: Merging Biological Self-Assembly and Polymer Synthesis," *Angew. Chem. Int. Ed.* **44**, 6986 (2005).

Berthier, D., Buffeteau, T., Legar, J.-M., Oda, R., Huc, I. "From Chiral Counterions to Twisted Membranes," *J. Am. Chem. Soc.* **124**, 13486 (2002).

Bong, D. T., Clark, T. D., Granja, J. R., Ghadiri, M. R. "Self-Assembling Organic Nanotubes," *Angew. Chem. Int. Ed.* **40**, 988 (2001).

Brizard, A., Oda, R., Huc, I. "Chirality Effects in Self-assembled Fibrillar Networks," *Top. Curr. Chem.* **256**, 167 (2005).

Bushey, M. L., Nguyen, T.-Q., Zhang, W., Horoszewski, D., Nuckolls, C. "Using Hydrogen Bonds to Direct the Assembly of Crowded Aromatics," *Angew. Chem. Int. Ed.* **43**, 5446 (2004).

Cornelissen, J. J. L. M., Rowan, A. E., Nolte, R. J. M., Sommerdijk, N. A. J. M. "Chiral Architectures from Macromolecular Building Blocks," *Chem. Rev.* **101**, 4039 (2001).

De Loos, M., Feringa, B. L., Van Esch, J. H. "Design and Application of Self-Assembled Low Molecular Weight Hydrogels," *Eur. J. Org. Chem.* 3615 (2005).

Dörfler, H.-D. "Chirality, Twist and Structures of Micellar Lyotropic Cholesteric Liquid Crystals in Comparison to the Properties of Chiral Thermotropic Phases," *Adv. Coll. Inter. Sci.* **98**, 285 (2002).

Estroff, L. A., Hamilton, A. D. "Water Gelation by Small Organic Molecules," *Chem. Rev.* **104**, 1201 (2004).

Feiters, M. C., Nolte, R. J. M. "Chiral Self-Assembled Structures from Biomolecules and Synthetic Analogues," *Advances in Supramolecular Chemistry* **6**, 41 (2000).

Fuhrhop, J.-H., Wang, T. "Bolaamphiphiles," *Chem. Rev.* **104**, 2901 (2004).

Fujiki, M. "Optically Active Polysilylenes: State-of-the-Art Chiroptical Polymers," *Macromol. Rapid Commun.* **22**, 539 (2001).

Fukushima, T. "π-Electronic Soft Materials Based On Graphitic Nanostructures," *Poly. J.* **38**, 743 (2006).

Hirst, A. R., Smith, D. K. "Two-Component Gel-Phase Materials: Highly Tunable Self-Assembling Systems," *Chem. Eur. J.* **11**, 5496 (2005).

Jonkheijm, P., Van der Schoot, P., Schenning, A. P. H. J., Meijer, E. W. "Probing the Solvent-Assisted Nucleation Pathway in Chemical Self-Assembly," *Science* **313**, 80 (2006).

Keizer, H. M., Sijbesma, R. P. "Hierarchical Self-Assembly of Columnar Aggregates," *Chem. Soc. Rev.* **34**, 226 (2005).

Knoester, J. "Collective Nonlinear Optical Properties of Disordered J-Aggregates," *Adv. Mater.* **7**, 500 (1995).

Koshima, H. "Generation of Chirality in Two-Component Molecular Crystals from Achiral Molecules," *J. Mol. Struc.* **552**, 111 (2000).

Kuroda, S. "J-aggregation and Its Characterization in Langmuir-Blodgett Films of Merocyanine Dyes," *Adv. Coll. Inter. Sci.* **111**, 181 (2004).

Nandi, N., Vollhardt, D. "Effect of Molecular Chirality on the Morphology of Biomimetic Langmuir Monolayers," *Chem. Rev.* **103**, 4033 (2003).

Oda, R., Huc, I., Schmutz, M., Candau, S. J., MacKintosh, F. C. "Tuning Bilayer Twist Using Chiral Counterions," *Nature* **399**, 566 (1999).

Rebek, J. Jr. "Simultaneous Encapsulation: Molecules Held at Close Range," *Angew. Chem. Int. Ed.* **44**, 2068 (2005).

Sato, S., Iida, J., Suzuki, K., Kawano, M., Ozeki, T., Fujita, M. "Fluorous Nanodroplets Structurally Confined in an Organopalladium Sphere," *Science* **313**, 1273 (2006).

Shimizu, T., Hato, M. "Self-Assembling Properties of Synthetic Peptidic Lipids," *Biochim. Biophys. Acta* **1147**, 50 (1993).

Shimizu, T., Masuda, M., Minamikawa, H. "Supramolecular Nanotube Architectures based on Amphiphilic Molecules," *Chem. Rev.* **105**, 1401 (2005).

Smith, H. E. "Chiroptical Properties of the Benzene Chromophore: A Method for the Determination of the Absolute Configurations of Benzene Compounds by Application of the Benzene Sector and Benzene Chirality Rules," *Chem, Rev.* **98**, 1709 (1998).

Terech, P. "≪Living Polymers≫ in Organic Solvents: Bicopper (II) Tetracarboxylate Solutions," *Nuovo Cimento* **16D**, 757 (1994).

Terech, P., Weiss, R. G. "Low Molecular Mass Gelators of Organic Liquids and the Properties of Their Gels," *Chem. Rev.* **97**, 3133 (1997).

Timmerman, P., Prins, L. J. "Noncovalent Synthesis of Melamine–Cyanuric/Barbituric Acid Derived Nanostructures: Regio- and Stereoselection," *Eur. J. Org. Chem.* **17**, 3191 (2001).

Yonezawa, Y., Yamaguchi, A., Kometani, N. "Current Topics in Photochemistry of Sensitizing Dyes," *Nippon Shashin Gakkaishi* **66**, 307 (2003).

Zemb, T., Blume, A. "Self-Assembly: Weak and Specific Intermolecular Interactions at Work," *Curr. Opin. Coll. Inter. Sci.* **8**, 1 (2003).

Zimmerman, S. C., Corbin, P. S. "Heteroaromatic Modules for Self-Assembly Using Multiple Hydrogen Bonds," *Structure and Bonding* **96**, 63 (2000).

PART II

NANOTECHNOLOGY

8

IMPLICATIONS OF SELF-ASSEMBLY FOR NANOTECHNOLOGY

8.1. GENERAL CONCEPTS AND APPROACH TO NANOTECHNOLOGY

The widely accepted definition of nanotechnology is "the manipulation of atoms or molecules on a nanometer scale of ~1–~100 nm." A more detailed description could be "the ability to manipulate/assemble nanometer-scale building units into integrated systems and the subsequent finding of novel properties/phenomena on a nanometer scale." In practice, systems with the upper limit of the length scale up to ~μm are usually considered to fit this definition. Since it was first conceptually envisioned by Richard Feynman in 1959 (Feynman, published in 1961), this revolutionary idea has been shaping a new trend in modern science/technology in expectation of its promising societal implications in a wide range of disciplines, including materials, medicines, environments, energy, communications, and more (Ball, 1998; Whitesides, 1998; Roco et al., 1999; Siegel et al., 1999; "Nanotechnology: A Special Report," 2000; National Nanotechnology Initiative, 2000; Roco and Bainbridge, 2001; Stupp et al., 2002; Schwarz et al., 2004). Unlike two earlier megatrends in the science/technology community, information technology and biotechnology, nanotechnology is truly being formed/developed with an inter- and multidisciplinary character (Figure 8.1).

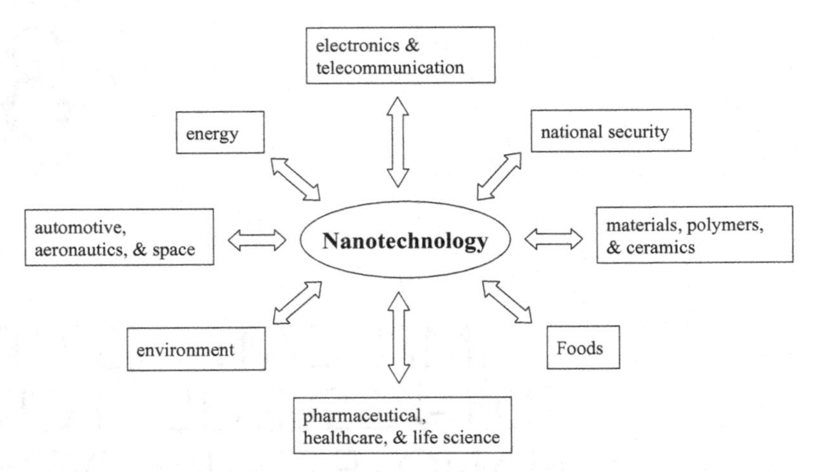

Figure 8.1. Societal and economic impacts of nanotechnology.

The two mainstream approaches to the issues of nanotechnology are *top-down* and *bottom-up*. The top-down approach adapts the concept that the construction of nanostructured systems that eventually consist of the length scale of atoms or molecules should begin with bulk materials. By removing the excess portion of bulk materials by physical, chemical, and mechanical means, the desired nanostructured systems can be constructed. Hard lithography using optical laser-light, electron-beam, or x-ray are typical examples. The bottom-up approach, on the other hand, starts with the individual building units that are the objects of the nanometer scale, such as atoms, molecules, polymers, and colloids. By assembling these building units with the needed controllability, the desired nanostructured systems can be obtained. Figure 8.2 represents the scheme of these two approaches.

Depending on the specific system to be constructed, *hybrid* approaches that try to integrate these two methods are also being sought. Figure 8.3 shows the schematic representation of these hybrid approaches. In the "top-down-after-bottom-up" approach, the rough quasi-nanostructure is assembled first, and the detailed desired structure is obtained by carving out by the top-down method of construction. The "bottom-up-after-top-down" approach is the exact opposite of this process. The rough quasi-nanostructure is obtained by the top-down approach first, and the desired structure is assembled on the base of that structure. Soft lithography such as microcontact printing and dip-pen nanolithography using the AFM (atomic force microscopy) probe are good examples.

The purpose of Part II of this book is to describe the implications of the self-assemblies from Part I for issues of nanotechnology. This first chapter of Part II is designed to address the primary questions: (1) why self-assembly is so critical to the progress of nanotechnology and (2) how it can be correlated with nanotechnology. It covers in detail the following issues:

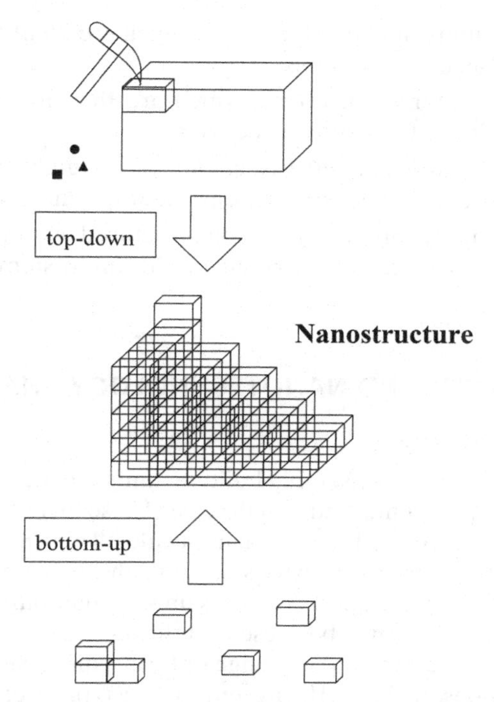

Figure 8.2. Bottom-up versus top-down approach for the preparation of nanostructures.

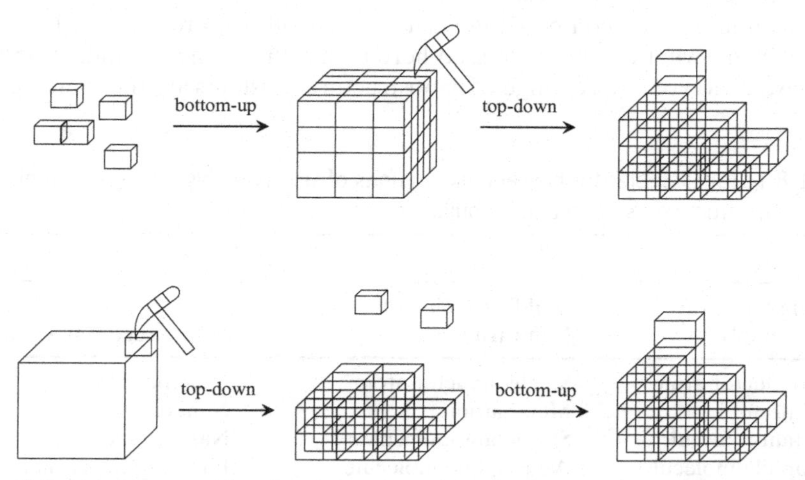

Figure 8.3. Hybrid approaches for the preparation of nanostructures: (a) top-down after bottom-up and (b) bottom-up after top-down.

1. The building units of nanotechnology, which are identified as the same as those of self-assembly systems
2. The principal interaction forces, which are the intermolecular/colloidal forces from the self-assembly processes
3. The concept of *spontaneous* association of building units as opposed to *one-by-one manipulation* for the construction of nanostructured systems
4. The self-assembly process for nanostructured systems, which can be deduced from the general scheme of the multi-stepwise self-assembly process

8.2. SELF-ASSEMBLY AND NANOTECHNOLOGY SHARE THE SAME BUILDING UNITS

In Chapter 1, varieties of self-assembly building units are identified and classified that cover a wide range of units and length scale. These include atoms, molecules, polymers, colloidal objects, and biological molecules. Table 8.1a lists some typical examples. As described in the previous section, the nanostructured system in nanotechnology is "an integrated system assembled/manipulated from nanometer-scale building units." As will be presented throughout Part II, the nanostructured system, in reality, covers a wide range of forms, dimensions, structures, and types of functionalities. Table 8.1b presents some typical examples. Here, the building units are primarily the basic and essential components for the construction of those systems. Simple systems such as quantum dots or nanocrystals consist mainly of one or two types of semiconductor or metal atoms. On the other hand, more complex systems such as nanodevices or model membranes can have multiple numbers of components such as molecules, polymers, and colloidal objects. Thus, what can be addressed here is that the primary building units for the construction of a wide range of nanostructured systems are the same building

TABLE 8.1. (a) Examples for typical building units of self-assembly, and (b) examples for a typical nanostructured system and its building units.

(a)	(b)	
Building Units for Self-Assembly	Building Units for Nanostructure	Nanostructured Systems
Semiconductor atom	Semiconductor atom	Quantum dot
Metal atom	Metal atom	Nanocrystal
Surfactant molecule	Surfactant molecule	Nanocomposite
Amphiphile molecule	Amphiphile molecule	Film/multilayer nanodevice
Amphiphilic polymer	Amphiphilic polymer	Polymeric nanocomposite
Colloidal object	Colloidal sphere, rod, etc.	Photonic device
Lipid, DNA, RNA	Lipid, DNA, RNA	Model membrane

units as are found in self-assembly systems. They also share the same range of length scale and origins.

This constitutes the first fundamental issue connecting self-assembly systems with nanostructured systems: the sameness of the building units and the sameness of the length scale. This then leads us to the second issue: the sameness of the operational force. This will be examined in the next section.

The sameness of the building units and operational force for both self-assembly and nanostructured systems is crucial, not only for the bottom-up approach that directly deals with the individual building units, but for the top-down approach as well, which will eventually be needed to create nanoscale structures. Of course, it is valid for the hybrid approaches, as well. At least, individual building units or the creation of a nanoscale structure will have to be dealt with at some point during the processes. An example can be found in the lithographic methods for the creation of patterned or fabricated circuits on a base of self-assembled monolayers.

8.3. SELF-ASSEMBLY AND NANOTECHNOLOGY ARE GOVERNED BY THE SAME FORCES

The main types of operational forces for the formation of nanostructured systems are intermolecular and colloidal forces. Since the building units and the length scale are the same for self-assembly as for nanostructured systems, it is only natural to arrive at the notion that the same operational forces are responsible for the interaction between the building units for both systems. All types of forces presented in Chapter 2 are operational with the exact same principles and correlation.

Also, the concept of force balance for self-assembly that is presented in Figure 1.2 is extensively valid for the formation of nanostructured systems. The attractive forces are always responsible for the initial assembly of the building units and repulsive opposition forces are responsible for the subsequent force balance. The same types of directional/functional forces are operating with the same principles as for self-assembly. Though the hydrogen bond is the most common force for nanostructured systems, the strong covalent bond or metal–ligand complexation is often found to be a directional/functional force. Typical examples include the thiol–metal or silyl–silica bonds for the formation of self-assembled monolayers, silanol condensation for the formation of nanostructured silica, and so forth.

8.4. SELF-ASSEMBLY VERSUS MANIPULATION FOR THE CONSTRUCTION OF NANOSTRUCTURES

The next issue is what makes self-assembly so advantageous for the construction of nanostructured systems. Figure 8.4 shows the comparison of a quantum corral with a surface micelle. In 1993 (Crommie et al., 1993), IBM researchers announced

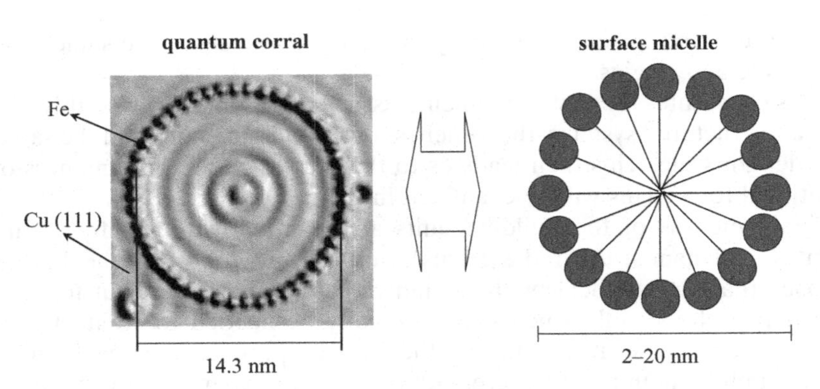

Figure 8.4. Quantum corral [from Crommie et al., *Science* **262**, 218 (1993); reprinted with permission from AAAS] and surface micelle.

that they were able to construct corral-shaped iron nanostructures by manipulating 48 iron atoms on the surface of copper. The diameter of this structure was 14.3 nm; they also later demonstrated the construction of a variety of derived versions of this type of structure. Without doubt, this work is clearly symbolic in the sense of the ability to manipulate individual atoms at will. The key to this issue, however, lies in its *process*. Each atom had to be manipulated *one-by-one* by the tip of an STM (scanning tunneling microscope), which requires time-consuming and labor-intensive efforts.

The very beauty of self-assembly comes from this point of view of the *process* of formation of nanostructures. The building unit of this surface micelle is the common surfactant molecule. And this structure is formed by *spontaneous assembly* of that molecule once the right condition is set, in most cases, at room temperature. Also, the size, which can range ~2–~20 nm, and the structure show striking similarity with those of the quantum corral. This leads to the key idea that self-assembly could be the avenue to ensuring a time-and-effort-saving, all-at-once process for the formation of nanostructures. As described throughout Part I, a variety of self-assembly processes and self-assembled structures are available. Also, since they are mainly governed by the controllable intermolecular and colloidal forces, the structures and sizes are very tunable. Thus, self-assembly can provide a variety of possibilities for the formation of nanostructures with fine tunability.

8.5. SELF-AGGREGATES AND NANOTECHNOLOGY SHARE THE SAME GENERAL ASSEMBLY PRINCIPLES

Self-assembled systems, in many cases, are certain types of nanostructured systems. Table 8.2 shows some typical examples of the relationship between

TABLE 8.2. Some examples of the relationship between the self-assembly processes from Figure 1.3 and typical nanostructured systems.

Primary Building Units	Self-Assembly Processes	Nanostructured Systems
Semiconductor, metal atom	Primary only	Quantum dot, nanoparticle
	Primary + secondary	Giant nanoparticle
	Primary + secondary + tertiary	Nanoparticle aggregate
Surfactant, polymer molecule	Primary only	Film, monolayer, multilayer
	Primary + secondary	Nanocomposite
	Primary + secondary + tertiary	Nanocomposite, nanostructured material
Colloidal object	Primary only	Colloidal crystal, film
	Primary + secondary	Colloidal composite
	Primary + secondary + tertiary	Photonic device
DNA, RNA	Primary only	Monolayer, lock-and-key system
	Primary + secondary	Model membrane
	Primary + secondary + tertiary	Sensory device, bio-nanocomposite

the concept of the multi-stepwise self-assembly process from Figure 1.3 and the formation of nanostructured systems. The very same general self-assembly process can be directly or, in some cases, indirectly adapted for nanotechnology with a whole range of building units that are identified in Chapter 1. An additional aspect of nanotechnology is that often the formation of nanostructured systems involves self-assembly with multiple types of building units. For example, the construction of biosensors usually requires the combination of bio-mimetic assembly units with the formation of self-assembled monolayers, which is two-dimensional molecular self-assembly (self-assembly at interface). The construction of nanocomposites often requires both primary self-assembly and secondary self-assembly processes along with molecular and inorganic or polymeric building units. Also, a variety of nanostructured systems are (or need to be) constructed under an external force or through self-assembly in a confined system. Assembly of colloidal crystals for phonic devices and arrangement of quantum dots or rods in the assembly of quantum-size nanoelectronic devices, such as field-effect diodes, are typical examples. Table 8.3 summarizes the comparison of self-assembled systems with nanostructured systems.

TABLE 8.3. Relationship between the self-assembled systems from Table 1.2 and nanostructured systems.

Self-Assembly	Self-Assembled Systems	Nanostructured Systems
Atomic	Epitaxial film,Quantum dot	Nanofabricated system (2D), nanodevice
Molecular	Micelle, reverse micelle liquid crystal, bilayer, microemulsion, emulsion	Nanostructured inorganic, nanocomposite, hybrid material (organic/inorganic), porous material (micro-, meso-, macro-), nanoparticle, nanocrystal
Polymeric	Micelle, liquid crystal, emulsion	Nanostructured inorganic, nanocomposite, hybrid material (organic/inorganic), porous material (meso-, macro-)
Colloidal	Colloidal crystal, suspension	Photonic device, photovoltaic device
Biological	DNA, RNA, protein antibody–antigen, ligand-acceptor	Nanofabricated system (2D, 3D), nanoparticle, nanocrystal nanodevice
Interfacial	Surface micelle, Langmuir-Blodgett film, self-assembled monolayer	Nanostructured film/multilayer, nanodevice

8.6. CONCLUDING REMARKS

The implications of self-assembly for nanotechnology have been briefly described in this chapter. Again, it is clearly intuitive for the understanding and progress of nanotechnology that self-assembly (or self-organizing phenomena) and nanoscale events share:

1. The same building units
2. The same working length scale
3. The same major forces of formation

Most of the processes for the formation of nanostructured systems are solely or mainly involved with intermolecular/colloidal forces and their interplay, and can be explained by the general self-assembly scheme proposed in Figures 1.2 and 1.3. The rest of the chapters in Part II will present individual subjects related to nanotechnology based on the concept described here.

REFERENCES

Ball, P. *Made to Measure: New Materials for the 21st Century* (Princeton University Press: 1998).

Crommie, M. F., Lutz, C. P., Eigler, D. M. "Confinement of Electrons to Quantum Corrals on a Metal Surface," *Science* **262**, 218 (1993).

Feynman, R. P. "There Is Plenty of Room at the Bottom," *Miniaturization* (Reinhold: 1961).

"Nanotechnology: A Special Report," *C&EN* 25 (October 16, 2000).

National Nanotechnology Initiative: The Initiative and Its Implementation Plan. National Science and Technology Council: www.nano.gov (2000).

Roco, M. C., Bainbridge, W., eds. *Societal Implications of Nanoscience and Nanotechnology* (NSF: March 2001).

Roco, M. C., Williams, R. S., Alivisatos, P., eds. *Nanotechnology Research Directions: IWGN Workshop Report—Vision for Nanotechnology R&D in the Next Decade* (National Science and Technology Council: 1999).

Schwarz, J. A., Contescu, C. I., Putyera, K., eds. *Dekker Encyclopedia of Nanoscience and Nanotechnology* (Marcel Dekker: 2004).

Siegel, R. W., Hu, E., Roco, M. C., eds. *Nanostructure Science and Technology: Worldwide Study on Status and Trends* (Kluwer Academic Publishers: 1999).

Stupp, S. I. et al., eds. *Small Wonders, Endless Frontiers: A Review of the National Nanotechnology Initiative* (Committee for the Review of the National Nanotechnology Initiative, National Academic Press: 2002).

Whitesides, G. M. *Nanotechnology: Art of the Possible* (Technology Review: November/December 1998).

9

NANOSTRUCTURED MATERIALS

In Chapter 3, we discussed three unique properties of micelle solutions that make *micellar catalysis* possible. They are the nanometer size of micelles which generates the extremely high water–oil interfacial area in aqueous solution, the binding affinity of ionic species on the micelle surface, and the solubilization capability of liphophilic compounds within the micelle interior. These were also the key concepts for a variety of traditional applications for surfactants and micelle solutions.

We also looked at surfactant self-assembly for the formation of micelles and its underlying phenomena throughout Chapter 3. If we remember the basic viewpoint of *nanotechnology* from Chapter 8, that it is "the manipulation of atoms and molecules on a nanometer scale," we can find unique implications of micelles for its development. First, the size of micelles ranges on the same nanometer scale. Second, its ability to provide a nanometer-scale reaction space with enhanced reaction rate can be used as a well-defined *nanoreactor*. Third, our understanding of the surfactant self-assembly process can be expanded to the *manipulation* (self-assembly) of more complex and functional systems. Finally, the versatility of micelle structures can add a rich spectrum of structural possibilities to nanometer-scale materials.

Self-Assembly and Nanotechnology: A Force Balance Approach, by Yoon S. Lee

In this chapter, we will first discuss the role of a variety of self-assemblies in the preparation of nanostructured bulk material. Different types of nanostructured materials including nanocomposites, microporous materials, mesoporous materials, and macroporous materials will be introduced. The concept of a *soft-template* (or *structure-directing agent*) for surfactant, polymer, or colloidal self-assembled aggregates such as micelles, emulsions, or colloidal crystals will have special attention. Also, it will be shown how the picture of force balance for self-assembly can be applied to follow the structural changes of nanostructured materials and how the general scheme of self-assembly can become useful in understanding the formation and evolution of nanostructured materials.

Characterization of nanostructured materials sometimes requires thoughtful attention owing to their kinetically stable characteristic and the length scale of their structures. This will be discussed later in this chapter. Novel and better physical and chemical properties that make these materials exciting will be touched on through the chapter. Finally, we will try to envision the potential applications of these materials and to address the possible challenges for their future development.

9.1. WHAT ARE NANOSTRUCTURED MATERIALS?

Nanoscale materials can be defined as materials with at least one of their dimensions in the range of the nanometer scale. Therefore, a variety of films that actually can have a centimeter and even larger scale of width and length can be defined as nanoscale materials as long as their height (thickness) is on the nanometer scale. It may not be easy to define nanostructured materials here in accordance with this. It could mean materials whose entire structures are on the nanometer scale. It could also mean materials whose critical structural features are within the range of the nanometer scale. But this actually can be a good guideline for the difference between *nanostructured materials* and *nanoparticles*. Generally, the former can be categorized as materials whose inner critical structural features are in the range of nanometers. This can include the size of pore (or void space) for porous materials, the size of unit cell for those with crystal-like symmetry, and the size of repeating unit (or component) for hybrid or composite materials. The latter category is rather strict. It refers to materials all three of whose dimensions are in the range of the nanometer scale. All particles, regardless of shape, can be referred to as nanoparticles as long as all three of their dimensions are in the range of the nanometer scale. This can include small dust particles, aerosols, colloidal objects, quantum dots, small metal and semiconductor particles, and so on.

Figure 9.1 shows the schematic representation for the three different types of nanostructured materials. *Natural clay-type nanocomposites* are inorganic materials whose inorganic frameworks are regularly arranged, thus creating the pore (or void space) inside the materials. The sizes of these pores are in the range

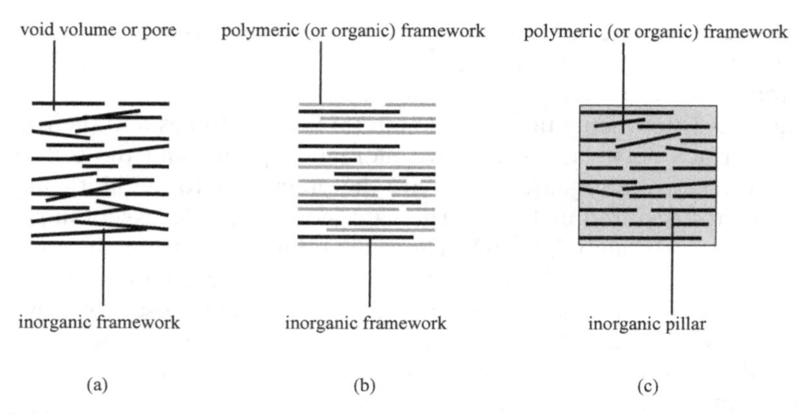

Figure 9.1. Schematic representation of the different types of nanostructured materials: (a) natural clay-type nanocomposite, (b) preparative polymeric (or organic)-inorganic nanocomposite, and (c) synthetic polymeric (or organic)-inorganic hybrid.

of the nanometer scale. The frameworks are usually aligned two-dimensionally which makes them exhibit sheetlike characteristics. Typical examples are bentonite, montmorilonite, and liponite.

Preparative organic–inorganic nanocomposites are composite materials that were formed by the regular arrangement of the organic (or polymeric) framework and inorganic framework (or pillar) on a nanometer scale. For most cases, there is no strong covalent bond involved in the formation of these types of materials. Physical bonds such as electrostatic attraction or hydrogen bonds are the dominant forces. The third type is *organic (or polymeric)-inorganic hybrid materials*. Their structural features are very similar to those of preparative nanocomposites. They are also arranged on a nanometer scale. But the organic frameworks and the inorganic pillars are bonded strongly, usually by covalent or ligand bonds, though the actions of weak intermolecular forces are strongly involved during the assembly processes for the formation of this structure.

In Figure 9.1b and c, the removal of organic (or polymeric) components generates porous inorganic materials whose pores can be replicated on a nanometer scale by the organic (or polymeric) components.

9.2. INTERMOLECULAR FORCES DURING THE FORMATION OF NANOSTRUCTURED MATERIALS

Intermolecular and colloidal forces play a critical role throughout the entire process for the formation of nanostructured materials. All possible intermolecular forces have the ability to affect this process at some degree, and their interplay is usually cooperative and complicated. Though this issue will be addressed throughout this chapter, in this section we will discuss the conceptual picture of those forces and the interplay between them.

Figure 9.2 represents the possible intermolecular forces and their interplay at the interface between the reactive inorganic species and the self-assembly units. The reactive inorganic species are the precursors to the inorganic framework of nanostructured materials. They are mostly anionic, except in rare occasions such as extremely low pH. The self-assembly units are the organic (or polymeric) components. Bear in mind that this is a schematic representation. There is no clear *macroscale* interface formed between these two components like the oil–water interface. Both components are in solution and the interface is formed at the contact region between the inorganic species and the self-assembly unit (or self-assembled aggregates).

Whether the self-assembly unit in Figure 9.2 is a single molecule (a), a self-assembled aggregate such as a micelle (b), an emulsion (or microemulsion) (c), or a colloidal particle (d), there is always the attractive force between the two components (i). This is the major driving force that brings two components close together; thus, it is responsible for the formation of nanostructured materials. When the self-assembly unit is cationic, force (i) is the direct electrostatic attraction. When it is anionic, this force is the mediated-electrostatic attraction, which

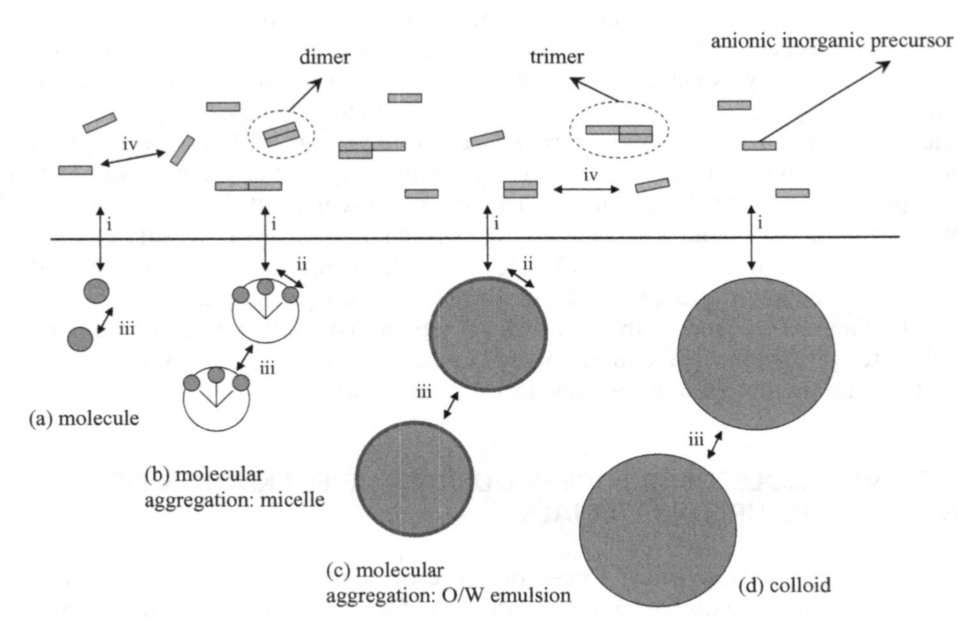

Figure 9.2. Interplay between intermolecular/colloidal forces at the interface between reactive inorganic species and self-assembly units or self-assembled aggregates.

is possible through the cationic species (or layer) between the two components. Hydration force is the major player when the self-assembly unit is nonionic. Force (ii) is the force that is induced as a result of force (i). As the two components come close together, the force balance within the self-assembled aggregates such as micelles or emulsions is forced to change. A shift toward the new balance is inevitable. And this will change the morphology of the self-assembled aggregates and also critical reaction conditions such as charge density, which will eventually affect the nanostructure formation. For the micelle, this occurs on the level of the entire micelle; for the emulsion, it mainly does so on the level of the self-assembled layer that stabilizes the emulsion droplets.

Force (iii) is also induced as a result of force (i). Assume that the self-assembly units are at stable condition before the interaction (either thermodynamically or kinetically). After the interaction between the two components, the reactive inorganic species that are confined on the surface of organic components can induce this attractive interaction. This is usually condensation. When the inorganic components are nonreactive, force (iii) will be electrostatic repulsion. This will stabilize the entire system, which will result in no formation of nanostructures.

Force (iv) is the force between the inorganic species. For many cases of inorganic precursors, a variety of structural geometries exist in solution. These include ring structure, cagelike structure, dimer, trimer, and so on. Understanding the effect of these different structural factors of inorganic species on the self-assembly unit is important in deriving a better explanation of the physical and chemical properties of nanostructured materials formed. Force (iv) is responsible for this to some degree. It can be a rather strong force such as condensation, or can be a weak force such as hydrophobic or van der Waals as well.

9.3. SOL–GEL CHEMISTRY

Sol–gel chemistry has been widely studied over the past few decades (Brinker, 1989; Pierre, 1998). It was perhaps one of the most valuable concepts in the development of a wide range of industrial materials, such as ceramics, porcelains, semiconductors, aerogels, xerogels, and so forth. This sol–gel process of inorganic precursors (usually metal alcoxides) is also at the heart of the chemistry of the formation of nanostructured materials. This section will briefly describe this issue. As shown in Figure 9.3, the sol–gel process is a two-step process of hydrolysis and condensation (Pouxviel et al., 1987; Turner and Franklin, 1987; Boonstra and Bernards, 1989). Alcoxide precursors are first hydrolyzed in the aqueous solution.

Note that aluminum alcoxide is trifunctional.

This hydrolyzed form is the main *reactive inorganic species* in the above section. It usually is anionic except in the case of extremely low pH. Depending on the solution pH, the ratio of anionic group to hydroxide group (the degree of

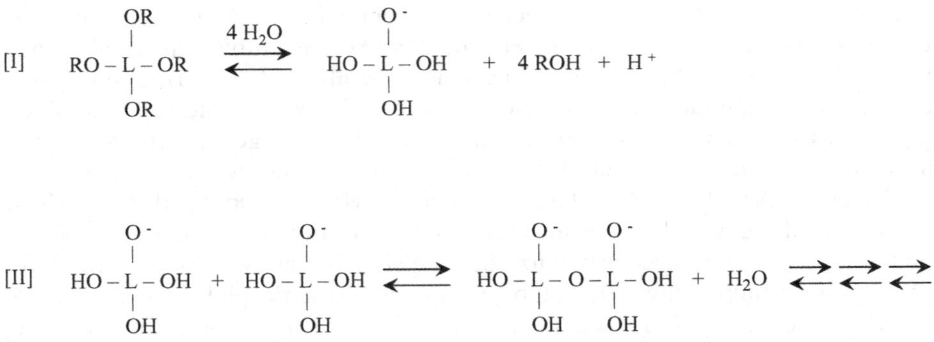

$$[\mathrm{I}] \quad \mathrm{RO-\underset{OR}{\overset{OR}{|}}{L}-OR} \; \underset{}{\overset{4\,H_2O}{\rightleftharpoons}} \; \mathrm{HO-\underset{OH}{\overset{O^-}{|}}{L}-OH} + 4\,\mathrm{ROH} + \mathrm{H^+}$$

$$[\mathrm{II}] \quad \mathrm{HO-\underset{OH}{\overset{O^-}{|}}{L}-OH} + \mathrm{HO-\underset{OH}{\overset{O^-}{|}}{L}-OH} \rightleftharpoons \mathrm{HO-\underset{OH}{\overset{O^-}{|}}{L}-O-\underset{OH}{\overset{O^-}{|}}{L}-OH} + \mathrm{H_2O} \rightleftharpoons \rightleftharpoons \rightleftharpoons$$

L = Si, Al, Ti, Zr, Ce, etc.

R = Me, Et, *n*-Pr, *i*-Pr, *t*-Bu, etc.

Figure 9.3. Sol-gel chemistry for the formation of nanostructured materials.

deprotonation) is changed (Babushkin et al., 1985). At this state, the solution can be considered as the solution of this monomeric form or, more realistically, a suspension of the monomeric form with possible higher-order species. The name *sol* comes from this state. The second step is the condensation between these hydrolyzed forms and possibly between the other forms, too. Continuation of this process provides bulk inorganic precipitations. Thus, this product inevitably has a high degree of cross-linked or networked morphology with a large amount of hydrogen-bonded water inside. The name *gel* represents this state. Usually, the first and second steps cannot be clearly separated, and occur with some degree of overlap. This makes kinetic control so critical for the sol–gel process.

Also, this is the important difference between these inorganic gels and other common types of gels, such as organogels and hydrogels. Both organogels and hydrogels are formed from organic, polymeric, biologically originated, or biologically derived gelling agents. Usually they possess (or are designed to possess) functional groups with hydrogen-bond capability, such as the carboxylic acid group, amide group, and hydroxide group. With the presence of aqueous or organic solvents, they form highly networked or cross-linked morphology with a large amount of water (hydrogels) or organic molecules (organogels; details in Section 7.5).

Aerogels and xerogels are formed by the drying of these inorganic gels, hydrogels and organogels. Water and organic solvent are removed from the inside while the framework retains its morphology. This creates highly porous but low-density materials.

9.4. GENERAL SELF-ASSEMBLY SCHEMES FOR THE FORMATION OF NANOSTRUCTURED MATERIALS

There are four categories of self-assembly processes that are directly responsible for the formation of nanostructured materials. Figure 9.4 shows the schematic representations. Type (a) is the *co-self-assembly* of organic self-assembly units with inorganic species. One step usually completes the self-assembly process. Both the intermolecular forces between the components and the reactions between the inorganic species interplay to form self-assembled aggregates. Depending on the degree of the reaction of the inorganic species (such as condensation), these organic–inorganic composites can possibly evolve into other structures.

Type (b) is called the *cooperative self-assembly* process. It is a multistep process and the most common type. The inorganic species interact with the already existing self-assembled aggregates, such as micelles, and these inorganic species–covered self-assembled aggregates self-assemble into secondary self-aggregates through interparticle (inter-self-assembled aggregate) interactions. More details on this process will be given in Section 9.6.

Type (c) is rather straightforward. *Organic self-assembly units* are self-assembled into the given structures first. This then acts as a *soft-template*. The inorganic species can be introduced into the interstitial regions between them, and react to form the nanostructures.

Type (d) begins with somewhat different starting self-assembly units. The organic and inorganic components are already bonded into one unit: an *organic-*

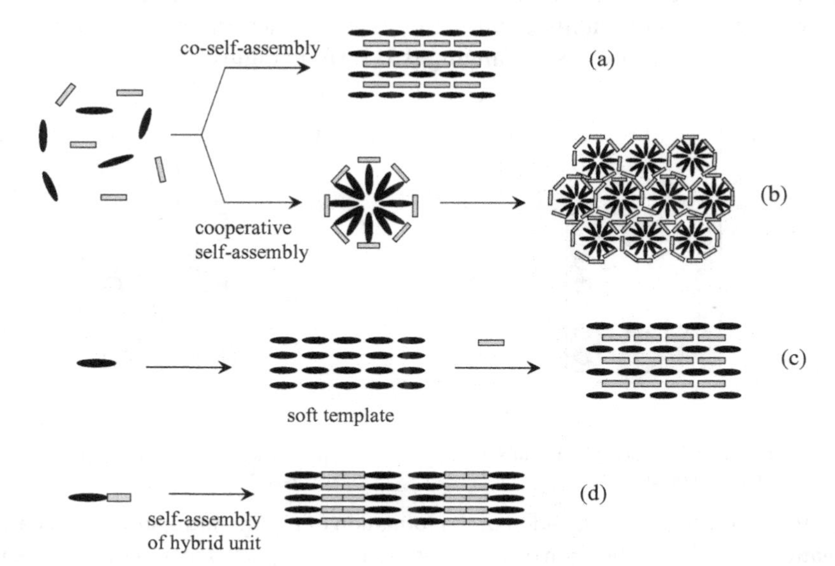

Figure 9.4. Different self-assembly schemes for the formation of nanostructured materials.

inorganic hybrid self-assembly unit. Usually, one step completes the self-assembly process while further reaction (such as condensation) can affect the structures formed. This type of self-assembly is also governed both by the intermolecular forces between organic parts and by the reaction between inorganic parts.

All four of these processes are subject to constant changes in the structures formed due to the constant changes in the intermolecular forces and the continuation of the reaction during the whole process. Therefore, the final nanostructured materials that are prepared from these schemes are usually highly kinetically stable structures.

9.5. MICRO-, MESO-, AND MACROPOROUS MATERIALS

This section will describe the schematic view of the formation of porous inorganic materials and their classifications. Details of the self-assembly issue will follow later in this chapter. By IUPAC nomenclature, the size term *micro* refers to the size range below 2 nm, *meso* (or *nano*) refers to 2–20 nm, and *macro* is assigned to the size range from 20 nm to μm. A widely accepted classification for porous materials is based on the size of their pores. Figure 9.5 shows the representative illustrations.

Microporous materials thus can be referred to as materials with a pore size range below 2 nm. Almost all synthetic zeolites and most of the natural zeolites belong to this category. For typical synthetic zeolites, regularly patterned inorganic frameworks are formed around self-assembled organic molecules. Removal of these organic components creates the exact replica of the inorganic pores. In practice, they have pore size ranging 0.5–1.5 nm (Figure 9.5a).

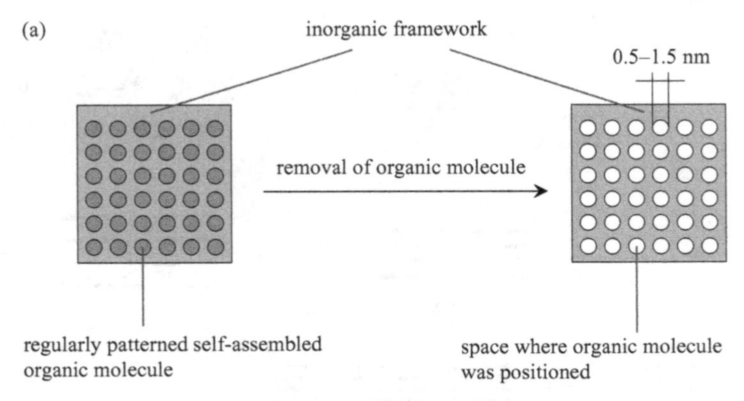

Figure 9.5. (a). Schematic illustration of the formation of zeolitic microporous materials. (b). Schematic illustration of the formation of meso- (or nano-) porous materials. (c). Schematic illustration of the formation of macroporous materials. (d). Schematic illustration of the formation of large macroporous materials.

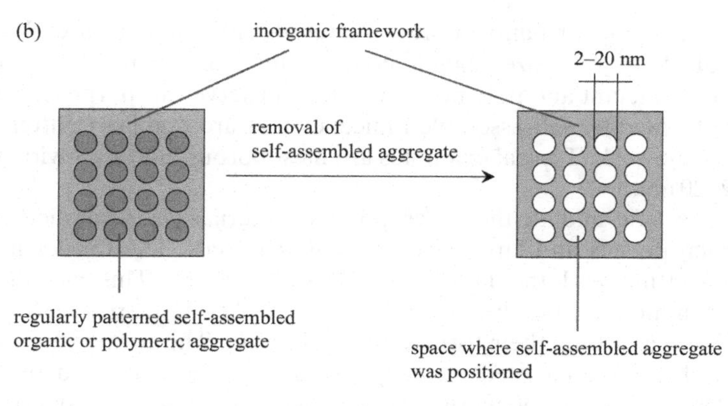

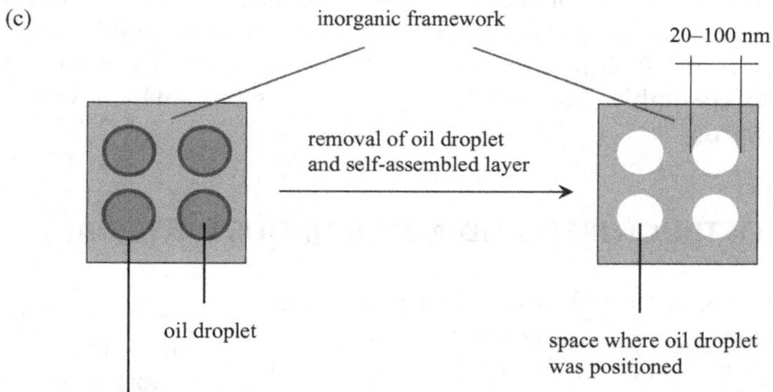

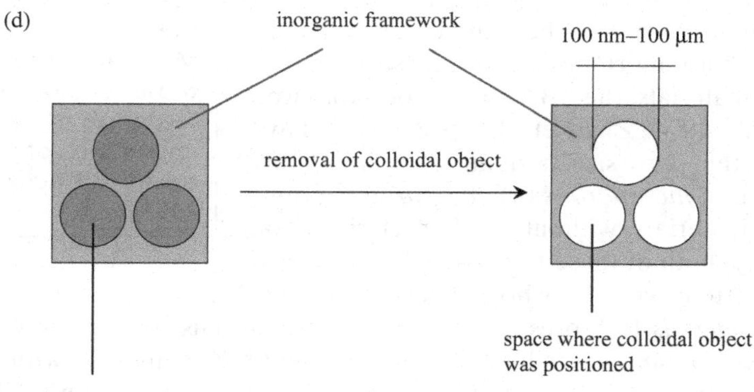

Figure 9.5. *Continued*

Similarly, meso- (or nano-) porous materials (Figure 9.5b) are obtained by the removal of the nanosize organic components from their composite forms. Most of the surfactant and polymeric micelles can serve this purpose. Nanocomposites are formed by self-assembled micelles that are regularly patterned with inorganic framework. Typical examples are mesoporous materials with pore size range of 2–20 nm.

In Figure 9.5c and d, the macroporous materials are classified into two groups: macroporous and large macroporous materials. The former have pore size of 20–100 nm and the latter have 100 nm–100 μm. This classification is not very common in the literature. However, it will help clarify the self-assembly issue and thus their formation mechanism. The former one is usually formed by the self-assembled oil droplets that can be stabilized by the self-assembled surfactant or polymeric layers. In the latter case, colloidal objects are self-assembled with inorganic framework. As can be imagined from the nature of oil droplets and colloidal objects, the former case can have a lot more complicated self-assembly issues than the latter cases, which can be rather straightforward.

9.6. MESOSTRUCTURED AND MESOPOROUS MATERIALS

The first self-assembly-based tailor-made mesostructured inorganic materials were discovered in the early 1990s (Beck et al., 1992; Kresge et al., 1992) by scientists from the Mobil Corporation.

> The preparation of silica with almost the same mesostructure using a very similar method actually appeared in the literature long before the claims made by Mobil. But it seems that those authors did not recognize the enormous importance of their discovery (Di Renzo et al., 1997).

The Mobil scientists named these new classes of materials *mesoporous materials* (M41S), surely because the pore size of these materials ranges 2–4 nm. Another important family of mesostructured inorganic materials, the SBA series, soon followed (Huo et al., 1994; Zhao et al., 1998). They have enhanced the pore size, ranging 2–10 nm. Many resources use, *mesoporous* or *nanoporous* to refer to these materials without further clarification. Technically, both of these terms are the same. But to be consistent with the whole theme of this book, we will use the term *mesoporous materials* to represent these types of materials. Accordingly, the term *mesostructured materials* will be used to represent the composites with which we obtain mesoporous materials by removal of organic components. The term *nanostructured materials* that is used throughout this chapter of course covers the whole range of structural scale from micro to macro.

This section will be devoted to describing the details of the mesostructured and mesoporous materials, from synthesis to characterization and potential applications. The critical roles of surfactant and polymer micelles in the preparation of these materials will be discussed. The concept of the surfactant packing parameter is very useful in designing, tracking, and explaining the variety of mesostructures we discuss here.

9.6.1. Formation of Mesoporous Silica with Hexagonal Structure

Though there are a lot of details to be considered, it is perhaps fair to say that mesostructured inorganic materials are obtained by using surfactant or polymer liquid crystal structures as structuring directing agents or soft templates (Ying et al., 1999). Three typical liquid crystal structures of surfactant or polymers are hexagonal, cubic, and lamellar. Thus, three typical structures of mesostructured materials are hexagonal, cubic, and lamellar types of structures. Figure 9.6 shows the schematic representation.

Regardless of the head group type, almost all of the known surfactants are useful for this purpose, as long as they are water soluble and form micelles. By simple control of solution conditions (such as pH), we can easily induce the attraction between the inorganic species and the micelles (Figure 9.2b), whether they are ionic, nonionic, or zwitterionic (Huo et al., 1994; Polarz, 2004). For most of the amphiphilic polymers, the same principles are applied. Since they form nonionic micelles, the hydration interaction will be the dominant force to form the composites. First, we will discuss the detailed mechanism for the formation of the hexagonal type of mesostructured silica in Figure 9.7 (Firouzi et al., 1995; Lee et al., 1996 and 2000; Stein and Melde, 2001).

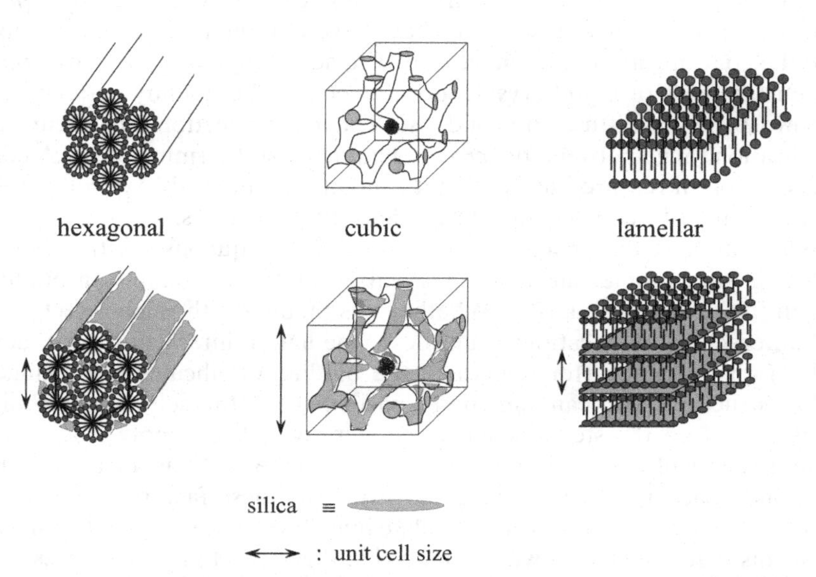

Figure 9.6. Three typical surfactant liquid crystal structures and the corresponding structures of mesostructured materials.

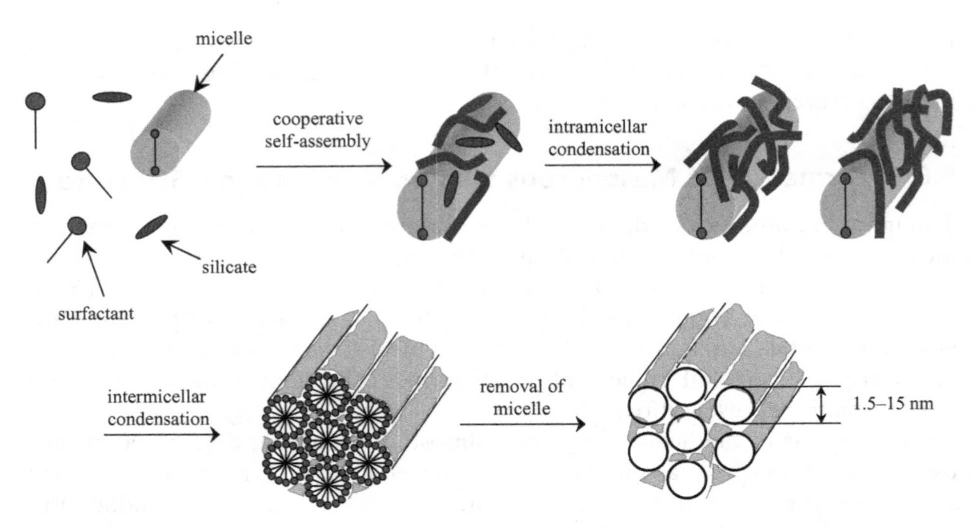

Figure 9.7. Mechanistic scheme of the formation of hexagonal mesoporous silica.

It has been known that the mesostructured silicas are formed only above the *cmc* of the participating surfactant. This is not a fully understood phenomenon even though the introduction of silicate anionic species that have strong counterion binding affinity (especially with cationic micelles) can reduce the *cmc* significantly. Also, the mesostructured silicas have not been successfully prepared above the concentration of liquid crystal. The actual aqueous interstitial regions between the units of liquid crystals do not seem to be volumy enough for the inorganic species to diffuse and condense. The actual reaction thus occurs at the concentration range between the *cmc* and liquid crystal forming point. A considerable amount of research shows that it is involved with multistep self-assembly with intra- and intermicellar silicate condensation reactions.

Spherical or rodlike micelles are present in the aqueous solution first. The addition of the anionic silicate species will promote the counterion binding of them on the surface of micelles. Whether this binding is through direct electrostatic interaction (for cationic micelles), mediated interaction (for anionic micelles), or hydration (for nonionic), the binding of silicates will change the micellar structures and can sometimes participate in the self-assembly of surfactant. Therefore, this step is actually a cooperative self-assembly process. These silicate species will then undergo condensation on the micellar surfaces (intramicellar condensation). For the hexagonal structure, these factors are believed to induce the transition of spherical or rodlike micelles into long, wormlike micelles. This means that micellar growth is induced in this step of primary self-assembly. In this sense, both the surfactant and silicate can be considered to be the primary self-assembly building units. Micelles with the silicate species on their surfaces then align into the hexagonal sites through the intermicellar attraction that is mainly the condensation between the micelles. This is the secondary self-assembly process, and the surfactant–silica composites are formed in this step. The micelles are the secondary self-assembly building unit and this nanocomposite (mesostructured silica) is the secondary self-aggregate. Mesoporous silica is obtained by the removal of the micelles, which can be done in a variety of ways: calcination, solvent extraction, supercritical fluid extraction, and so on.

The pore size, geometry, and structure of mesoporous silicas are determined by the size, geometry, and structure of the micelles that are removed. Though there is often some degree of shrinkage during the surfactant removal steps, the diameters of the micelles solely determine the size of pores on a nanometer scale. The micelles are also aligned along the two-dimensional hexagonal symmetry, which determines the hexagonal symmetry of the pores. And, of course, the spherical shape of the micellar cross-section determines the spherical shape of the individual pore.

Polymer micelles can be used with the same mechanistic principles for the synthesis of mesoporous silicas. Since the size of micelles formed from most of the polymers is larger than that of surfactant micelles, mesoporous silicas with much larger pore size are usually obtained.

These striking structural features make these mesoporous silicas unique in the field of nanotechnology. Not only are the sizes of the pores much larger than those of conventional zeolites (Which limits their applications to small single molecules), but they also can be finely tuned on the nanometer scale of 1.5–15 nm by simply choosing appropriate micelles. Polydispersity of pore size is very narrow. Wide ranges of mesoporous silicas show 300–1,200 m^2/g of extremely high surface area and 0.3–2.5 cm^3/g of extremely large pore volume. This provides a wide variety of new application possibilities that did not exist before (Section 9.9). Now we will describe the role of self-assembly in the structural control of these materials.

9.6.2. Structural Control of Mesostructured and Mesoporous Materials

Now let us look at how the concept of force balance for surfactant self-assembly can be used to control the structures of mesostructured materials and mesoporous materials. In Chapter 3, we discussed the details of the surfactant packing parameter and its application to track the structural change of micelles. Also, we noted in Chapter 4 (Exercise 4.1) that this concept can be directly applied in the formation and change of surfactant liquid crystal structures. In the previous subsection, we have seen, in the case of the hexagonal structure, that the structural features of mesostructured materials are almost solely determined by those of surfactant liquid crystals. Therefore, in principle, we should be able to follow and explain the structural changes that are very common during the synthesis of mesostructured materials. We also should be able to *design* the structures of the mesostructured materials by the empirical design of the surfactant packing parameter.

Figure 9.8 shows the different ways of designing the g value (surfactant packing parameter) of surfactant micelles. Path (a) is by use of the concept of micellar solubilization. The g value of 1/2 represents the spherical micellar unit that is the assembly unit of the hexagonal liquid crystal structure. Hexagonal mesostructured silica from Figure 9.7 can be largely obtained by setting this micellar condition prior to the silicate condensation reaction. Now let us assume that organic molecules are solubilized near the surface of this spherical micelle in such a way as to have increased g value close to 2/3. Since g value of ~2/3 is

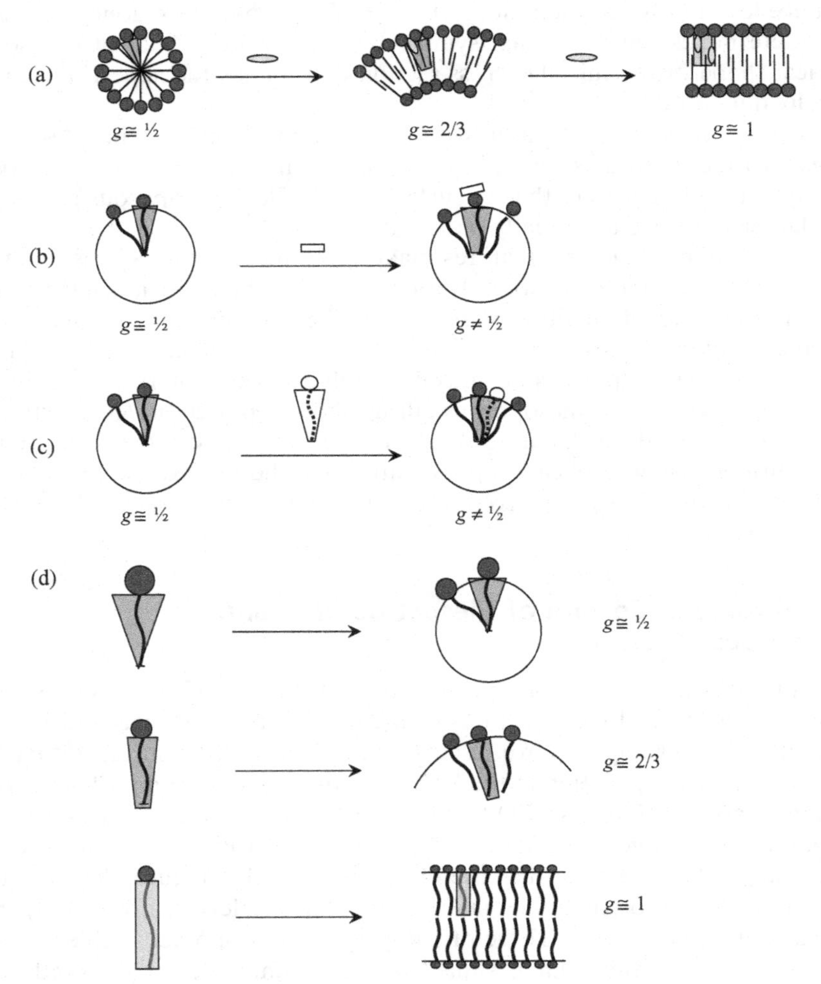

Figure 9.8. g value design for the structural control of mesostructured composite: (a) by micellar solubilization with lipophilic molecule, (b) by counterion binding, (c) by mixed micelle with co-surfactant, and (d) by direct structural design of surfactant molecule.

for the repeating unit of cubic liquid crystals, we should be able to obtain mesostructured materials with cubic structures by setting this micellar packing condition. Further micellar solubilization of the same solubilizate can provide the packing condition with *g* value close to 1. Reaction at this condition will likely create the lamellar type of mesostructured materials. The *design* of the *g* value is not exactly quantitative. Depending on the other reaction parameters, even exactly the same *designed g* value can sometimes result in different structures. But in most cases it has been shown that it is possible to control the structures of the mesostructured materials with reasonable accuracy with the understanding of molecular properties of surfactant and solubilizate.

The same principles apply for paths (b), (c), and (d) in Figure 9.8: Path (b) is by controlling the degree of the counterion binding prior to the reaction, which can create a different micellar packing condition from the original one. Possible binding competition between the counterion and the silicate species that actually undergo the reaction can make the structural prediction somewhat complicated. Path (c) is by the formation of mixed micelles prior to the reaction. As discussed in Chapter 3, the addition of the second surfactant or co-surfactant to the host micelles can create micelles with a different packing condition. This process also can be reasonably predicted by following the molecular geometries and properties of those surfactants. Path (d) is by directly designing the surfactant molecule. When the molecular structure of one surfactant favors the spherical packing, the same surfactant with a smaller head group will favor the curved packing (increased *g* value) due to the decreasing of the head group area a_0. The surfactant with a further decreased head group size will then favor the planar type of packing.

It has been shown that the design of the packing condition of micelles is quite powerful for the formation of mesostructured materials with controllable structures. It can serve not only to design the structures of the mesostructured materials but to track the structural evolution after the reaction as well. Figure 9.9 summarizes the relationship of the *g* value of the micelles to the surfactant liquid crystals and to the structures of the mesostructured materials obtained. This is a valid relation not just for mesostructured silicas but for mesostructured materials of other inorganics as well. These include titania, alumina, ceria, zirconia, aluminosilica, titanosilica, other transition metal oxides, and many more (Polarz, 2004). The symbols, H, C, and L represent hexagonal, cubic, and lamellar, respectively. Subscripts I and II represent the normal and reverse states, respectively. Two cubic regions, between the normal micelle and normal hexagonal, and between the inverse micelle and inverse hexagonal, are usually very narrow and undetectable in many cases. Thus, just C has been used for both cases. The two most popular series of known materials of MCM and SBA are shown in the figure with their symmetries. The number in parentheses is the year when they were first reported. Blank lines are for the structures of the surfactant liquid crystals whose counterpart mesostructured materials have not been reported as of 2007. Some structures such as cubic with *Im*3*m* symmetry have been reported, but with some controversy, and thus have not been included here.

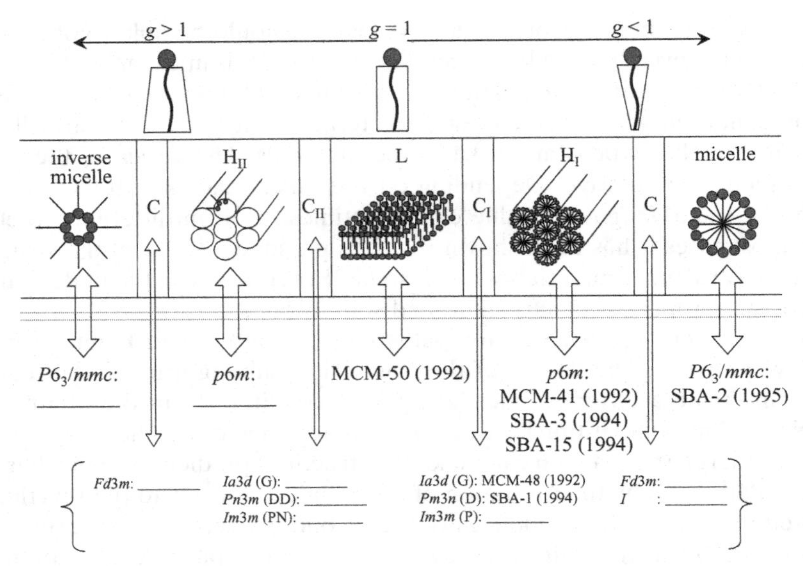

Figure 9.9. Summary of the relationship between micellar packing geometry and the structures of surfactant liquid crystals and the structures of the mesostructured materials obtained.

Careful precautions have to be always in place for this exercise in micellar packing design, because all three subparameters of v, a_o, and l almost always affect the g value in a coupled way. For example, changing the chain length of surfactant changes not only the l value, but also the v value and a_o value. The constant change in the condition for the charge density on the micelle surface makes this even more complicated. This type of change can be caused by the constant hydrolysis and condensation of inorganic species such as silicate, binding competition with other counterions, and the existence of other forms of species such as dimer, trimer, and so on. It also strongly affects the physicochemical properties of the final materials, such as surface acidity and stability (both hydrothermal and mechanical). This issue will be described in detail in the next subsection.

9.6.3. Epitaxial Analysis at the Micelle-Silica Interface

Growth of solid films whose surface patterns are governed by the epitaxy or crystal structure of the substrate is called *epitaxial film growth* (details in Chapter 11). It is common in the area of vapor phase depositions, especially for the preparation of functional semiconductor films. The key is the epitaxial (or geometrical) matching between the films deposited (semiconductor) with the substrate. This geometrical constraint serves as another major factor along with the bonding possibility in determining the physicochemical properties of the film formed (Ohring, 2002). Many functions in biological systems are also known to be governed by the epitaxial type of interactions. In this section, we will adopt this very concept onto the silica–micelle interface of the mesostructured silica. It provides a nice explanation for why all of the mesostructured silicas that have been reported so far show only amorphous silica walls. On the other hand, zeolites that are widely synthesized with similar quaternary ammonium salts show highly crystalline silica wall structures. It also serves as an excellent model to track and explain the structural changes during and after the synthesis of mesostructured silicas.

Figure 9.10a shows the schematic illustration of the micelle surface. To be simple but clear, the micelle surface is projected as a two-dimensional surface. Ignorance of the surface curvature should not cause that much error. The area we are focusing on is a couple of molecules (headgroups) wide, which can be assumed to be much smaller compared with the entire surface of the micelle. The most commonly used quaternary ammonium cationic surfactant, cetyltrimethylammonium bromide (CTAB) (or cetyltrimethylammonium chloride: CTAC), is selected as the model system in this figure. With the assumption of close-packing of the head group on the micelle surface, it should be reasonable to deduce that the surface has a *pseudo-hexatic phase*. This means that the head groups are arranged into the hexagonal lattice of symmetry. Thus, the parameters for the unit cell, d_{100} and a_{DD} values, for this hexagonal symmetry can be obtained from the head group size of the cetyltrimethylammonium cation, as shown in the figure. The scheme on the right-hand side of Figure 9.10a is for the most common silicate species in aqueous solution: monomer, dimer, and bridged-type trimer, from top to bottom, respectively (Iler, 1979). The light-gray filled circle represents the monomer. These three types of silicate species are known to be active participants in the silicate condensation reaction of most of the ordinary reaction systems. The open circle on the left-hand side of the figure is for the surfactant head group and is scaled to the size of the silicate species. The dark-gray filled circle in the upper-left corner of the figure is for the counterion (bromide or chloride ion; their Pauling diameters are 3.92 and 3.62 Å, respectively). The relative size of this counterion is scaled to the bromide ion.

Now let us superimpose (or build) the three types of silicate species on top of the micelle surface; this is *epitaxial matching*. Since there is a strong electrostatic attraction between the cationic head group and anionic silicate monomer, it can be reasonably assumed that the position of the silicate monomer should be located near the center of the head group. As shown in the lower-left corner of Figure 9.10, this simple geometrical picture shows that each silicate monomer is too far away to be contacted and thus to react on the micelle surface. To form the dimer or trimer on the micelle surface, the additional silicate monomer should be placed between the monomers that are already on the micelle surface. This is energetically unfavorable. Simple geometrical calculation can give the distance between the adjacent anionic groups that are the binding sites on the

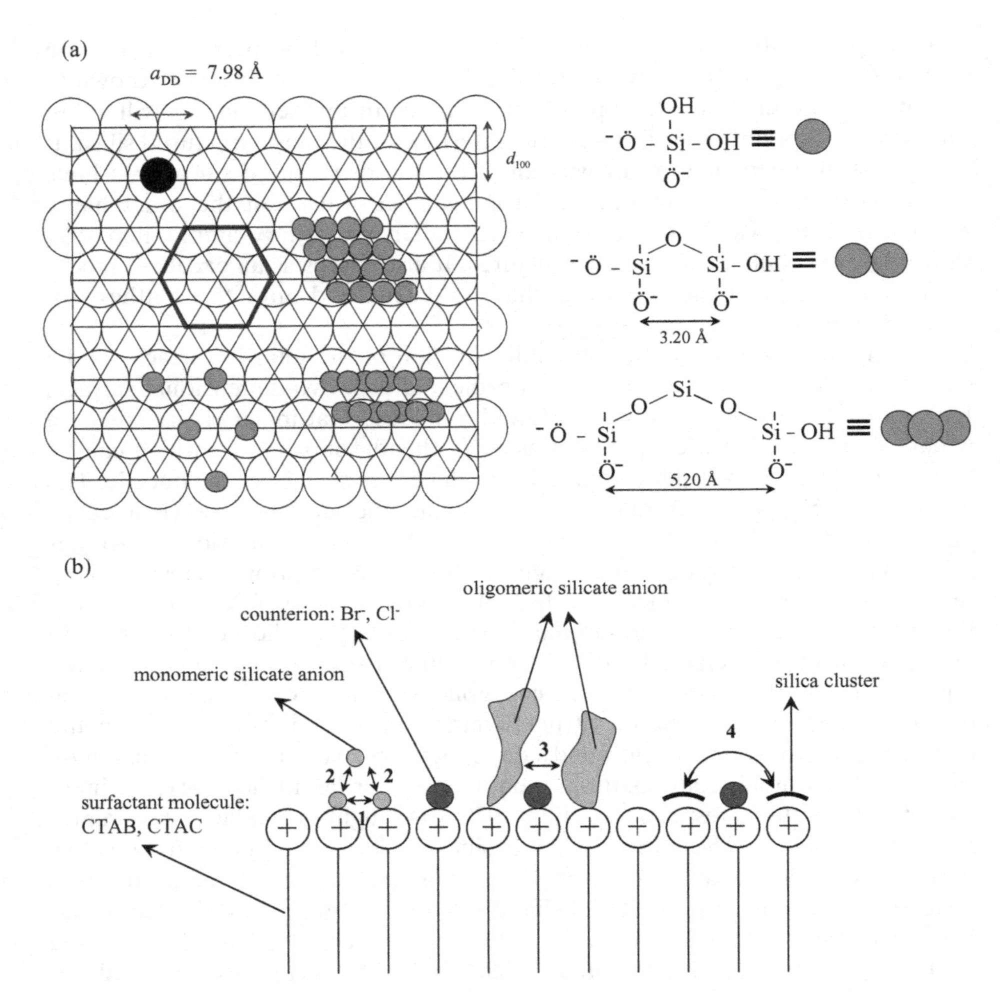

Figure 9.10. Epitaxial analysis of the formation of mesostructured silica.

micelle surface. As shown in the figure, these are 3.20 Å and 5.20 Å for the dimer and trimer, respectively. Neither of these values is matched with the head–head distance of 7.98 Å for the cetyltrimethylammonium cation. Neither of the configurations can fulfill the simple geometrical (epitaxial) matching. The next silicate monomers will be attracted on the bare cationic charge of the other head groups. This analysis suggests that the monomeric form of the silicate species may not play a major part in the formation of mesostructured silicas, at least in the early stages of the reaction. Rather, the oligomeric form of silicate anions that can be preformed in the silicate solution (Swaddle et al., 1994) may have the better geometrical advantage to be a major player.

EXERCISE 9.1

The cross-sectional area of a close-packed hydrocarbon chain in an alkane crystal is 21 Å^2. Given this, show that it may not be possible for silicate monomers to condense into either of the two configurations (dimer, trimer) in Figure 9.10a on any typical surfactant micelles.

Solution: To be geometrically matched with the distance of 3.20 Å between the two adjacent silanol groups of directly condensed silicate anions (dimer), a_{DD} should be 3.20 Å, too. This leads the radius of the surfactant head group to be 3.20/2 = 1.6 Å; then this leads the area of the head group to be $\pi r^2 = 8.04$ Å^2. By the same approach, in the case of 5.20 Å of bridged silicate trimer, the radius of the surfactant head group should be close to 5.20/2 = 2.6 Å; then this leads the area of head group to be $\pi r^2 = 21.2$ Å^2. The cross-sectional area of the close-packed hydrocarbon chain in an alkane crystal, 21 Å^2, is the minimum possible value of the area of the surfactant head group. This can be obtained only when the micelle is assumed to be a disk shape with rounded edge with infinite micelle size (infinite aggregation number). All surfactant molecules in real systems should have a much larger head group area than 21 Å^2. Thus, it is not possible to achieve satisfactory geometrical matching for either the dimer or the trimer configurations with any of the typical micelle surfaces.

Figure 9.10b is designed to better clarify this aspect. It is the side view of scheme (a). Process **1** is for the formation of the dimer through the direct condensation of silicate monomers, and **2** is for the formation of the trimer through the third silicate monomer. Again, this simple geometrical analysis shows that neither of these processes is favorable; the silicate monomers are too far away to react to each other.

> Remember that the sizes of head groups, counterions, and silicate species are scaled.

Also, the counterion on the micelle surface can be a big factor. The degrees of counterion binding of bromide and of chloride anion on a cetyltrimethylammonium micelle are ~0.75 and ~0.70, respectively, at room temperature. Preoccupation of the micelle surface by the counterions can hinder silicate (direct or bridged) condensations *physically*.

The oligomeric forms of the silicates, on the other hand, have a size that is comparable to that of the surfactant molecules and are much bigger than the counterions. They also exist in a variety of shapes. Thus, once they are attracted onto the micelle surface, they may have a much better chance to react without being constrained by the geometrical factor. The condensation may occur directly between the oligomers or even over the counterions; this is process **3**. Thus, this process can overcome both the geometrical constraint and the hindrance problem of counterions.

Since the oligomeric forms of silicate species are amorphous and since the condensation between them occurs mainly on the surface of those silicate species, it would take a considerable amount of reaction time and high reaction temperatures to create the mesostructured silica with crystalline silica. This reaction condition is not a unfavorable for the synthesis of zeolites. But this is not a benign condition for the synthesis of mesostructured silica. The organic components, surfactant molecules, should stand throughout the reaction to create the characteristic structures of the mesostructured silica. But, for all of the cases reported, the surfactant molecules were decomposed well before this condition was reached, even with highly thermally stable fluorinated surfactants. This might be the reason why the silica walls of all of the mesostructured and mesoporous silicas reported so far are amorphous, rather than crystalline as in zeolitic silicas.

Finally, let us add one more consideration from Figure 9.10. The formation mechanism of zeolitic silicas is still controversial. But, for the cluster–cluster model (Kirschhock et al., 1999 and 2002), the silicate clusters that are formed around the quaternary ammonium head groups (just like the clusters that are formed around the short chain tetraalkylquaternary ammonium cation for the formation of zeolitic silica) will experience strong hindrance mainly by the counterion; this is process **4**. And this will hinder the formation of networked silica clusters, which prevents the formation of crystalline silica along the surface of micelles.

EXERCISE 9.2

Estimate the degree of the counterion binding of the silicate species for the formation of mesostructured silica prepared using cetyltrimethylammonium chloride (CTAC). In most cases, this synthesis is satisfactory on the molar ratio of Si/CTAC from 1.5 to 3.5. Use the experimental value of 1–2 nm as the thickness of the silica wall, which can make roughly 10–17 layers of silicate monomers.

Solution: Let us assume that we have spherical micelles in the reaction solution with CTAC and assume that the aggregation number is 75. Also, assume that we are working at a concentration of 100 mM (~3.5 wt. %) of CTAC. *Cmc* of ~1 mM is negligible at this condition. This gives us $\sim[(6.02 \times 10^{23}) \times 0.1]/75 = \sim 8.03 \times 10^{20}$ micelles in this solution. Considering the Si/CTAC = 1.5–3.5, the total amount of Si should be $1.2 \times 10^{21} - 2.8 \times 10^{21}$. Next, while 10–17 silicate layers consist of the whole silica walls, two end sides are contacted with the micelle surfaces; thus, let us assume that 8–15 layers are inside the silica walls. Thus, 20–12% of total Si can be available for the binding with the head groups, that is, $\sim 1.4 \times 10^{20} - 5.6 \times 10^{20}$ Si. This means that the degree of counterion (silicate) binding can be estimated as 0.17–0.70. On the other hand, this means that ~83—30% of the cationic head groups on the micelle surfaces can be free from the silicate ion binding.

9.6.4. Charge Matching at the Micelle–Silica Interface

This subsection deals with the issue of charge matching at the silica–micelle interface of mesostructured silicas. Again, cationic cetyltrimethylammonium surfactant is chosen as the model system. Figure 9.11 shows the schematic illustration. Suppose the surfactant molecule has a certain value of *g* as a result of the interaction of its head group with the deprotonated form of silanol groups from the surface of silica (left-hand scheme of the figure). This silica is the inorganic component of the mesostructured silica. And let us suppose that for some reason the degree of the silanol group condensation has been changed. This will change the density of the deprotonated silanol groups at this interface, which will then change the charge matching possibility with the head group. The force balance will be changed, mainly because of the change of the electrostatic repulsion between the head groups. This will then change the *g* value of the surfactant molecule into a different value from the original one (right-hand scheme of the figure).

This change in the *g* value may not seem likely to guarantee the change in the curvature of the silica component inside the mesostructured silica. But, surprisingly enough, whole structures of mesostructured silicas are easily and delicately changed, for most of the cases reported so far, in response to the change in *g* value. And the structural symmetries of the mesostructured silicas such as hexagonal, cubic, and lamellar are well tracked by the appropriate *g* value of the surfactant molecule. This concept of charge matching is versatile and handy; it provides a meaningful avenue to track and explain not only the constant change in the structures during the reaction (changes of pH, temperature, etc.) but the structural changes by postsynthetic treatments (such as hydrothermal treatment and surface modification) as well.

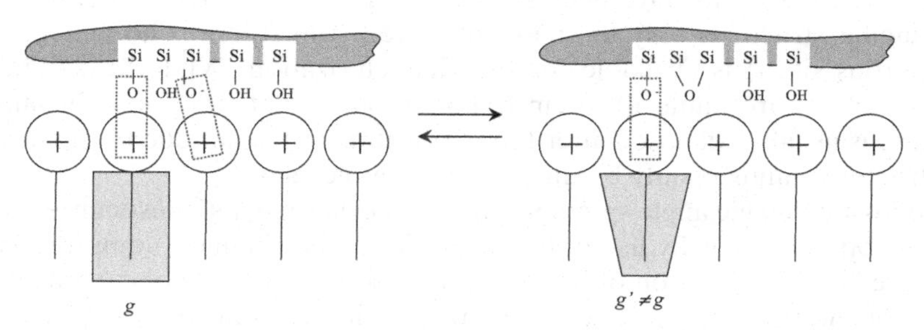

Figure 9.11. Issue of change of charge matching at the interface of silica–cationic micelle.

9.6.5. Characterization of Mesostructured and Mesoporous Materials

A full description of the characterization of mesostructured and mesoporous materials is beyond the scope of this book. This subsection will briefly describe some of the important issues in characterizing these mesoscale materials.

Three basic parameters for the three typical structures (hexagonal, cubic, and lamellar; Figure 9.6) are their geometrical symmetry (unit cell type), unit cell size, and pore size (for mesoporous materials). Two widely used techniques for determining these parameters are transmission electron microscopy (TEM) and powdered x-ray diffraction (XRD). While the former technique can provide a direct, distinguishable image for each of the structures, the latter can provide the quantitative information to determine the structural parameters. These two techniques work in a complementary way with each other. For example, when the TEM image happens to be taken on the side of the *c*-axis (dimensionless long axis) of the hexagonal structure, the image can look like the lamellar type. XRD peak analysis can clarify the difference. Also, it is not possible to determine the different cubic structures with only TEM images. Three known cubic symmetries, *Pm*3*n, Ia*3*d*, and *Im*3*m*, for mesostructured materials can be identified only through the assignment of XRD peaks.

Even though modern scattering and diffraction techniques offer a much wider range of q (scattering vector, 2θ for x-ray techniques) available, the length scale of these mesoscopic structures is located in the lower limits of the techniques. This is why the XRD peaks are much broader than the usual XRD peaks for crystalline minerals and zeolites. Thus, another complementary experiment using small-angle neutron scattering (SANS) and small-angle x-ray scattering (SAXS) is often required to expand the q-range even lower. The contrast matching technique is very useful in studying the mesostructured and mesoporous materials in detail.

The *conventional* x-ray diffraction theory (Bragg diffraction) is directly applied to assign the XRD peaks of the mesoscopic structures. Dimensionality, unit cell type, and unit cell size can be calculated from it. Excellent literature is available for the details of the analysis of diffraction and scattering data (Förster et al., 2005). Unit cells have been shown in Figure 9.6 for the comparison. For the lamellar structure, only the mesostructured type (there is no lamellar-type mesoporous silica) is available for the characterization. While hexagonal and cubic structures are stable after removal of the surfactant molecules, the lamellar structure is usually collapsed after the removal of surfactant due to the fact that this structure stands mainly by the surfactant molecules.

Solid-state magic angle spinning (MAS) nuclear magnetic resonance (NMR) spectroscopy is especially useful for determining the chemical composition and the degree of condensation of inorganic species. For example, the condensation degree of the silicate species can be easily deduced from the ^{29}Si MAS NMR spectra. BET (Brunauer-Emmett-Teller) or BJH (Barrett-Joyner-Halenda) sorption techniques are critical in determining the other critical physical parameters for mesoporous materials: surface area, pore volume, and pore size distribution. They are also useful for studying condensation (gas or liquid) phenomena inside mesopores.

9.7. ORGANIC–INORGANIC HYBRID MESOSTRUCTURED AND MESOPOROUS MATERIALS

This section will focus on the formation and properties of organosilica mesostructured and mesoporous materials. We discussed surfactant–silica mesostructured composites in the previous section. In those types of materials, the silica framework and the organic component (mainly surfactant) were physically bonded by intermolecular interactions. However, the organosilica in this section is the type of material whose organic components are covalently bonded with a silica framework (Asefa et al., 2000; Sayari and Hamoudi, 2001; Inagaki, 2004). We defined this as *synthetic organic–inorganic hybrid material* in Section 9.1.

Figure 9.12 shows the synthetic scheme of typical organosilicas. The whole mechanism is almost the same as the sol–gel process described in Section 9.3. And the mesostructures are templated (or directed) by the surfactant (or polymer) micelles exactly as shown in Figure 9.7. The critical difference is the structural features of the silicate precursors. They have two trifunctional silicon moieties covalently bonded with organic groups in between. The degree of deprotonation is of course strongly dependent on solution pH. Ethene, methyl, ethyl, phenylene, and biphenylene are among the organic groups reported to form the organosilicas successfully. These *functional* organic components remain intact after removal

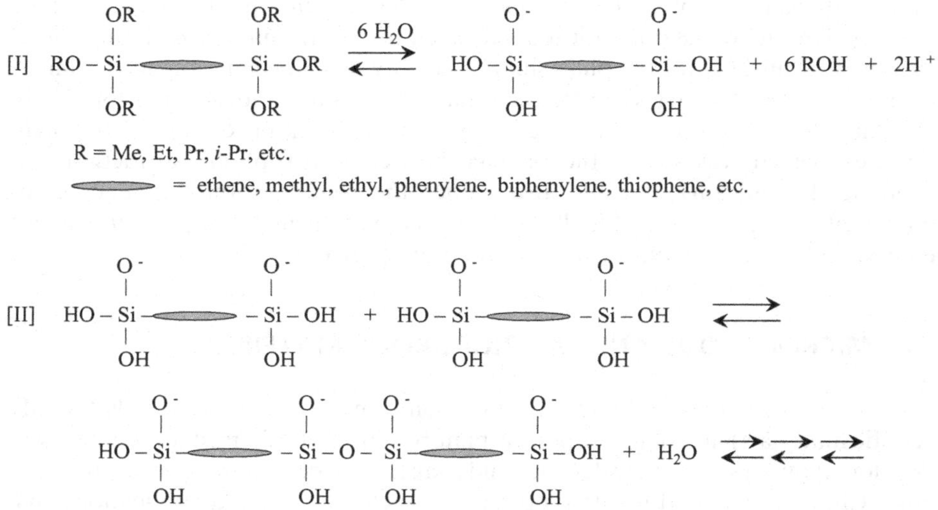

[I] $(RO)_3Si$–▬–$Si(OR)_3$ $\xrightleftharpoons{6\,H_2O}$ $HO\text{–}Si(O^-)(OH)$–▬–$Si(O^-)(OH)\text{–}OH$ + 6 ROH + $2H^+$

R = Me, Et, Pr, *i*-Pr, etc.

▬ = ethene, methyl, ethyl, phenylene, biphenylene, thiophene, etc.

[II] $HO\text{–}Si(O^-)(OH)$–▬–$Si(O^-)(OH)\text{–}OH$ + $HO\text{–}Si(O^-)(OH)$–▬–$Si(O^-)(OH)\text{–}OH$ $\rightleftharpoons$

$HO\text{–}Si(O^-)(OH)$–▬–$Si(O^-)(OH)\text{–}O\text{–}Si(O^-)(OH)$–▬–$Si(O^-)(OH)\text{–}OH$ + H_2O $\rightleftharpoons\rightleftharpoons\rightleftharpoons$

Figure 9.12. Synthetic scheme of the formation of organosilica.

of the surfactant micelles from the mesostructured organosilicas (to prepare mesoporous organosilicas). With the notion that the precursors are trifunctional (three possible hydrolysis and condensation sites on each silicon), the organosilicas obtained here might be considered as the silsesqueoxane type of mesostructured material.

One of the best features of these organosilicas is, of course, the functional organic groups that are strongly (covalently) bonded within the silica framework. This aspect provides the possible avoidance of the postsynthetic surface treatment for many cases, which is often necessary for the functionalization of mesoporous silicas for further applications. But that sometimes requires additional steps of reaction that can cause destabilization of the framework or even structural transition/collapse. Organosilicas also show better structural regularity for some cases (such as the crystalline-like wall structure for the benzene group) (Inagaki et al., 2002) compared with the mesoporous silicas. This is believed to promote better hydrothermal stability for organosilicas. Maybe this is possible because only the monomeric forms of organosilicate that happen to have the right geometrical feature are involved in the interaction with the surfactant head groups. For example, the distance between the two silanol groups on the hydrolyzed precursor [1,4-*bis*(triethoxysilyl)-benzene], 7.6 Å (Inagaki et al., 2002), is well matched with the head–head distance, 7.5 Å, of octadecyltrimethylammonium cation on its micelle surface (Lu et al., 1993) that was used as a template for this particular benzene-functionalized organosilica.

Preparation of organosilicas is quite a facile way to incorporate organic functionality within the silica framework. However, since this method inevitably requires the development of the proper organosilicate precursors, the choice of functional groups might be somewhat limited. Also, it shows limited structural diversity. Most of the mesostructures reported so far show hexagonal geometry even at quite different synthetic conditions. The scheme for the force balance described in Figure 9.8 does not seem to work for the formation of organosilicas, as it does beautifully for the plain silicas. Possibly this is because the fixed dimensions of bi-trifunctional groups (fixed distance between silanol groups) could help the silanol–head group interaction dominate the packing design of template micelles (which was set by the balance between hydrophobic attraction and head–head repulsion). It could also reduce the condensation frequency compared with silicate anions. Finally, thermal stability at high temperatures is not guaranteed, especially on the side of the organic groups.

9.8. MICROPOROUS AND MACROPOROUS MATERIALS

This section describes in detail the formation of microporous materials (mainly zeolitic materials) and of macroporous materials. Zeolite chemistry has been well developed over the past few decades and numerous excellent resources are available (Chon et al., 1996) for its synthesis, characterization, and applications. We will not explore this vast area, which would only dilute the point of this book.

The aspects of self-assembly in the formation of zeolites will be the focus of this section.

Preparation of macroporous materials in a controllable manner is an important field of study, too. This can be especially critical for applications in submicron-scale devices such as phonic devices, capillary-derived devices, and microelectromechanical systems (MEMSs). Proper design of self-assembly for emulsions and colloids is the key. We will stick to the issue of self-assembly in this section.

9.8.1. Co-Self-Assembly for the Formation of Microporous Materials

Natural zeolite was discovered more than two-and-a-half centuries ago. Over 150 naturally formed zeolites have been identified. As the name *zeolite* (*stone-to-boil*) implies, it has been extensively used in the areas of catalysis, sorption, and separation (Davis, 1991; Davis and Lobo, 1992). Synthetic zeolites began to be discovered decades ago. And now we have more than 80 different classes of zeolites that are commercially available. Since these synthetic zeolites can be obtained with much narrower pore size distribution than the natural ones, the application possibilities of modern zeolites have been broadened and are much more sophisticated.

Note that aluminum alcoxide is trifunctional.

Figure 9.13 shows the schematic representation of the formation of synthetic zeolitic materials.

OR | RO – L – OR | OR + R' | R' – N^+ – R' | R' —hydrolysis, condensation, electrostatic binding→

L = Si, Al, Ti

R = Me, Et, *n*-Pr, *i*-Pr, *t*-Bu

R' = Me, Et, *n*-Pr

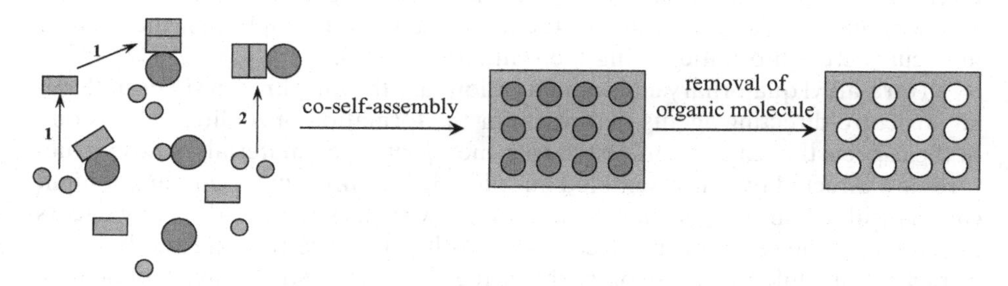

Figure 9.13. Schematic representation of the formation of zeolitic materials.

Co-self-assembly of individual organic molecules with inorganic species (Figure 9.4a) is the key to control over critical structural parameters, including pore size and pore size distribution. Alkoxides of silicon, aluminum, and titanium are among common inorganic precursors and tetra-alkyl quaternary ammonium cations are typical organic components. Final zeolites can be obtained as single components such as pure siliceous or multicomponents such as titanosilica, aluminosilica, and so on.

The inorganic precursors are hydrolyzed first, and then these hydrolyzed forms co-self-assemble with organic cations to form microstructured composites. Removal of the organic cations produces the microporous zeolites. Numerous studies have been preformed to reveal the true molecular-scale mechanism for this co-self-assembly step. Two well-established ones are the cluster–cluster route (process **1**) and the monomer–cluster route (process **2**). For the case of pure siliceous zeolite, well-defined oligomeric silicate species are formed in the solution first (induced by organic cations), and then these clusters are aligned in stepwise and directional ways. As this process progresses, the whole system is self-assembled into the microstructured composite. This is the typical cluster–cluster route (Kirschhock et al., 1999 and 2002). On the monomer–cluster route, the silicate species directly grow into clusters around the surface of the organic cations (Kinrade et al., 1998; Knight and Kinrade, 2002). For both routes, the silica nanoclusters are considered to be the critical building blocks in the formation of crystalline siliceous zeolite. Tetra-alkyl quaternary ammonium cations with short chains of methyl, ethyl, and propyl are the only known organic components that can produce crystalline siliceous zeolites. Hence, it is quite unlikely that there can be a significant self-assembly tendency between the organic components by strong hydrophobic interaction. The main driving force for this co-self-assembly process must be the silanol group condensation. The organic cations in the obtained composites are highly organized and well-structured with a nanometer scale (microstructured), which strongly implies that this driving force is balanced with the opposition force. Possibly, the repulsive or steric interaction between the organic cations is responsible for this. The sites where these assembled organic cations were located solely determine the pore size and geometry of the final zeolites. Since the organic cations also experience the dynamic change during the synthetic process, these types of organic molecules are commonly referred to as *structure-directing agents*. The terms *template, pseudo-template*, or *soft-template* are also used for this, though these seem to imply that the organic molecules are quite static during the synthetic process.

Tetra-alkyl quaternary ammonium cations are the only known *single* molecular structure-directing agents that work for the formation of zeolites. This is one of the aspects that can be contrasted with mesostructured materials, whose structures are directed by self-assembled micelles. Tetra-alkyl quaternary ammonium cations with chains longer than propyl do not work as structure-directing agents for zeolites. The reason is not clear; possibly it is because they are too bulky to attract silica building units, or perhaps they have too strong hydrophobicity, which can cause phase separation. As a major part of the effort to enlarge the pore size of zeolites (which is critical for many applications), researchers have long been trying to employ the longer-chain cations as structure-directing agents. They have also tried the asymmetrical cations that have one (or two) long alkyl chain with three (or two) methyl (or ethyl) groups. These are *the* quaternary ammonium surfactants that form the micelles, and mesostructured materials have been discovered at this point.

9.8.2. Emulsions for the Formation of Macroporous Materials

This and the following subsections show the preparation of macroporous materials through employing somewhat bigger self-assembled aggregates: emulsion and colloidal crystal.

Figure 9.14 shows the schematic illustration of the preparation of macroporous materials using oil-in-water emulsion. As discussed in Chapter 4, this type of emulsion has submicron-size oil droplets dispersed in aqueous solution. The self-assembled surfactant (or polymer) layers at the oil–water interface stabilize this structure. The reactive inorganic precursors can be dissolved at the aqueous region, and then are attracted onto the surface of oil droplets by electrostatic (for ionic surfactant) or by hydration (for nonionic surfactant) interaction. Condensation between the inorganics (silanol condensation for silicates) forms the macrostructured composites, and the subsequent removal of the organic components provides the final macroporous materials. For this method, it has been

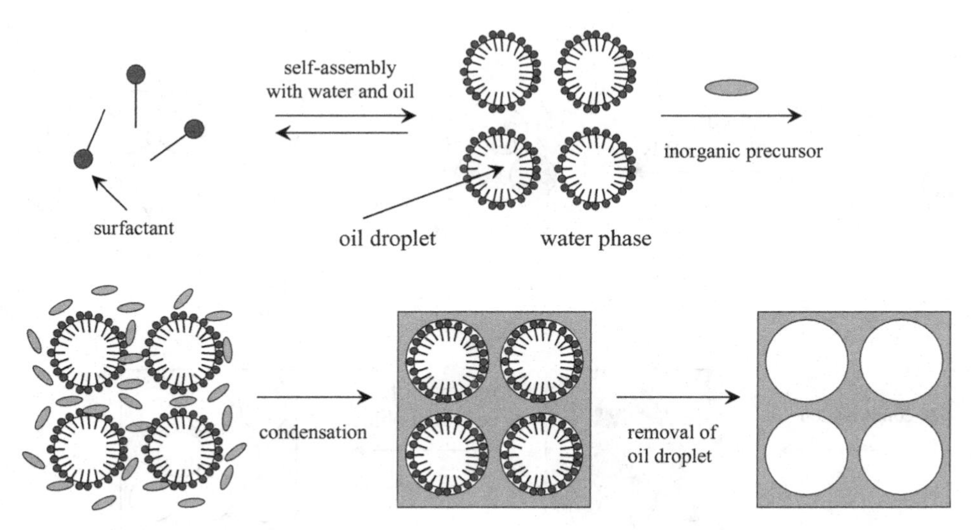

Figure 9.14. Mechanism of the formation of macroporous materials through an emulsion template.

known that the structure of the oil droplets is often deformed throughout the process, possibly because of the change in the charge matching on the surface of the surfactant layer. This often causes quite a large pore size distribution. But, unlike in cases of mesostructured materials, the structure of the oil droplets does not usually undergo a significant structural transition such as changing into entirely different morphological geometries. This is why they are considered to be the *soft template* of the self-assembly scheme (Figure 9.4c).

9.8.3. Colloidal Self-Assembly for the Formation of Macroporous Materials

Figure 9.15 is the schematic illustration of the formation of large macroporous materials. Though it is similar to the above case of emulsion, this process is rather straightforward. Crystal-like colloidal self-assembled aggregates are first formed through colloidal self-assembly. The polystyrene latex sphere is one of the most widely used colloidal particles for this. The inorganic precursors are then dissolved and condensed in the interstitial aqueous region to form the composite. Removal of the colloids provides the macroporous materials. The range of the pore size that can be obtained from this method is usually larger than from the emulsion-based process. Since there can be almost no structural changes in the size of the colloidal particles, the pore size distribution can be very narrow. The basic self-assembly scheme is the same as in the emulsion case (Figure 9.4c). Since the colloidal particles are harder than oil droplets, this is sometimes referred to as *hard template*.

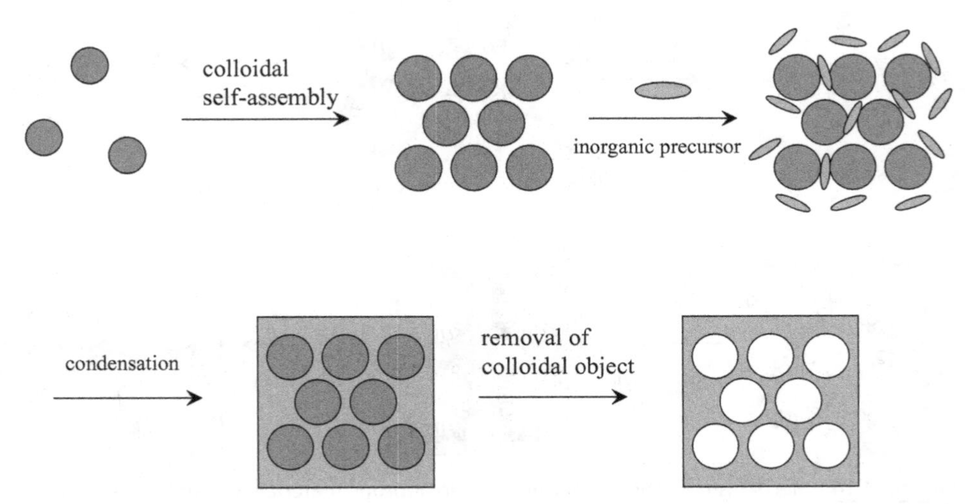

Figure 9.15. Mechanism of the formation of macroporous materials through a self-assembled colloidal template.

9.9. APPLICATIONS OF NANOSTRUCTURED AND NANOPOROUS MATERIALS

As discussed in the previous section, microporous zeolitic materials have been applied in a wide range of fields. However, the applications of mesoporous and macroporous materials are still in their infancy. This section describes the important application possibilities of these materials. Some of the applications are in the trial stage of practice while others are still in the stage of concept development. These include drug-delivery vehicles (Han et al., 1999), catalysts (Corma, 1997) and catalytic support (Corriu et al., 2004), size-selective catalysts (Davis, 2002), exotemplates (hard, secondary templates) (Schüth, 2003), functional devices (Davis, 2002), sensory devices (electrochemical, bio, optical, surface acoustic wave, etc.) (Walcarius et al., 1998; Scott et al., 2001), environmental sorbents (Feng, 1997), green chemistry (Macquarrie, 2002), and model nanospaces for confined gas or liquid (Schreiber et al., 2001). Three key structural prerequisites for all of these possible applications are their well-defined tunable pore size and shape, well-defined tunable geometry, and possibility for easy surface modification.

Figure 9.16 shows some schematic representations. Scheme (a) is for the concept of size-selective separation. Since the pore sizes of mesoporous and macroporous materials can be obtained with a wide range of nm—μm with very narrow distribution, even much larger molecules, including polymers and biological materials, can possibly to be separated with high selectivity. This can be a significant improvement compared with the zeolitic materials that have separation capability only for molecules smaller than a few nm.

Scheme (b) is for separation based on selective adsorption. The inner surface of mesoporous and macroporous materials can be easily modified with functional groups such as silanols. This then can be decorated with secondary functional groups such as ligands for metal ions or biofunctional groups for biological molecules. Big pore size and large pore volume provide enough room for these functional groups and for the adsorbates at the same space. The large surface area is critical for the efficiency of this type of practice. By employing the concept of the synthesis of the organosilicas, the functional groups can be impregnated within the silica framework, too. Surface modification is especially useful for the separation of heavy metal ions from wasted water, soil remediation, and catalytic reactions (Feng, 1997). Not only can the porous sorbents adsorb/hold large volumes of hazardous adsorbates, but they are environmentally benign as well.

> Their chemical compositions are common inorganics, just like ordinary soils or sands.

Scheme (c) is for the use of mesoporous and macroporous materials as *hard templates*. While "hard template" as used in Section 9.8.3 implies the role of a direct template, the same name here refers to the secondary template role. The porous materials can be used to synthesize the secondary nanostructured or nanosized materials. In this sense, it can also be called *secondary template* or

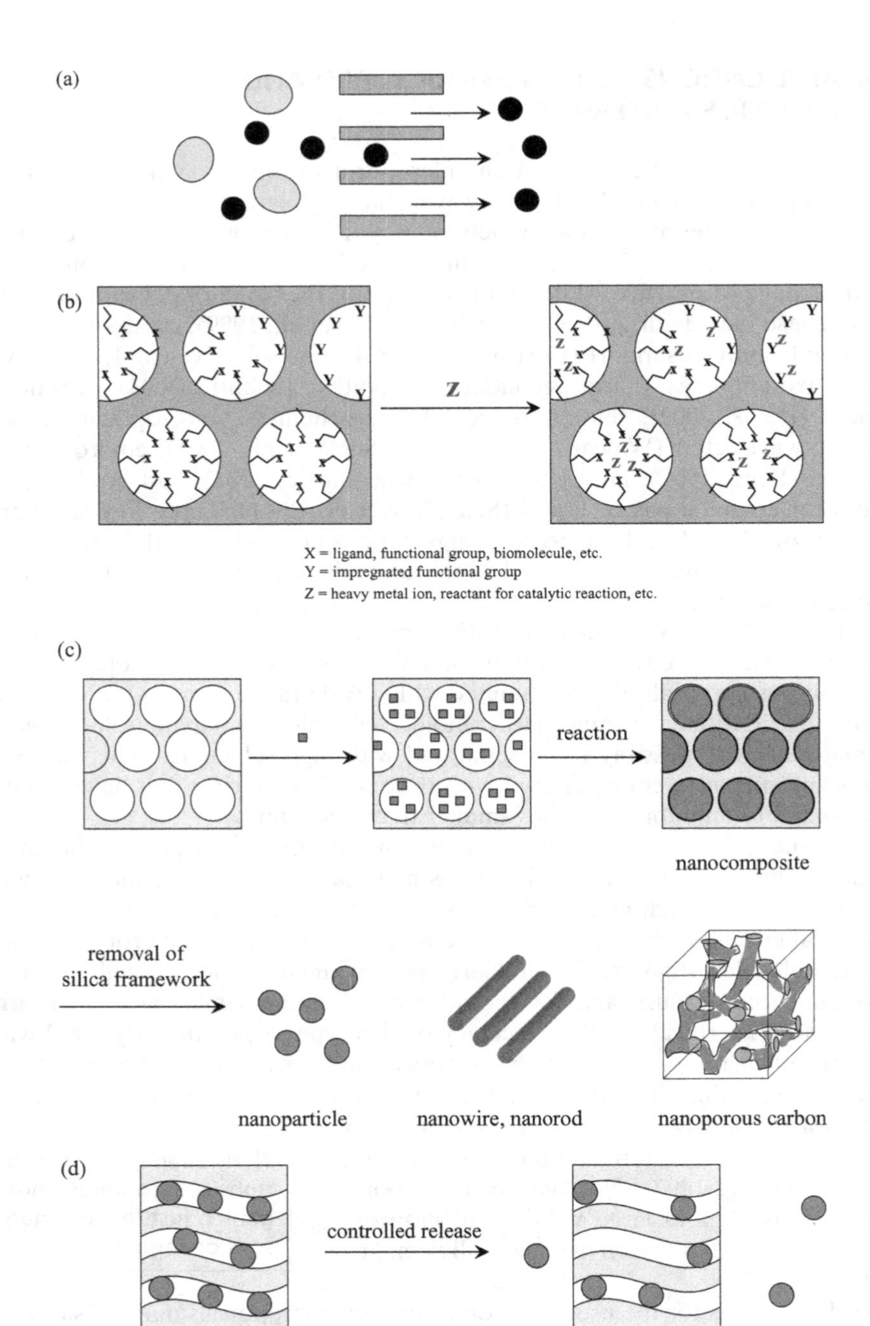

Figure 9.16. Applications of nanostructured and nanoporous materials.

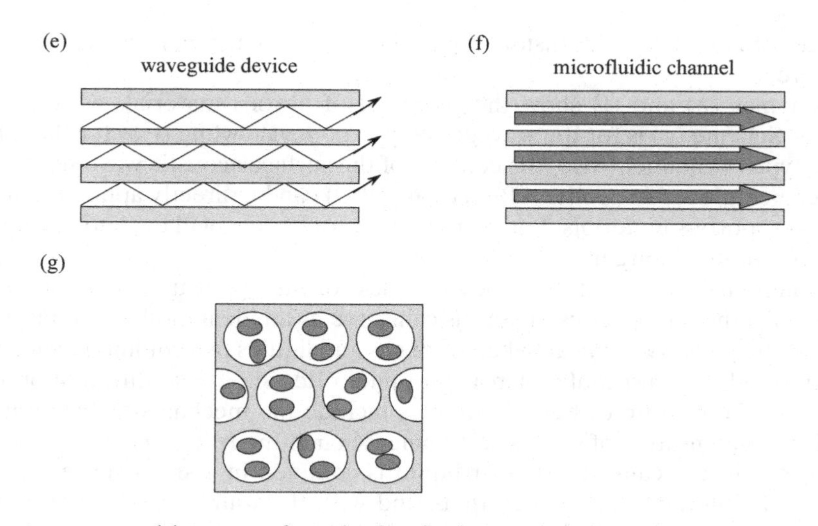

Figure 9.16. *Continued*

exotemplate (Schüth, 2003) (as opposed to *endotemplate* for the soft templates such as single molecules for zeolite preparation and micelles for mesoporous materials).

Special types of nanocomposites, for instance, nanoparticle-nanostructured inorganic composites (Konya et al., 2002; Xu et al., 2002), can be obtained using this concept. It will be possible to array nanoparticles of semiconductors or metals in a controllable and orderly manner in this way. This exercise is important because unique, novel physicochemical properties are expected from this type of nanocomposite. And the structural parameters are the key to success the control of spatial geometry, the control of interparticle distance, and the control of density distribution.

Nanoporous carbon was successfully obtained by filling the organic materials within the silica mesopore, subsequent carbonization, and removing the silica framework (Schüth, 2003). Also, nanoparticles, nanorods, and nanowires have been prepared by a similar procedure (Schüth, 2003).

Scheme (d) shows the application for the controlled release. The substances that we want to release in a controlled manner can be confined inside the meso- or macropore in composite forms. The release can be triggered by the change in the solution conditions, such as pH or salt concentration. The high surface area and large pore volume of porous materials make them ideal hosts that can confine large amounts of substances in relatively small amounts of the host. Adjustment of the pore geometry and size will provide the possibility of control of diffusion rate. For example, three-dimensional cubic geometry will enforce much slower diffusion than the two-dimensional hexagonal pore structure.

Schemes (e) and (f) show the potential of porous materials as functional devices. Scheme (e) is for the waveguide type of device while (f) is for the microfluidic type of channel. Also, the concept of the surface acoustic wave sensor that has been employed in zeolites (Bein et al., 1989) can be directly applied to meso- and macroporous materials. Larger pore sizes, of course, will expand its application even to much larger molecules.

Finally in this section, but not as the last of the application possibilities, in Scheme (g) the nanoporous materials can serve as an ideal model system for the study of the physicochemical behavior of gas and liquids in a confined nanospace. Gas or liquids that are confined in a nanospace often show very different or novel properties from those of bulk. Examples include the mechanism change in the capillary condensation of vapors and changes in melting/freezing points of liquids. This is not just because the gas or liquid is confined in a small space, but also because its interaction (between them and with the wall of the host) is now at the nanometer scale, which is often a few molecules across. Thus, the discrete character of matter such as molecular layering or cavitation becomes more prominent over the continuum of description of the state. Also, the effect of an external field such as an electric field or magnetic field becomes comparable to molecular-level interactions. Studies on this issue can be a valuable avenue to understanding important biological phenomena including the behavior and implication of thin water film or confined water within biological systems. Controllable pore size at the nanometer scale, controllable pore structure, and controllable surface properties are the essential prerequisites to select an ideal nanospace for this study, all of which should be possible with the nanoporous materials described in this chapter.

9.10. SUMMARY AND FUTURE ISSUES

We have discussed nanostructured materials and their relation to self-assembly in this chapter. Also, the interplay of intermolecular and colloidal forces during the formation of these materials was described. The self-assembly process plays an important role not only as the facile route for the preparation of nanostructured materials but as a practical tool to design/control their physiochemical properties as well. For the preparation of microporous zeolitic materials, co-self-assembly of single organic molecules with silicate precursors was the key step. For the preparation of mesostructured materials, cooperative self-assembly of surfactant micelles with inorganic precursors was critical. Emulsion droplets and colloidal self-assembled aggregates can be successfully employed for the preparation of macrostructured materials.

A large part of this chapter was especially devoted to mesostructured materials. With the picture of force balance, the implication of micelles and micellization for the formation of mesostructured materials and the concept of charge matching at the micelle–inorganic interface were developed. It was a simple but clear picture, which led us to the practical scheme that the surfactant packing parameter, *g*-value, becomes a quantitative tool to track and even to design the structures of a variety of mesostructured materials.

There are plenty of possibilities for more sophisticated and complex applications of nanostructured and nanoporous materials. But this, of course, will require identifying key directions and addressing proper strategies. Some of those issues will be proposed here.

1. Hierarchically Ordered Structure. Hierarchical structures could be identified as nanostructures with multipore systems, multiple geometrical symmetries, derived external features (such as ring, ribbon, helix, and tube), and so on. This will significantly expand the scope of possible applications. This goal could be approached (1) by the transcription of the features of biological self-assembly, (2) by developing combined molecular and colloidal self-assembly processes, and (3) by employing the external field.

2. Stability. Neither hydrothermal nor mechanical stabilities for most nanostructured materials are quite satisfactory. This is an especially important issue for cost-effective uses and advanced applications at extreme conditions. For example, the recycle rate of mesoporous inorganics for use as sorbents or catalysts in solution is currently very low. Effective uses as sensory or functional devices usually require durability at high temperatures and pressures for an extended period of time. The wealth of synthetic knowledge from inorganic, organic, and polymer chemistry will have to be employed more actively in the synthesis of nanostructured materials. And its effect on the underlying self-assembly issue will have to be properly addressed.

3. Polymorphism. Like many of the natural and synthetic solid materials, nanostructured materials are mostly kinetically stable materials. This brings up the issue of polymorphism. Constant change in the structures is not uncommon during the synthesis of nanostructured materials, and in many cases we can end up with a mixture of different geometries. Studies show that construction of the general phase diagram does not seem to be possible for most of the symbolic reaction systems, such as those for MCM-41 and SBA-15.

It is common that even slight changes in physical or chemical conditions such as molar ratio between organic and inorganic components can induce totally different phase behavior.

The new paradigm of force balance we have described in this chapter is more reliable to provide a reasonable prediction of morphology than the conventional phase diagram approach. It will certainly be worth it to address this concept in more complicated systems.

4. *Structure-Functionality Relationship.* Efforts to correlate the atomic or molecular structure with its activity or property are a well-established exercise in the broad disciplines of science and technology. For example, the *structure–property* relationship of surfactant molecules with their interfacial properties such as foamability, detergency, or emulsificability is very useful for surfactant technology. Atomic structures on solid surfaces are strongly correlated with their catalytic activity. Organic chemists have long been developing the *structure–reactivity* relationship. This includes the Hofmeister series and the Woodward-Hoffmann symmetry rule. Also, the extensively studied *structure–activity* relationship in modern biology (such as bioinformatics) is valuable for the further development of that area, including the discovery of novel drugs and the design of delivery tools. It might require a much greater accumulation of knowledge to establish these types of models in the area of nanostructured materials. However, the well-established relationship of their structural parameters to their specified "functionalities" will help us extract the fundamental principles and so apply them efficiently for further advancement in this area.

Finally, all the nanostructured and nanoporous materials described in this chapter are needed in another scientifically and technologically important form (film). This will be presented in Chapter 11 with other classes of nanostructured films.

REFERENCES

Asefa, T., Yoshina-Ishii, C., MacLachlan, M. J., Ozin, G. A. "New Nanocomposites: Putting Organic Function 'Inside' the Channel Walls of Periodic Mesoporous Silica," *J. Mater. Chem.* **10**, 1751 (2000).

Babushkin, V. I., Matveyev, G. M., Mchedlov-Petrossyan, O. P. *Thermodynamics of Silicates* (Springer-Verlag: 1985).

Beck, J. S., Vartuli, J. C., Roth, W. J., Leonowicz, M. E., Kresge, C. T., Schmitt, K. D., Chu, C. T. W., Olson, D. H., Sheppard, E. W., McCullen, S. B., Higgins, J. B., Schlenker, J. L. "A New Family of Mesoporous Molecular Sieves Prepared with Liquid Crystal Templates," *J. Am. Chem. Soc.* **114**, 10834 (1992).

Bein, T., Brown, K., Frye, G. C., Brinker, C. J. "Molecular Sieve Sensors for Selective Detection at the Nanogram Level," *J. Am. Chem. Soc.* **111**, 7640 (1989).

Boonstra, A. H., Bernards, T. N. M. "Hydrolysis-Condensation Reactions in the Acid Step of a Two-Step Silica Sol-Gel Process, Investigated with ^{29}Si NMR at –75 °C," *J. Non-Cryst. Solids* **108**, 249 (1989).

Brinker, C. J., Scherer, G. W. *Sol-Gel Science: The Physics and Chemistry of Sol-Gel Processing* (Academic Press: 1989).

Chon, H., Woo, S. I., Park, S.-E., eds., *Studies in Surface Science and Catalysis*, Vol. 102 (Recent Advances and New Horizons in Zeolite Science and Technology) (Elsevier: 1996).

Corma, A. "From Microporous to Mesoporous Molecular Sieve Materials and Their Use in Catalysis," *Chem. Rev.* **97**, 2373 (1997).

Corriu, R., Mehdi, A., Reyé, C. "Nanoporous Materials: A Good Opportunity for Nanosciences," *J. Organometal. Chem.* **689**, 4437 (2004).

Davis, M. E. "Zeolites and Molecular Sieves: Not Just Ordinary Catalysts," *Ind. Eng. Chem. Res.* **30**, 1675 (1991).

Davis, M. E., Lobo, R. F. "Zeolite and Molecular Sieve Synthesis," *Chem. Mater.* **4**, 756 (1992).

Davis, M. E. "Ordered Porous Materials for Emerging Applications," *Nature* **417**, 813 (2002).

Di Renzo, F., Cambon, H., Dutartre, R. "A 28-Year-Old Synthesis of Micelle-Templated Mesoporous Silica," *Microporous Mat.* **10**, 283 (1997); reference therein.

Feng, X., Fryxell, G. E., Wang, L.-Q., Kim, A. Y., Liu, J., Kemner, K. M. "Functionalized Monolayers on Ordered Mesoporous Supports," *Science* **276**, 923 (1997).

Firouzi, A., Kumar, D., Bull, L. M., Besier, T., Sieger, P., Huo, Q., Walker, S. A., Zasadzinski, J. A., Glinka, C., Nicol, J., Margolese, D., Stucky, G. D., Chmelka, B. F. "Cooperative Organization of Inorganic-Surfactant and Biomimetic Assemblies," *Science* **267**, 1138 (1995).

Förster, S., Timmann, A., Konrad, M., Schellbach, C., Meyer, A., Funari, S. S., Mulvaney, P., Knott, R. "Scattering Curves of Ordered Mesoscopic Materials," *J. Phys. Chem. B* **109**, 1347 (2005).

Han, Y.-J., Stucky, G. D., Butler, A. "Mesoporous Silicate Sequestration and Release of Proteins," *J. Am. Chem. Soc.* **121**, 9897 (1999).

Huo, Q., Leon, R., Petroff, P. M., Stucky, G. D. "Mesostructure Design with Gemini Surfactants: Supercage Formation in a Three-Dimensional Hexagonal Array," *Science* **268**, 1324 (1995).

Huo, Q., Margolese, D. I., Ciesla, U., Feng, P., Gier, T. E., Sieger, P., Leon, R., Petroff, P. M., Schüth, F., Stucky, G. D. "Generalized Synthesis of Periodic Surfactant/Inorganic Composite Materials," *Nature* **368**, 317 (1994).

Iler, R. K. *The Chemistry of Silica: Solubility, Polymerization, Colloid and Surface Properties, and Biochemistry* (Wiley: 1979).

Inagaki, S. "FSM-16 and Mesoporous Organosilicas," *Stud. Surf. Sci. Catal.*, Vol. 148 (Mesoporous Crystals and Related Nano-Structured Materials), pp. 109–132 (Elsevier: 2004).

Inagaki, S., Guan, S., Ohsuna, T., Terasaki, O. "An Ordered Mesoporous Organosilica Hybrid Material with a Crystal-like Wall Structure," *Nature* **416**, 304 (2002).

Kinrade, S. D., Knight, C. T. G., Pole, D. L., Syvitski, R. T. "Silicon-29 NMR Studies of Tetraalkylammonium Silicate Solutions: 1. Equilibria, ^{29}Si Chemical Shifts, and ^{29}Si Relaxation," *Inorg. Chem.* **37**, 4272 (1998); "2. Polymerization Kinetics," *ibid.*, 4278.

Kirschhock, C. E. A., Ravishankar, R., Jacobs, P. A., Martens, J. A. "Aggregation Mechanism of Nanoslabs with Zeolite MFI-Type Structure," *J. Phys. Chem. B* **103**, 11021 (1999).

Kirschhock, C. E. A., Ravishankar, R., Van Looveren, L., Jacobs, P. A., Martens, J. A. "Mechanism of Transformation of Precursors into Nanoslabs in the Early Stages of MFI and MEL Zeolite Formation from TPAOH-TEOS-H_2O and TBAOH-TEOS-H_2O Mixtures," *J. Phys. Chem. B* **103**, 4972 (1999).

Kirschhock, C. E. A., Ravishankar, R., Verspeurt, F., Grobet, P. J., Jacobs, P. A., Martens, J. A. "Reply to the Comment on 'Identification of Precursor Species in the Formation of MFI Zeolite in the TPAOH-TEOS-H_2O System,'" *J. Phys. Chem. B* **106**, 3333 (2002).

Knight, C. T. G., Kinrade, S. D. "Comment on 'Identification of Precursor Species in the Formation of MFI Zeolite in the TPAOH-TEOS-H_2O System,'" *J. Phys. Chem. B* **106**, 3329 (2002).

Konya, Z., Puntes, V. F., Kiricsi, I., Zhu, J., Alivisatos, A. P., Somorjai, G. A. "Nanocrystal Templating of Silica Mesopores with Tunable Pore Sizes," *Nano Lett.* **2**, 907 (2002).

Kresge, C. T., Leonowicz, M. E., Roth, W. J., Vartuli, J. C., Beck, J. S. "Ordered Mesoporous Molecular Sieves Synthesized by a Liquid-Crystal Template Mechanism," *Nature* **359**, 710 (1992).

Lee, Y. S., Surjadi, D., Rathman, J. "Effects of Aluminate and Silicate on the Structure of Quaternary Ammonium Surfactant Aggregates," *Langmuir* **12**, 6202 (1996): "Compositional Effects and Hydrothermal Reorganization of Mesoporous Silicates Synthesized in Surfactant Solutions," *ibid.* **16**, 195 (2000).

Lu, J. R., Simister, E. A., Thomas, R. K., Penfold, J. "Structure of an Octadecyltrimethylammonium Bromide Layer at the Air/Water Interface Determined by Neutron Reflection: Systematic Errors in Reflectivity Measurements," *J. Phys. Chem.* **97**, 6024 (1993).

Macquarrie, D. "Micelle-Templated Silicas as Catalysts in Green Chemistry," *Handbook of Green Chemistry and Technology*, pp. 120–149 (Blackwell Science: 2002).

Ohring, M. *The Materials Science of Thin Films: Deposition and Structure*, 2nd ed. (Academic Press: 2002).

Pierre, A. C. *Introduction to Sol-Gel Processing* (Kluwer Academic Publishers: 1998).

Polarz, S. "Ordered Mesoporous Materials," *Encyclopedia of Nanoscience and Nanotechnology*, Vol. 8, pp. 239–258 (American Scientific Publishers: 2004).

Pouxviel, J. C., Boilot, J. P., Beloeil, J. C., Lallemand, J. Y. "NMR Study of the Sol/Gel Polymerization," *J. Non-Cryst. Solids* **89**, 345 (1987).

Sayari, A., Hamoudi, S. "Periodic Mesoporous Silica-Based Organic-Inorganic Nanocomposite Materials," *Chem. Mater.* **13**, 3151 (2001).

Schreiber, A., Ketelsen, I., Findenegg, G. H. "Melting and Freezing of Water in Ordered Mesoporous Silica Materials," *Phys. Chem. Chem. Phys.* **3**, 1185 (2001).

Schüth, F. "Endo- and Exotemplating to Create High-Surface-Area Inorganic Materials," *Angew. Chem. Int. Ed.* **42**, 3604 (2003).

Scott, B. J., Wirnsberger, G., Stucky, G. D. "Mesoporous and Mesostructured Materials for Optical Applications," *Chem. Mater.* **13**, 3140 (2001).

Stein, A., Melde, B. J. "The Role of Surfactants and Amphiphiles in the Synthesis of Porous Inorganic Solids," *Surfactant Science Series*, Vol. 100 (Reactions and Synthesis in Surfactant Systems), pp. 819–851 (Marcel Dekker: 2001).

Swaddle, T. W., Salerno, J., Tregloan, P. A. "Aqueous Aluminates, Silicates, and Aluminosilicates," *Chem. Soc. Rev.* **23**, 319 (1994).

Turner, C. W., Franklin, K. J. "Studies of the Hydrolysis and Condensation of Tetraethylorthosilicate by Multinuclear (^{1}H, ^{17}O, ^{29}Si) NMR Spectroscopy," *J. Non-Cryst. Solids* **91**, 402 (1987).

Walcarius, A., Despas, C., Trens, P., Hudson, M. J., Bessiere, J. "Voltammetric In Situ Investigation of a MCM-41-Modified Carbon Paste Electrode: A New Sensor," *J. Electroanal. Chem.* **453**, 249 (1998).

Xu, W., Liao, Y., Akins, D. L. "Formation of CdS Nanoparticles within Modified MCM-41 and SBA-15," *J. Phys. Chem. B* **106**, 11127 (2002).

Ying, J. Y., Mehnert, C. P., Wong, M. S. "Synthesis and Applications of Supramolecular-Templated Mesoporous Materials," *Angew. Chem. Int. Ed.* **38**, 56 (1999).

Zhao, D., Feng, J., Huo, Q., Melosh, N., Fredrickson, G. H., Chmelka, B. F., Stucky, G. D. "Triblock Copolymer Syntheses of Mesoporous Silica with Periodic 50 to 300 Angstrom Pore," *Science* **279**, 548 (1998).

10

NANOPARTICLES: METALS, SEMICONDUCTORS, AND OXIDES

Nanoparticles and carbon nanotubes are perhaps the two most popular subjects in the area of nanotechnology; they have fascinated not only the scientists and engineers in this field but the general public as well. Their unique size and shape, and their novel properties, which have never been discovered in the microscopic and macroscopic worlds, provide us enough incentive to be excited about them.

The history of nanoparticles reaches back to the ancient world. More than 2,500 years ago, the ancient Egyptians were able to apply a variety of pigments with a wide color spectrum. The source of this color was the metal nanoparticles they put in the sols. The Chinese and Europeans in the later Middle Ages also developed pigments by the similar methods. They used some of metal sols for medical purposes, too. And the varied range of colorful ornamentation was due to the action mainly of gold nanoparticles of various sizes. The popular "black ink" used in Asian calligraphy was the sols of carbon nanoparticles. The seventeenth through nineteenth centuries saw the continuation of this trend.

In the twentieth century, synthetic varieties of nanoparticles began to be developed, and a variety of interesting and novel properties have been revealed from them. We now rediscover the value of these tiny particles and recognize

Self-Assembly and Nanotechnology: A Force Balance Approach, by Yoon S. Lee

their enormous potential in a variety of applications. An increasing number of commercial applications can be already found in the market in the areas of electronics, photonics, optics, pharmaceuticals, environmental, materials, and consumer products.

This chapter focuses on nanoparticles consisting of metal, semiconductors, and oxides. The topic of carbon-based nanoparticles (mainly carbon nanotubes and fullerenes) is enormous and fills numerous books. Thus, these will not be described here. There are many aspects in common between these two groups, especially on the issues of quantum size effect, surface atom effect, and their assemblies. But their relationship with self-assembly for synthesis is quite different. The synthesis of the former group is strongly dependent on the confinement effect provided by the self-assembly of organic building units or their self-assembled aggregates. The synthesis of the latter group, on the contrary, is solely the result of the chemical reaction between carbon atoms catalyzed mainly by metal atoms. Those unique nanostructures (nanotubes and fullerenes) happen to be the most stable products at the given reaction conditions.

This chapter is organized to cover the range from the basic concept and facts of nanoparticles to their application potential and future issues. The key point that is addressed throughout the chapter is the issue of intermolecular forces and self-assembly for the synthesis and size/shape control of nanoparticles, and for their applications.

10.1. WHAT ARE NANOPARTICLES?

Nanoparticles can be defined as particles with at least one of their three-dimensional sizes in the range of 1–~100 nm. This is between the size of atoms or molecules and bulk materials. Within this size range, they can usually consist of 10–10,000 atoms.

First, let us clarify the terms. Various terms that describe the nanosized particles have been adopted in the literature. The term is sometimes based on just the size, sometimes based on the physical property of the particles, and sometimes based on the shape of the particles. The term *nanoparticle* is obviously the most widely used one, and it is safe to say that this term encompasses most of the nanosized particles regardless of their physical property, size, and shape. Also, nanoparticles can be in either an amorphous or crystalline state. *Nanoparticle* refers to both states. When the crystalline state is taken into account, the term *nanocrystal* is preferred.

The terms *cluster* or *nanocluster* are also frequently used. Some of the literature uses these words as equivalent to or as replacements for *nanoparticle*. They are sometimes used to refer even to the aggregates of colloidal particles. This seems not to be a problem, as long as they are meant to represent the particles of the nanometer range. However, *cluster* is originally adopted to define the size range of atomic aggregates whose optical transition is not dependent on the number of atoms. As we will discuss in detail later in this chapter, the optical properties of nanoparticles are strongly dependent on their size. Hence, once the optical properties show variation in changes in size, the term *cluster* had better not be used to replace the term *nanoparticle*.

Based on the materials, nanoparticles can be composed of semiconductive, pure metallic, metal oxide, organic, polymeric, and biological components. Table 10.1 lists some examples. Figure 10.1 presents the different types of nanoparticles. A variety of different shapes of nanoparticles have been identified. This includes, but is not limited to, sphere, prism, cube, tetrapod, branched, triangle, hexagon, and pentagon. Their shapes can also be tube, rod, needle, and hollow sphere. Carbon nanotubes and peptide nanotubes are typical examples of the tube-type of nanoparticles. Fullerene-based particles and a variety of organic, inorganic, polymeric, or biological nanosized capsules (such as vesicles) can be classified as typical hollow sphere types. By composition, acorn, core-shell, and alternative types have been widely synthesized.

TABLE 10.1. Typical examples of various nanoparticles.

Composition	Nanoparticles
Pure metal	Au, Ag, Pd, Pt, Cu, Co, Ni, Ru, etc.
Bimetal	Fe-Co, Co-Ni, Pd-Au, etc.
Alloy	FePt, CoPt, PdNi, PtRu, etc.
Semiconductor	GaAs, CdTe, CdSe, CdS, ZnSe, AgBr, etc.
Oxide	SiO_2, Al_2O_3, TiO_2, CeO_2, Fe_3O_4, ZrO_2, ZnO, SnO_2, etc.

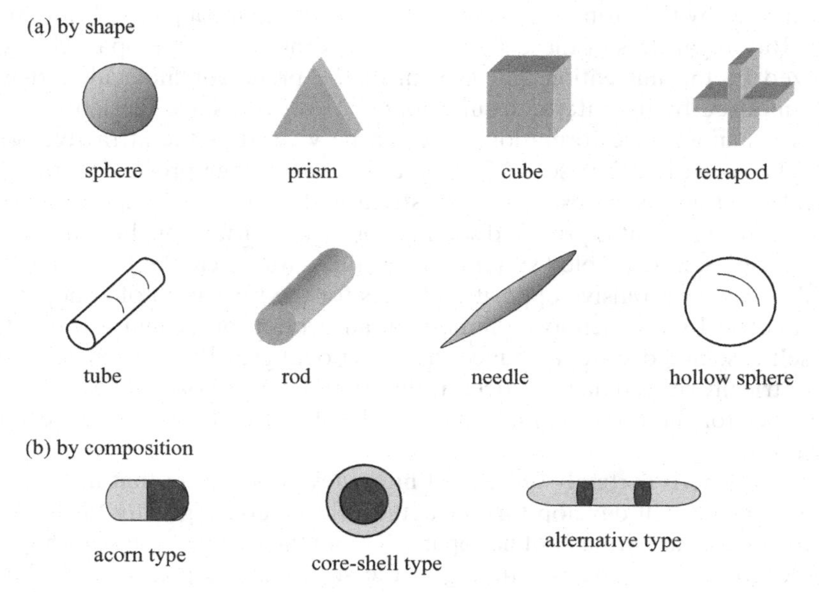

Figure 10.1. Different types of nanoparticles.

The most straightforward reason why nanoparticles are so important for current and future science and technology developments is the fact that their mesoscopic properties are at the intermediate state between atoms or molecules and bulk materials. For the most representative example, two gold atoms have the same properties, and two gold bulks have the same properties, too. This is always the case, regardless of their volume, size, and shape. But two gold nanoparticles do not share the same properties; they can be greatly different, based on their size and shape. Their melting temperature, color, electrical conductivity, and semiconductivity are all different. Also, bulk gold is non- (or barely) reactive inert metal, but gold nanoparticle is highly reactive and shows great catalytic reactivity, for example, for surface oxidation and surface epoxidation. Below a certain diameter of ~3 nm, its metallic property converts into semiconductive. This brings up the fascinating point that a variety of properties of gold nanoparticles can be tuned by their sizes and shapes. Excellent literature is available that reviews solely gold nanoparticles (Daniel and Astruc, 2004).

10.2. INTERMOLECULAR FORCES DURING THE SYNTHESIS OF NANOPARTICLES

In Figure 1.2 in Chapter 1, the universal concept of force balance for the self-assembly process was shown, that is, the balance of attractive force with repulsive force that is influenced by directional force. This picture can be directly adopted for the synthesis of nanoparticles that is guided by the self-assembly process or by self-assembled systems. As will be seen in the next section, the synthesis of nanoparticles by this principle can be categorized into a couple of different subroutes. But, regardless of these subroutes, the formation of nanoparticles is first initialized by the nucleation and growth of the precursor molecules. It is not always induced by the intermolecular forces. However, since these are the sole forces for nanoparticle formation, they can be viewed as the attractive driving force. The growth of the "seed" formed at the early stage proceeds until a geometrical constraint is imposed, which restricts and quenches further growth. This geometrical constraint is provided spatially or on the surface by the self-assembly process or by self-assembled systems. Thus, these forces can be treated as detrimental effects or repulsive opposition forces for the formation of nanoparticles. Unsymmetrical facts such as asymmetrical adsorption or asymmetrically developed self-assembled systems induce the directional growth, and so can result in asymmetrically shaped nanoparticles rather than spherical ones. This is the directional force for the formation of nanoparticles. Figure 10.2 shows the schematic illustration.

This picture is certainly quite a bit more developed from that in Figure 1.2. However, as we will develop further details, this exercise greatly facilitates the unified view of the synthesis of nanoparticles and their changes in size/shape. For example, whenever there is a tuning of the size of space provided by the self-assembled systems, the size of the nanoparticles obtained shows a direct relation-

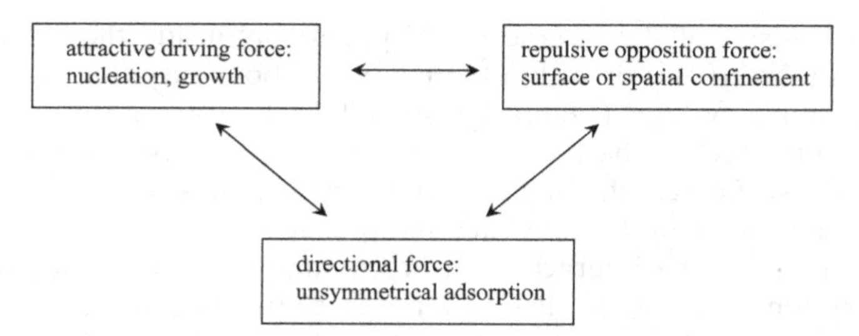

Figure 10.2. Force balance for the synthesis of nanoparticles.

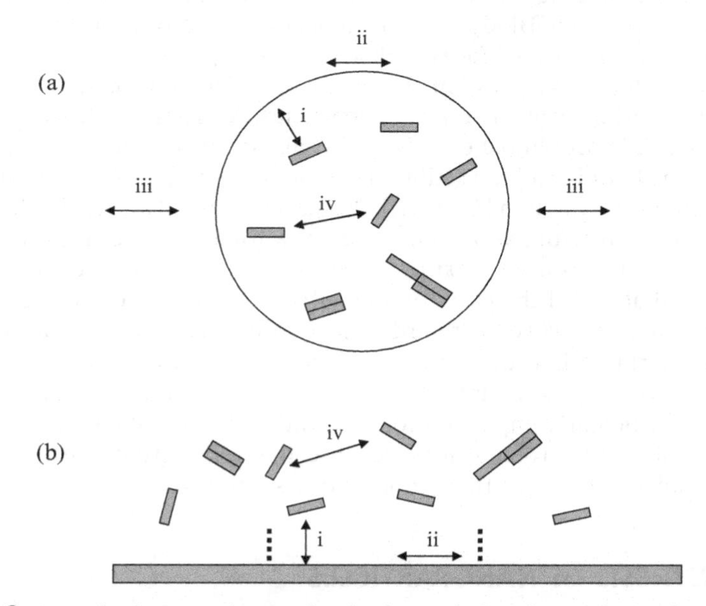

Figure 10.3. Interplay between intermolecular forces during the synthesis of nanoparticles: (a) inside self-assembled aggregates and (b) on the surface of self-assembled substrates.

ship with it; a larger space induces bigger nanoparticles. This can be understood as the lesser (more precisely, "late" in this case) opposition force gives more room to the driving force until they are balanced. Also, introduction (or presence) of adsorbent that is adsorbed on the specific (or different) site of the nanoparticles being formed makes them asymmetrically shaped. Thus, this is a directional force. Figure 10.3 presents more details from two representative situations of the intermolecular forces in the synthesis of nanoparticles. For the synthesis of nanoporous materials (Section 9.2, Figure 9.2), the nanoscale confinement was achieved as a result of the interaction of inorganic precursor with self-assembled aggregates. Therefore, it was the self-assembled aggregates that were confined in the nanoscale. For nanoparticles, nanoscale confinement is achieved as a result of the interaction of precursors inside the self-assembled aggregates (Figure 10.3a) or within the nanoscale boundary that is provided by the self-assembled systems (Figure 10.3b). While the former case is the spatial confinement imposed by the nanoscale space of the self-assembled aggregates, the latter is the surface confinement imposed by the nanoscale area that is forced by the interplay between the substrate and precursors.

Force type (i) is the interaction of precursors with the surface of the self-assembled systems. This is the driving force that attracts and retains the precursors inside the nanoscale space or on the surface. For case (a), this is the counterion binding on the inner surface of the self-assembled aggregates. For case (b), this is the counterion binding on the surface of soft substrates such as the Langmuir monolayer or Langmuir-Blodgett (LB) films, or the adsorption (chemisorption or physisorption) on hard surfaces such as metals or polymers. Force type (iv) is the attractive interaction between the precursors. This acts as the driving force for the nucleation and growth of the nanoparticles. A strong condensation process and rather weak hydrophobic or van der Waals forces are among those forces.

Whenever there is an interaction of foreign components (such as precursor) with a self-assembled system [force (i)], the self-assembled system itself is subject to a change in its structure as a result of the new force balance. This is force type (ii). For case (a), this is mostly expressed as the change in the packing geometry. For case (b), that is still the case for soft substrates such as the Langmuir monolayer or LB films. However, for hard surfaces, this force is negligible. The strain on the hard surfaces is orders of magnitude stronger than the intermolecular forces. Force type (iii) is the interaction between the self-assembled aggregates. This force is particularly important for the synthetic route of nanoparticles using reverse micelles or microemulsions because this force greatly affects the size, shape, and polydispersity of the nanoparticles synthesized.

10.3. SYNTHESIS OF NANOPARTICLES

Synthesis of nanoparticles can be performed in a variety of ways (Cushing et al., 2004; Daniel and Astruc, 2004; Masala and Seshadri, 2004; Burda et al., 2005). But the majority of the modern advances have been made through the marriage of synthetic techniques from traditional colloidal chemistry with the demands from materials sciences. And the self-assembly process of amphiphilic molecules lies at the heart of it. Even for the recently developed gas-phase techniques such as atomic layer deposition, the knowledge of colloidal force is crucial for the satisfactory control of the nanoparticles produced by these routes.

Figure 10.4 outlines the self-assembly schemes for the synthesis of nanoparticles. Besides the fact that the spatial or surface confinement of the precursor is key, these schemes are comparable to those for the nanoporous materials in Figure 9.4. Process (a) is cooperative self-assembly. Amphiphile molecules and inorganic or metallic precursors are self-assembled together through cooperative but nonselective interaction. This results in the formation of a reverse micelle

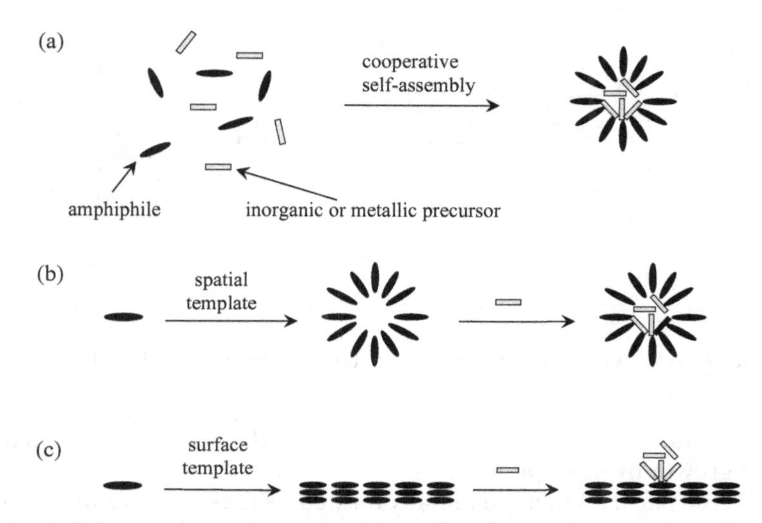

Figure 10.4. Different self-assembly schemes for the synthesis of nanoparticles.

type of aggregate where the nanoparticles are being formed inside of it. The amphiphiles adsorbed on the surface of nanoparticles provide the layers that protect from unwanted coalescence and at the same time spatial confinement for the growth of nanoparticles. Process (b) is the case where the space for the reaction *nanospace* is provided first. This is mostly either by reverse micelles or microemulsion. Precursors are then transferred into the nanospace and reacted. Process (c) is for the surface confinement. The self-assembled monolayers or films act as surface templates for the synthesis of nanoparticles.

Regardless of the synthetic methods described below, nanoparticles are prepared within the two- or three-dimensional space that provides the geometrical confinement or constraint. This is sometimes a real space such as microemulsion and mesoporous template, and sometimes an imaginary one such as a soft or hard epitaxial layer. But, regardless of whether it is real or imaginary, that space is imposed as a result of the interplay of the intermolecular forces: between the components of self-assembly building units and between the self-assembled units and components of nanoparticles being prepared. The size/shape and even the physicochemical properties of the nanoparticles are determined by this cooperativity of intermolecular forces for self-assembly.

10.3.1. Direct Synthesis: Confinement-by-Adsorption

This is the process wherein the synthesis of nanoparticles is achieved by the direct reaction of the precursors and the subsequent kinetic confinement by the adsorption of organic layers on the surface of growing nanoparticles. In this sense, this method is often called the *arrested precipitation method*. Figure 10.5 shows the schematic illustration. It can be understood as having four steps: decomposition

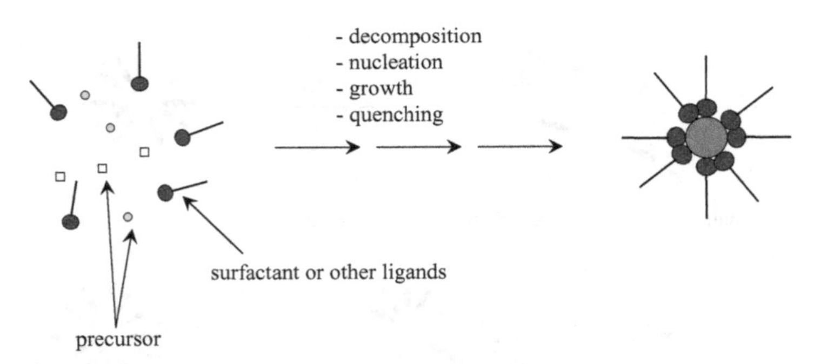

Figure 10.5. Synthesis of nanoparticles: type I—confinement-by-adsorption.

of precursor, nucleation of seed particle, growth of the seed particle, and quenching of the growth by adsorption.

Small seed particles that are initially formed by the decomposition of precursors have higher reactivity, and so grow faster. As the reaction goes further into nucleation and growth stages, the size of particles becomes bigger. The organic molecules that preexist with the precursors now can be adsorbed and become saturated on the surface of the particles, which quenches (stops or arrests) particle growth. In most cases, this final step can be controlled on the nanoscale level of the organic-particle complex, and the particle that is obtained also has a nanometer size, which means that a nanoparticle is synthesized. Decomposition of the precursor can be achieved by a variety of methods. This includes the sol–gel process, hydrothermal reaction (thermolysis), electrochemical process, sonication, irradiation by UV, IR, or laser, and a sudden jump in reaction condition such as pH, pressure, or temperature. Depending on the precursor, nucleation can be a homogeneous or heterogeneous type of process.

Control of the size and shape of the nanoparticles can be achieved by the control of various synthetic parameters. For example, the interplay between thermodynamics and kinetics during the nucleation, growth, and quenching steps can provide some degree of control in Ostwald ripening. The nonadsorbed species that can change the colloidal force, such as a double-layer or depletion force, will affect the force balance between the nanoparticles. This in turn will affect the coalescence between them, which can have a great effect on their sizes and size distribution. Also, the affinity of amphiphile molecules to any specific facet or specific composition of the nanoparticles can significantly affect their morphology. Instant inhomogeneity on the surface of the nanoparticles can create facets with different surface energy. The amphiphiles or capping agents can be preferentially adsorbed onto this facet, which can cause uneven growth of nanoparticles. This process produces nanoparticles with nonspherical structures.

Two typical organic layers that are widely used are amphiphiles and capping ligands. Not only do they arrest the growth of nanoparticles, but they protect them from the common coalescence problems as well. The difference lies in the way they interact with the nanoparticles. Amphiphiles (mostly surfactants or amphiphilic polymers) bind on the surface of nanoparticles mainly by physisorption of ion–ion, ion–dipole, or dipole–dipole interactions. But the capping agents that consist of functional groups and bulky or linear chains are mainly chemisorbed by complexation or covalent bonds. Thus, unlike amphiphiles, capping agents can have a stronger influence on the physicochemical properties of nanoparticles. Typical functional groups of capping agents include thiol, amine-, silane-, phosphine, and disulfide-.

10.3.2. Synthesis within Preformed Nanospace

On the previous adsorption-to-saturation route, the nanoparticles are confined by the adsorption of organic layers. For classification purposes, let us define this process as a *type I reaction*. This spatial confinement for the synthesis of nanoparticles can also be provided by the preformed nanoscale space. This is a *type II reaction*. Figure 10.6 presents the schematic illustration. The nanospace can be provided by self-assembled aggregates of surfactants (a), self-assembled aggregates of biological or bio-mimetic building units (b), branched dentritic polymers

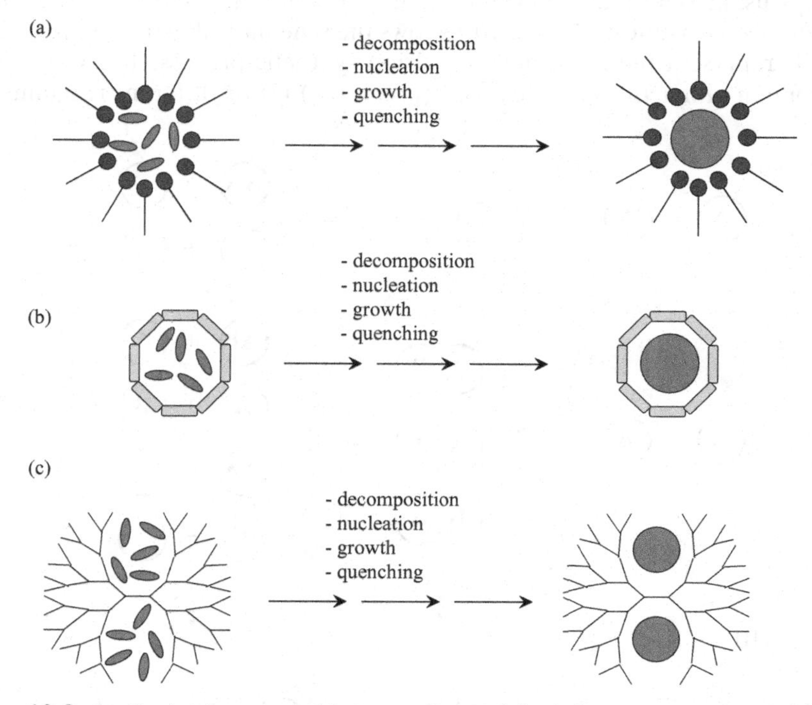

Figure 10.6. Synthesis of nanoparticles: type II—spatial confinement reaction. Within (a) self-assembled aggregate, (b) biological aggregate, (c) dendrimer, and (d) porous solid (Figure 9.16c).

(c), and presynthesized porous solids (d). Precursors undergo the same reaction steps of decomposition, nucleation, and growth, but mainly within the nanospace. Thus, the quenching of the reaction is the result of the predetermined spatial confinement (Vriezema et al., 2005).

10.3.2.1. Surfactant Self-Assembled Aggregates. A variety of self-assembled systems have structural features that can provide nanoscale space for the heterogeneous reaction. For example, normal micelles and oil-in-water (O/W) microemulsions retain the nanometer-sized oil phase that can be useful for synthesis in nonaqueous media. On the contrary, reverse micelles and water-in-oil (W/O) microemulsions provide aqueous environment. Most of the nanoparticle syntheses require the reaction in aqueous media. As mentioned in Chapter 4 (Section 4.4.2), the water molecules in the core of reverse micelles are mainly hydrated; thus, they cannot provide enough space and mobility for the proper reactions. Naturally, W/O microemulsion systems have mostly been used for the synthesis of nanoparticles (Lisiecki, 2005; Uskokovi and Drofenik, 2005). They are easy to prepare, their water pool sizes are easy to tune, and there are a variety of choices of surfactant and co-surfactant that can be employed.

Two different classes of reactions can be applied to the synthesis of nanoparticles using W/O microemulsion: multi-microemulsion synthesis and single-microemulsion synthesis. Figure 10.7 shows the schematic illustration. Routes (I) and (II) represent the multi-microemulsion synthetic process. The water pools that contain either the same reactant (precursor) (I) or different reactants (II)

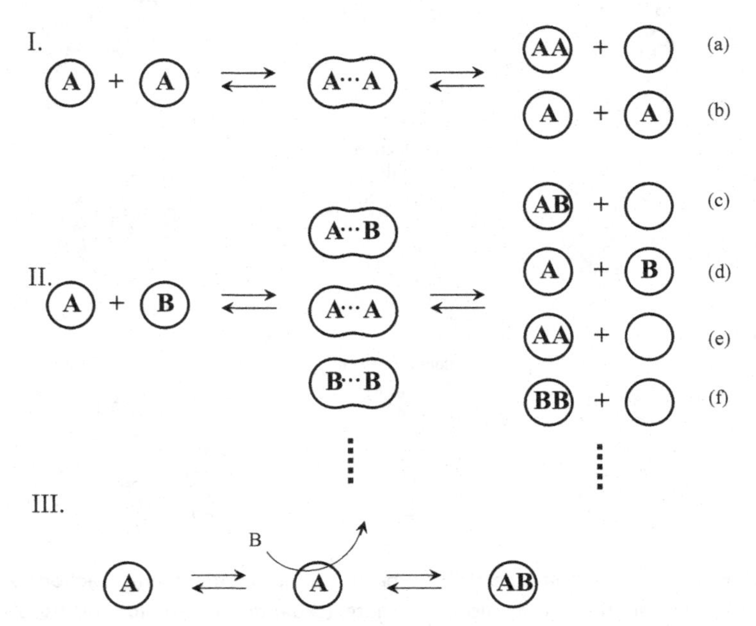

Figure 10.7. Different routes for the synthesis of nanoparticles using W/O microemulsion.

first undergo mixing and interchange processes. Decomposition, nucleation, and growth steps follow until the given space of water pool quenches further growth. The most desirable mechanisms for the synthesis of nanoparticles are (a) and (c). But there always can be the side processes of (b), (d)–(f), and more, and these are the main sources for the polydispersity of the size of the nanoparticles synthesized using microemulsion.

Route (III) is the single-microemulsion synthetic process. Steps from decomposition to growth are triggered by the diffusion of the second reactant into the water pool that contains the first reactant. Consecutive reactions of diffusion and growth can be achieved under certain conditions, which can be useful for the synthesis of multicomponent or multilayered nanoparticles.

Whether the synthesis is performed via the multi- or single-microemulsion route, the main factors that affect the physical properties of nanoparticles remain almost the same. They can be put into six categories:

1. The ratio of water to surfactant concentration is the primary factor in determining the size of water pool, as well as the size of nanoparticles synthesized.
2. The nature and structure of surfactant and co-surfactant, and their ratio, affects film stability at the oil–water interface.
3. Not only the precursors but the intermediate species can affect film stability.
4. The exchange rate of water and precursor contents inside the water pool can significantly affect the reaction kinetics.
5. pH is also a strong factor, since the chemical state of most of the precursors/intermediates can be greatly affected by it.
6. Changes in common oil phases such as cyclohexane, dodecane, hexane, and isooctane can greatly affect the rate of the reaction inside the water pool.

Size of Water Pool. As mentioned briefly in Section 4.4.4, the size of the microemulsion droplets usually changes with their simple composition factors. Based on the assumption of the spherical shape of droplets in W/O and O/W microemulsions, the size of the droplets increases linearly with the volume fraction of the dispersed phase (the phase of droplets) (Steytler et al., 2004). For example, at aerosol OT (AOT)-based W/O microemulsion, the size of the water pool can be tuned by simple control of the parameter, ω, which is the molar ratio of water to surfactant: R_w (nm) $\approx 1.5\,\omega$, where R_w is the radius of water pools (Pileni, 1993). At constant AOT concentration, the ω value increases as the water amount increases, so the size of the water pool increases. There is some limitation on the validity of this relation. The concentration range of the surfactant should be moderate and the optimum concentration of surfactant is dependent on the specific system component and also on temperature. For most cases, this relation works best in the ω range of 15–35. However, that is enough to explore the range of reaction conditions or formulation conditions for the synthesis of nanoparticles. As the size of the water pool increases in this range, the size of the nanoparticle synthesized linearly increases, too.

Film Stability. The self-assembled film of surfactants around the water pools restricts the size of the nanoparticles. Its stability is critical to preventing their coalescence during synthesis. It also affects the diffusion of the precursors into the water pool during the collision (mixing)-interchange process and therefore the kinetics of the overall reaction.

Counterion Binding. As the reaction progresses, both the composition of the precursors and their amounts are inevitably changed. Thus, there will be a change in the degree of counterion binding. This can change the microstructure of water pool and alter the structure of the nanoparticles synthesized.

Nonspherical Nanoparticles. The sphere is the most common structure obtained by the synthesis of nanoparticles using microemulsions. But rod-shaped, cube, and needle-shaped structures are often obtained even though the reaction system was aimed at the sphere. This is primarily caused by the change in the whole phase during the reaction. Also, when the interfacial film is not rigid enough, irregular growth of the particle can be generated. The degree of surfactant adsorption onto the surface of nanoparticles is not always symmetrical, which can change the structure of nanoparticles synthesized.

10.3.2.2. Bio-mimetic Self-Assembled Aggregates. The scheme in Figure 10.6b describes the synthesis of nanoparticles using self-assembled aggregates of biological or bio-mimetic building units. Precursors are confined inside the nanosized aqueous pool of the bio-self-assembled aggregates, and then undergo the decomposition, nucleation, and growth processes. The geometrical confinement by the bio-self-assembled aggregates quenches further growth, which produces nanoparticles. As described in Chapter 7, bio-self-assembly processes are hierarchical and directional, which provides a variety of "supramolecular self-aggregates" from a wide spectrum of building units. For nanoparticle synthesis, a well-defined aqueous nanospace is crucial. This can be provided by certain types of proteins and viruses that are self-assembled into cagelike structures.

A typical example can be found in the nanoscale cage structures formed by ferritin-based proteins such as ferritin and apoferritin. They are abundant proteins in nature, and self-assembled into cage (or hollow sphere)-like structures. This provides an inner space of aqueous environment with ~8 nm of inner diameter. Hydrogen bonding along with other intermolecular forces is strongly involved, so the self-assembled structures are relatively robust. Also, their walls are permeable to small ions and solvents. These conditions are suitable for synthesis of nanoparticles by the spatial confinement approach. Effective synthesis of iron sulfide nanoparticles using ferritin is well demonstrated. Also, manganese oxide nanoparticles have been synthesized from apoferritin.

Viruses are other self-assembled systems that can provide "nanoreactors." Viral capsids are self-assembled proteins with unique scaffold structures. Their inner space is also aqueous, and usually bigger than protein analogs. Their length can reach up to hundreds of nanometers. They are stable enough in a wide range of pH and temperatures and a wide range of chemical reactions. They also come with more structural variety. Besides the cagelike structures, rod-shaped, hollow-tube, and other higher-order structures are common for viral capsids. These features can provide more structural diversity in the nanoparticles synthesized. The most typical example is the tobacco mosaic virus (TMV). It has a long rod-shaped structure with ~300 nm length. Both its inner and outer surfaces can be used to obtain well-defined nanoparticles or their assembled structures. For example, its 4 nm diameter channel is permeable by small ions and solvent molecule and thus can be used to prepare well-defined nanowires with a very high length-to-diameter ratio.

Bio-self-assembled aggregates are usually more stable than surfactant self-assembled aggregates, so their sizes and shapes can be well retained in a wider range of reaction conditions. For a certain class of nanoparticles, this provides some unique advantages. Also, they can provide the heterogeneous nucleation sites that can serve as surface-confined space for the growth of nanoparticles. On the other hand, reactions with the bio-self-assembled aggregates are often highly selective, which means that the variety of nanoparticles (size, shape, and composition) can be quite limited compared with the microemulsion-based approach. This limitation can be overcome to some degree by the fact that their surfaces can be easily tuned by genetic modification.

The *biomineralization* process also generates well-defined nanoparticles (more precisely, bio-inorganics or bio-nanocrystals). In the sense that they are well-defined *nanosized particles*, they are also a family of *nanoparticles*. But, in the sense that their compositions are common minerals and their formation processes are highly selective mineralization rather than "planned" reactions, these two can be sometimes differentiated.

10.3.2.3. Dendritic Polymers. Dendritic polymers are a class of polymers that have a multigeneration branched structure. Poly(amidoamine) (PAMAM, starburst dendrimer) and poly(propyleneimine) are typical examples. These are not, by any means, self-assembled aggregates. But their unique structural features provide the space inside their molecular body. This space has a nanometer-ranged size. Also, when their end-groups are properly functionalized, the nanospace becomes aqueous. Precursors can be diffused into this space, and can undergo decomposition, nucleation, and growth. As in the previous cases, the geometrical confinement of this space quenches further growth (Scott et al., 2005). Figure 10.6c is the scheme.

10.3.2.4. Nanoporous Solids. The concept of this approach has been described in Section 9.9. As presented in Figure 9.16c, the nanospace provided by inorganic nanostructured (mesoporous) materials can be adopted for the synthesis of nanoparticles. The same reaction steps are taken as in the previous cases. The precise tunability of the nanospace size/morphology (via control of the self-assembly process) makes it possible to tune the size and shape of the nanoparticles synthesized. Spherical and rod-shaped ones have been synthesized along with nanowires and planar nanosheets. Since this nanospace is "hard," the space size–nanoparticle size relation can be much more predictable and straightforward than with the previous approaches. But that in turn can cause some limitations during the reactions, in terms of capillary condensation, wetting, and dewetting.

10.3.2.5. Directed Growth by Soft Epitaxy. Geometrical confinement, which is the primary requirement for the synthesis of nanoparticles, also can be provided by surfaces. While the nanoscale *real* space provided by a variety of self-assembled aggregates was the key aspect in the previous type II reactions, this surface confinement is provided by the *imaginary* space or area on self-assembled or prepatterned surfaces. In both cases, the interplay between the intermolecular and colloidal forces is key to determining the degree of confinement. Three typical routes are presented in Figure 10.8. The first two are for the synthesis of nanoparticles under the Langmuir monolayer (a) or on Langmuir-Blodgett (LB) film (b). Part (c) is for synthesis on hard prepatterned surfaces, and will be described in the next subsection.

Langmuir and LB films are the typical self-assembled films that are formed from a variety of surfactants or lipids (Chapter 6). First, the precursor species interact with these films mainly through the electrostatic attraction. But this interaction is highly specified by the ionic property, molecular configuration, and structure of these self-assembled films. Thus, nucleation and growth of the interacted precursors are also controlled by the properties of the film. For most cases, these films have unit (or cell) surface structures formed in the nanometer range, and the growth of the nucleated precursor is quite well limited within this range. This quenches further growth; thus, the particles are grown within the nanometer size. Figure 10.8d presents three typical unit structures. These approaches also can be referred to as *soft epitaxial* reaction routes.

The size and shape (also crystal orientation when nanocrystals are formed) of nanoparticles are greatly affected by the unit structures and sizes. And, as described in Chapter 6, they can be controlled with great precision: by the choice of amphiphile, by the control of physical conditions, especially surface pressure, by the combination of amphiphiles (mixed films), and so on.

An excellent example can be found in the preparation of lead sulfide nanocrystal under the mixed Langmuir monolayer of arachidic acid and octadecylamine. This shows the dramatic control of the structure and crystal orientation of nanocrystals by the simple control of the ratio between the amphiphiles and surface pressure (Fendler and Meldrum, 1995).

10.3.2.6. Directed Growth by Hard Epitaxy. Nanoparticles also can be directly formed on hard surfaces. When the precursors are deposited in the gas

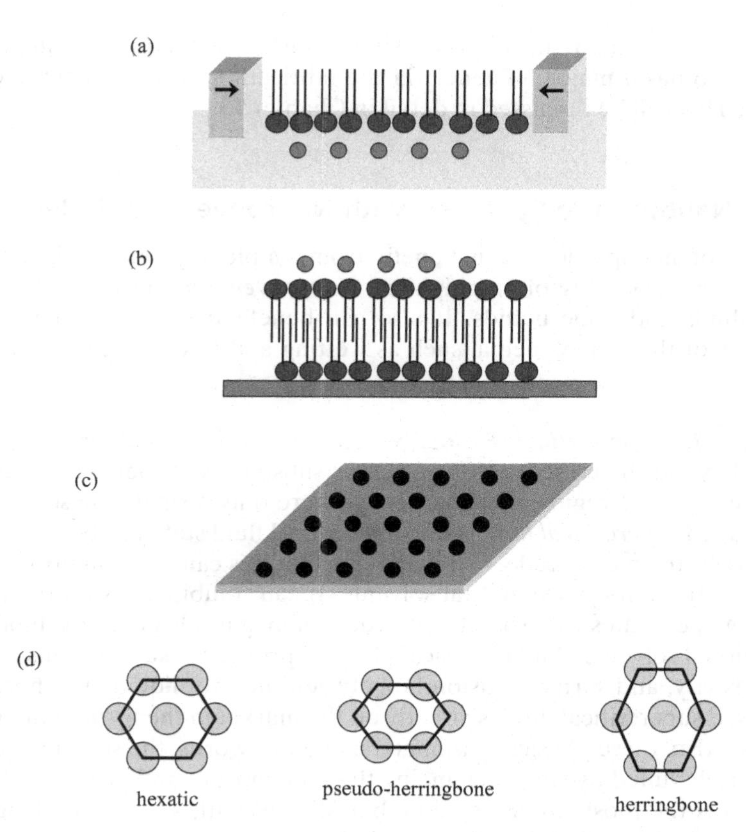

Figure 10.8. Synthesis of nanoparticles: type III—surface confinement reaction. (a) Below Langmuir monolayer, (b) on Langmuir-Blodgett film, and (c) on solid surface. (d) Three typical lattice structures on these surfaces.

phase on hard surfaces, the precursors can have more adsorption tendency on certain sites (or small areas) with high surface energy. The precursors are selectively adsorbed onto these sites. Then they undergo the decomposition, nucleation, and growth steps. Metal or semiconductor surfaces can be prepatterned for this process, or surfaces with inherited regular structures such as HOPG (highly oriented pyrolytic graphite) can be employed. Quenching of further growth is primarily determined by the intermolecular interaction between the precursor and the surface. But the size and shape of nanoparticles are also strongly controlled by experimental features such as beam condition, vacuum condition of the chamber, and deposition times. Gas evaporation, laser vaporization/pyrolysis, laser/ion beam sputtering, and atomic vapor deposition are among the commonly employed procedures for *epitaxial deposition*. The nanosized particles from this route also have other names, such as *nanoisland, surface island*, and just *island*. Synthesis of semiconductor nanoparticles, for instance, gallium arsenide, gallium silicide, or germanium silicide, on a silicon surface are typical examples. This method also has a major influence on the fabrication of nanoparticles on solid surfaces. This will be discussed in detail in Chapter 13.

10.3.3. Nanoparticle Synthesis with Nonconventional Media

Synthesis of nanoparticles can benefit from employing the newly developed media for self-assembly of amphiphiles. *Nonconventional media* include supercritical fluids and ionic liquids. It also can benefit from employing the novel properties of those new media, such as wetting and low surface and interfacial tensions.

10.3.3.1. Supercritical Fluids. When temperature and pressure are increased beyond the critical point of certain substances at their vapor–liquid coexistence region, a region is encountered where only a single-phase fluid exists. This is called *supercritical fluid*. This supercritical fluid state has been extensively studied over the past decade, and it appears that this can be, in many cases, a superior substitute for conventional solvents. It can solubilize a variety of organic and inorganic solutes with sometimes orders of magnitude greater solubility than conventional solvents. And its typical physical properties such as density, viscosity, diffusivity, and surface tension lie between those of liquid and those of gas. CO_2-based supercritical fluid is already on the market in the form of dry cleaning solvents. Also, these physical parameters, in other words, the solvent properties, can be easily tuned by simply changing the temperature and pressure of the system. CO_2 is the most studied system, but alkanes with short chain length have been extensively studied, too. These include methane, ethane, propane, pentane, and hexane.

The unique properties of the supercritical fluids can provide some unique advantages for the synthesis of nanoparticles in both the spatial confinement reaction using microemulsion and the arrested precipitation approach (Shah et al., 2004):

1. Supercritical fluids exist within a wide range of temperatures and pressures. Thus, they can provide accessibility to a wide range of high reaction temperatures and high pressures, which water-based reaction systems cannot. Temperatures ranging from room to more than 500 °C are often possible, which can be ideal for the synthesis of highly ordered nanocrystals. The discovery of heat-tolerant ligands, capping agents, or emulsifiers that are mostly fluorocarbon-based substances was crucial for this.
2. The water-in-CO_2 (W/C) microemulsion can serve as the reaction compartment just like in the case of W/O microemulsion. This provides additional ways to control the solvent (CO_2) solvation property and water pool size via simple control of temperature and pressure. Also, the diffusion of the precursors between water pools, which sometimes has a pro-

found impact on the property of the nanoparticles synthesized, can be simply tuned by the same process.
3. They are nonflammable and nontoxic and therefore environmentally sound solvents. This could have potentially huge implications, especially where nanoparticle synthesis is shifted toward the venue of industrial-scale production.
4. Enhanced wettability with controllable solvation can provide a unique opportunity for the assembly of nanoparticles, possibly by the manipulation of intermolecular forces between them; this is *controlled nanofabrication.*

One limitation on the use of supercritical fluid for now is the limited number of available ligands. This quite limits the variety of nanoparticles that can be synthesized, especially compared with water-based synthetic systems.

10.3.3.2. Ionic Liquids. An ionic liquid is an organic salt that has high polarity and low melting point. When metal cations are bonded with anions through ionic bonds, they form very stable crystalline salts that show high melting point. This of course is a result of strong ionic bonds. Typical salts such as NaCl and KCl are good examples. When organic cations are combined with anions, they are still seemingly bonded through ionic bonds. However, their high polarizability makes their ionic bonds much weaker than the conventional ones. Van der Waals force takes the significant portion of the interaction over ionic bonds. Thus, the salts formed have much weaker bonds than the conventional salts, which make them exist in liquid state even at room temperature. Some of them stay in liquid state at a wide range of temperatures: ~−100 °C–~400 °C. The name *ionic liquid* is due to this phenomenon (Wasserscheid and Keim, 2000).

A typical example of ionic liquids is the salt of 1-alkyl-3-methylimidazolium with larger anions such as hexafluorophosphate, tetrafluoroborate, and nitrate. These have low melting points, and show very unique physical properties when compared with the conventional organic and water-based solvent systems. These include high polarity, almost negligible vapor pressure, high ionic conductivity, and high thermal stability. Also, they are highly structured by a strong hydrogen-bonded network. These aspects make them unique solvent systems for synthesis of certain types of nanoparticles (Antonietti et al., 2004):

1. Their hydrogen-bonded networked structures, in some cases, provide spatial confinement within solvent systems for synthesis of nanoparticles. This means that there can be no need of ligands or capping agents to impose spatial confinement during the synthesis.
2. Their high thermal stability provides a wide range-of-temperature window, just as in the case of supercritical fluids. Again, this is important especially for the synthesis of highly crystalline nanoparticles such as silicon and germanium. Also, for ionic liquids, there is no need to impose high pressure to reach this condition.

3. The electrochemical approach is another promising route toward the synthesis of nanoparticles. But the conventional solvent systems very much restrict its usefulness because of their low conductivity. The unusually high ionic conductivity of ionic liquids can provide much greater versatility for this electrochemical approach.
4. Ionic liquids are nontoxic and, in most cases, are recyclable, which means they can be considered as highly environmentally benign solvent systems.

As with the supercritical fluids, there are some limitations. Self-assembled systems in ionic liquids are not well developed so far, which limits the spectrum of self-assembled spatial confinement opportunities as compared with conventional water-based systems. This reduces the versatility of the size/shape controls and the size ranges of the nanoparticles synthesized. Also, there are limited numbers of types and available structural varieties of ionic liquids themselves. This limits the possible compositional spectrum of nanoparticles.

10.4. PROPERTIES OF NANOPARTICLES

Nanoparticles have unique properties compared with their bulk counterparts. Many of these properties, including physical, chemical, optical, electrical, and magnetic, can be controlled by relatively simple tuning of their sizes, shapes, compositions, protecting organic ligands, and interparticle distance. These novelties are not just the result of the scale-down in their sizes or masses; they also come from the different routes that can vary depending on the type of materials (Burda et al., 2005). This section is divided into two subsections of quantum size effect and surface atom effect, where the novel properties originate mainly from the quantum confinement and from the energetics of the surface atoms, respectively.

10.4.1. Quantum Size Effect

Electronic, magnetic, and optical properties of semiconductors and noble metals are greatly changed when their size reaches to the nanometer scale. Quantum confinement and surface plasmon of conduction electrons are two key concepts in understanding these phenomena.

10.4.1.1. Optical Properties of Semiconductors. Suppose the semiconductor particles have their sizes smaller than the photon wavelength but still larger than the size of their crystal lattice. Then the de Broglie wavelength of their valance electrons becomes comparable to their diameter. This means that the free electrons can be confined (or trapped) in the nanoparticles; this is *quantum confinement*. This is the same picture as the classical quantum box, but with nanoparticles. The famous name *quantum dots* for nanoparticles stems from this.

This effect induces changes in band-gap energy and the quantization of the electronic energy level. Depending on the relation of the particle size to the Bohr exciton radius (the length scale that defines the electronic dynamics in bulk semiconductor materials), there is a difference in the detailed mechanisms of quantum confinement, but the fundamental concept of size-dependent optical transition remains the same. The smaller-sized nanoparticles have the bigger band gap, and therefore the higher energy transition. This means that the optical properties of the nanoparticles can be simply tuned by a change in their size. Figure 10.9 shows the schematic representation.

> This picture of quantum dots is the case for the confinement of the motion of electrons in all three dimensions. When the confinement comes to any two dimensions or any one dimension, the system is referred to as *quantum wire* and *quantum well*, respectively.

The *Kubo gap* is very small for the bulk materials; thus, not only its value itself but its dependency on the size of the materials can be negligible as well. The energy level can be considered as continuous. But for nanoparticles, the quantization of energy level (Kubo gap and band gap) is dependent on their size.

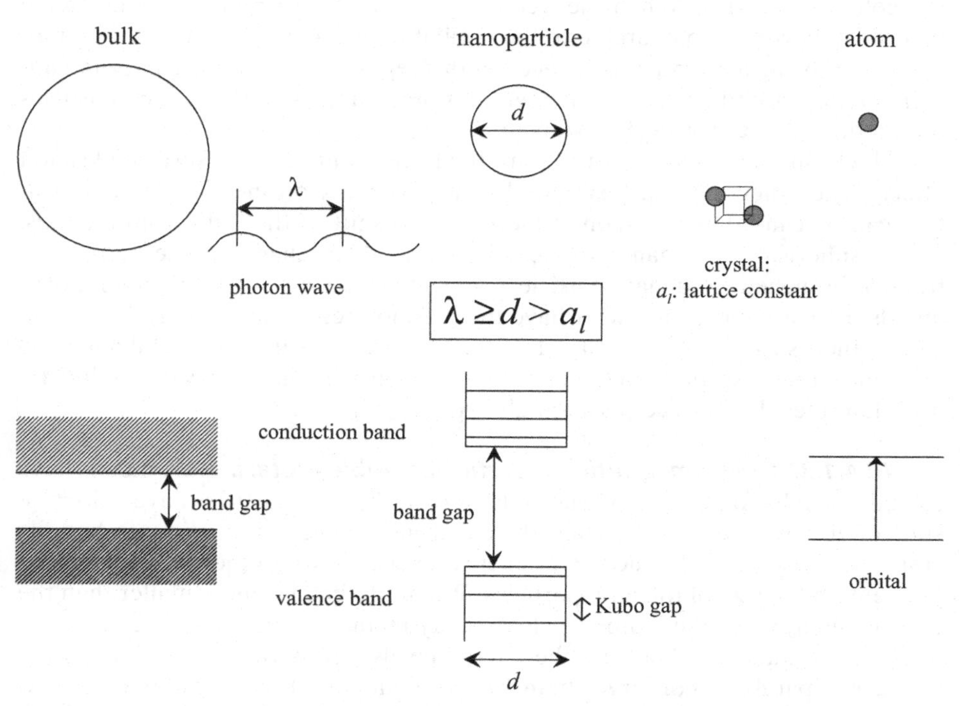

Figure 10.9. Schematic representation of the concept of quantum size effect of nanoparticles.

Since the band gap is increased as the size decreases in the range of ~2 nm—~20 nm of radius for most of the semiconductors (Bawendi et al., 1990; Gaponenko, 1998), there is a monotonic blue shift of absorption in the decreasing of the size. Furthermore, the selection rule allows electron-hole transition only with the same quantum numbers; thus, the broad feature of the bulk absorption spectrum becomes the spectrum with a number of distinct peaks that correspond to each transition for the nanoparticles. As the size of nanoparticles increases, the band of nanoparticles approaches to the value of bulk. For atoms or molecules, this type of transition occurs only between the atomic or molecular orbitals.

10.4.1.2. Optical Properties of Noble Metals. When the size of noble metals is reduced to the nanometer scale, the electrons near or at the surface become loose compared with those inside the core. Thus, when they interact with an incoming light wave, the electrons in the conduction band begin to polarize to one side of the surface by the action of the electric field of the wave. As the wave keeps oscillating between (+) and (−) (in respect to the direction of arbitrary coordination scale), the polarized electrons also begin to oscillate from one side of the surface (of the nanoparticle) to the other side. Diffusion of the nanoparticle (Brownian motion) is so much slower than the frequency of the light wave that the nanoparticle can be assumed to be fixed at its position, which means that this is a collective motion. When the frequency of the electric field of the incoming light wave becomes comparable to the oscillation frequency of this electron motion, very strong absorption is induced in this region. This is *surface plasmon absorption*. Surface plasmon is dependent not only on the size of the nanoparticles but on their shape as well (Liz-Marzan, 2006).

The brilliant colors of gold sols are the best example of the surface plasmon effect. When their sizes are increased, inhomogeneity on the electron polarization can be caused, which then induces the red shift of their absorption bands. While spherical gold nanoparticle sols show their characteristic red color, triangle-branched gold nanoparticle sols show blue color. Also, other noble metals such as copper and silver have the plasmon resonance at a visible region, too, so their sols also show brilliant color. Most of the transition metal nanoparticles have this absorption in the ultraviolet region. The theory developed by Mie well describes this surface plasmon absorption (Mie, 1908).

10.4.1.3. Electromagnetic Properties of Noble Metals. Bulk metals have a conduction band that is continuous. However, the nanoparticles have a discrete state of energy level, which means the electronic state is not continuous but discrete. The spacing of the successive energy levels, known as the Kubo gap, is dependent on the size of the nanoparticles. When this gap is much smaller than the thermal energy kT, the nanoparticles display metallic properties. The electrons can make a transition between the energy levels freely. But, when the gap becomes comparable to or larger than kT, the nanoparticles become nonmetallic. This happens when the diameter is below ~3—~5 nm for noble metal nanoparticles such as gold, platinum, silver, and palladium. The electrons cannot move freely between the quantum levels. This is why the transition from conductive metals to semiconductive nanoparticles and to nonconductive nanoparticles happens as their size decreases. Thus, the electrical conductivity and magnetic susceptibility of nanoparticles show the quantum size effect, which means these properties can be modulated by the control of the band gap, which is mainly determined by the size and shape of nanoparticles.

10.4.1.4. Electric Properties of Metals. As the size of the metal particles decreases to the nanometer scale, the Kubo gap becomes comparable to the thermal energy kT and even larger. This means the Coulomb energy, that is, the electrostatic energy, $e^2/2C$, becomes much larger than the thermal energy kT, where C is the capacitance. The Kubo gap is increased as the size of the nanoparticles becomes smaller, which means that C becomes smaller as the size of the particles decreases. Thus, Ohm's law for the bulk metals is no longer valid. The electron that enters into the nanoparticles no longer shows continuous flow (current) in response to the voltage of the system. The electron can be transferred only when the Coulomb energy is compensated by the external voltage entering into the nanoparticle. Therefore, there can be no current until this threshold value of the voltage is reached. This causes the electron to be transferred in a stepwise or jumpwise one-by-one way, instead of by continuous flow. This is *single electron tunneling*. It is also called the "Coulomb blockade." The next tunneling occurs only when the voltage provided fulfills this condition again. This is the reason for the *staircase* current-voltage characteristics that are typical for nanoparticles.

10.4.2. Surface Atom Effect

As shown in Figure 10.10, the ratio of atom at surface to bulk increases sharply as the size of any particles decreases. For spherical ones, the surface-to-bulk atom ratio is inversely proportional to their radius. For example, a 10 nm diameter of a spherical nanoparticle has ~10% of atoms at the surface. Atoms at the surface are much more active because they have fewer bonding numbers than bulk atoms. They have more uncompleted bond sites and dangling bonds. Also, this "imperfect" surface of nanoparticles means they can have surface defects that are even more active and can provide additional electronic states and reactivity.

When the size of materials decreases to the nanometer scale, the total surface area per unit mass is also sharply increased. For example, as the diameter of a spherical nanoparticle decreases to 1 mm, 1 μm, 10 nm, and to 1 nm, the total surface area available becomes 0.003, 0.03, 30, and 300 m^2/g.

These two factors are the key phenomena that derive the following typical surface atom effects for nanoparticles:

1. Surface Catalytic Properties of Metals and Oxides. Both of the typical kinds of catalysis are effective for nanoparticles. A *homogeneous* catalyst is

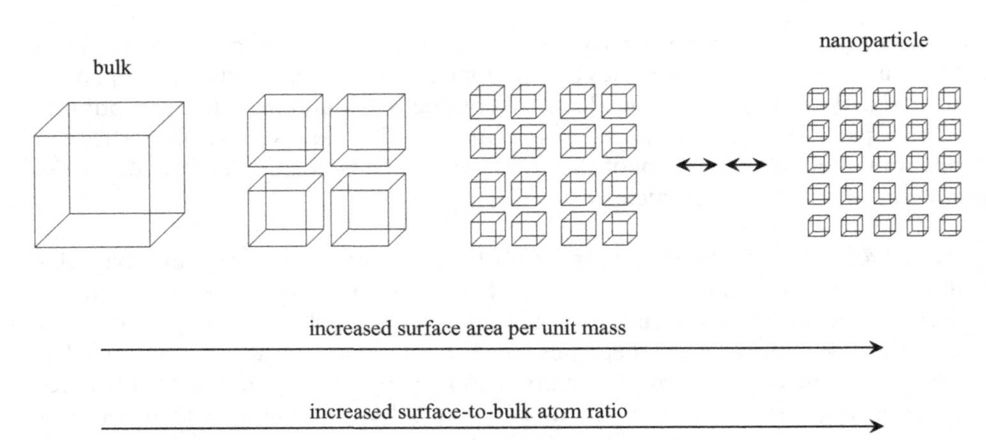

Figure 10.10. Schematic representation of the concept of surface atom effect of nanoparticles.

where nanoparticles are used alone as a catalyst. They can be also used in an anchored or supported form on substrates, where they show a catalytic effect in a synergistic way. This is a *heterogeneous* catalyst. A variety of catalytic reactions can be preformed with great efficiency, including hydrogenation, halogenation, oxidation, reduction, decomposition, electron-transfer reaction, full-cell reaction, and so forth. Different sizes, different shapes, different compositions, and different types of nanoparticles show different catalytic efficiency (Raimondi et al., 2005).

2. Melting Point Depression. Nanoparticles including semiconductors, metals, and oxides show melting points lower than their bulk counterparts. They also decrease as the size decreases. When solid materials melt, the total surface energy is decreased. This is because the high-energy surface atoms can be moved into the liquid, and therefore can minimize the surface energy. Since the nanoparticles have more surface atoms, the contribution of these surface atoms to the total energy of the system becomes larger than the bulk and even larger as the size of nanoparticles decreases. Decreased melting point can help in materials processing such as the increase of sintering efficiency.

3. Surface Functionalization. The high surface energy of nanoparticles can make their surface functionalization easy. This provides a variety of routes for properties modulation. Versatility in the anchoring of organic functional groups, biological ligands, and other molecules can provide a better spectrum for further applications. For example, metallic nanoparticles can show their plasmon absorption band coupled with electronic eigenstates of the same particles and/or energy levels of the surfactant or emulsifier molecules physisorbed or chemisorbed on their surface.

10.5. APPLICATIONS OF NANOPARTICLES

Given the variety of novel properties and their easy tunability, there are a wide range of applications of nanoparticles. This chapter will focus on their "lower-order application." This encompasses the direct application of nanoparticles with minor modifications, which include surface modification, the change in size/shape, and the formation of composite/hybrid types of materials. Among those examples are the primary type of sensory application, medical uses such as biological fluorescence tracer tags, imaging agents such as MRI (magnetic resonance imaging) contrast agent, and homogeneous catalysis.

Applications with major fabrication of nanoparticles, on the other hand, can be considered "higher-order applications." Examples include light-emitting diodes, low-bandwidth lasers, photovoltaic solar cells, display devices such as OLED (organic light-emitting diode), drug-delivery vehicles, nanoelectronics, photoelectronics, photonics, and heterogeneous catalysis. This topic will be described in Chapter 14.

10.5.1. Chemical and Biological Sensors

Figure 10.11 shows a schematic illustration of these types of applications. Ligand is attached covalently or sometimes physically onto the nanoparticle surface and the acceptor is attached onto the target substrate in a similar way. Targets can include chemical entities, biological subjects, and colloidal particles. The recognition between the ligand and the acceptor occurs when there is a chemical process such as complexation, biological complementarity, or geometrical matching. Detection is confirmed through the identification of the properties of the nanoparticles that are selectively attached onto the target substrates. For example, when the anthrax virus is targeted by using DNA-modified gold nanoparticles, the color of the sample is changed as a result of their aggregation.

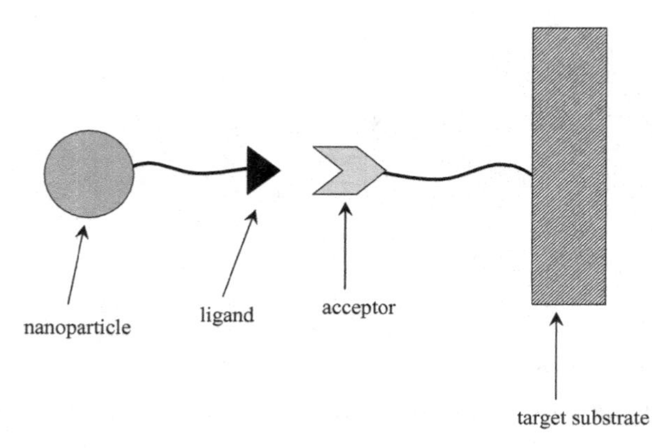

Figure 10.11. Schematic illustration of nanoparticles in chemical or biological sensory applications.

Also, nanoparticles with different sizes/shapes that are modified with different ligands can be applied as a noninvasive optical (fluorescence) marker both *in vitro* and *in vivo*. Those with one size/shape with a certain ligand will bind on the specific target that has the right acceptor; then the properties of the nanoparticles such as their given color can indicate the detection of that target.

10.5.2. Optical Sensors

Figure 10.12 shows a schematic illustration of these types of applications. The band gap (E_g) of the semiconductor nanoparticles increases as their size decreases. This induces the increase of the emission energy, meaning the blue shift as the size decreases. Since the degeneracy of the electronic structure is also delicately dependent on the shape and composition of the nanoparticles, the band gap can be even more precisely tuned by manipulating these factors, mainly because of the different surface polarization. Applications can range from simple optical sensors to fabricated devices such as photovoltaic cells and light-emitting diodes (LEDs).

The surface scattering from nanoparticles is dependent on the above structural factors, too. This leads to applications ranging from consumer products such as sunscreen to surface-enhanced Raman scattering (SERS).

The surface plasmon band, which is a broad absorption band of noble metal nanoparticles in aqueous solution, is located in the visible region. And the color

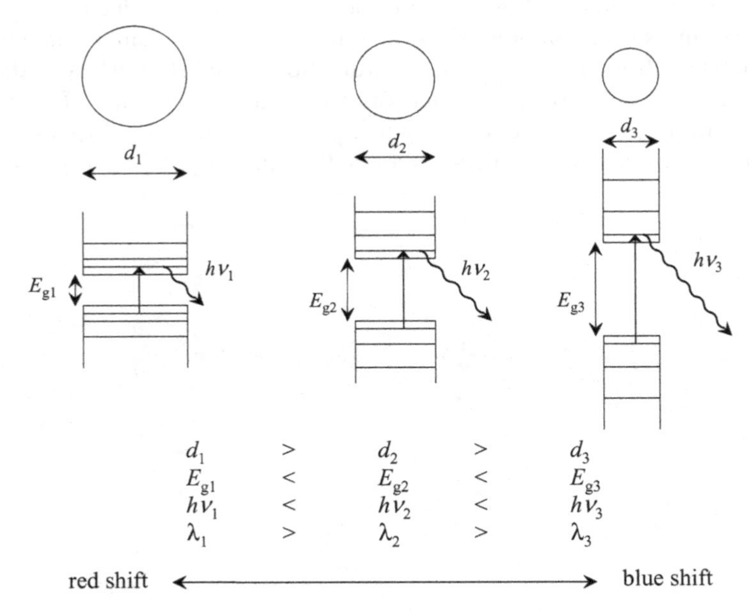

Figure 10.12. Schematic illustration of nanoparticles in optical sensory applications.

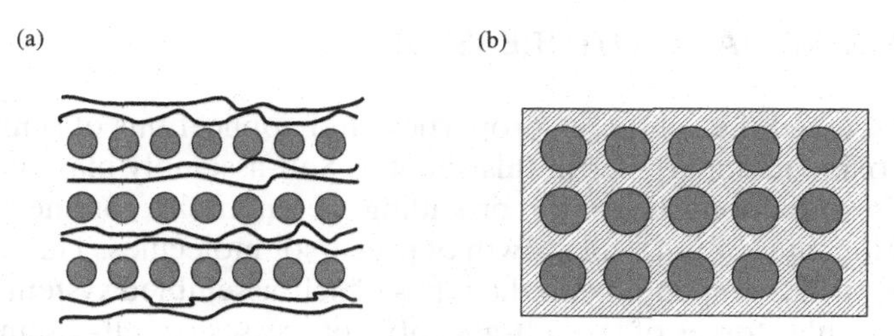

Figure 10.13. Scheme for (a) film-type and (b) bulk-type nanoparticle-based composite or hybrid materials.

of its sols is solely dependent on their size, shape, and composition. Applications such as biological marker as described in the previous subsection have great potential.

10.5.3. Nanocomposites and Hybrid Materials

The primary merit for the applications of the composites or hybrid materials is their noble or synergistic physicochemical properties. As shown in Figure 9.1, inorganic materials can form these types of materials with organic or polymeric materials. The same principles can be adopted for composites or hybrid materials with any types of nanoparticles. They can be either film-type or bulk-type as shown in Figure 10.13. When the nanoparticles are physically fabricated with organic or polymeric materials, they are called *composite* materials, while the *hybrid* materials are the chemically bonded ones.

10.5.4. Catalysis

Based on the surface atom effect, nanoparticles have (1) increased total surface area, (2) increased number of atoms accessible on the surface, (3) increased catalytic activity of those atoms, and (4) different (tunable) surface catalytic properties by the change in shape, size, and composition. These are the keys to their use as catalysts.

10.5.5. Functional Fluids

A typical example of these types of applications is *ferrofluids*. These are a suspension of ferromagnetic nanoparticles in solutions, so their physical properties such as viscosity and density are sensitive to external electric or magnetic field. Again, these properties can be tuned with great precision by the change in size, shape, and composition of the nanoparticles.

10.6. SUMMARY AND FUTURE ISSUES

We have discussed the synthesis, properties, and applications of semiconductor, metal, and oxide nanoparticles in this chapter. Self-assembly plays a crucial role in synthesis of nanoparticles by providing geometrical confinement during decomposition, nucleation, and growth of precursor molecules. This confinement is spatial or surface, depending on the types of self-assembled systems. By tuning the intermolecular forces of the self-assembly processes or self-assembled aggregates, the structural properties of the nanoparticles synthesized, including size, shape, and composition, can be easily controlled. The noble properties of nanoparticles can be precisely tuned by controlling these structural factors.

In the sense that there are plenty of elements that can be explored as a possible form of nanoparticles, and that there are plenty of new unexplored self-assembly systems available, further advances in nanoparticles look promising. Here are some of the future issues:

1. Structure–Property Relationship. Once the structure of the self-assembly building unit and the environmental conditions are acquired, the self-assembly process and the structure of the subsequent self-assembled aggregates can be obtained with a reasonable accuracy. A well-established relationship of the structures of the self-assembly building units (ligands, capping agents, and amphiphiles) to the physicochemical properties of the nanoparticles synthesized will provide better control of their structural/chemical features. This will also help expand the scope of the nanoparticles that can be synthesized through self-assembly.

For nanoparticle properties, the structure–property relationship between spherical properties, and also the relationship between nonspherical properties, will be among important future challenges. Even a slight alteration in the structure of the nanoparticles often ends up causing a dramatic change in their physical and chemical properties. Systematic studies and the construction of the generalized relationship for the properties of the nanoparticles across the wide range of their structures will provide a much better pathway for expediting their applications. The general phase diagram for surfactant self-assembly in Figure 4.10 would be a good example.

2. Force Balance within Confined Space or Surface. As in the synthesis of nanostructured materials in Chapter 9, there are constant changes in the intermolecular forces during the synthesis of nanoparticles. This is due to the constant change in the concentration of the precursors and to the constant evolution of the intermediate species. This strongly affects the force balance for the self-assembled aggregates, and therefore the packing geometry. For nanoparticle synthesis, this new balance will be in the confined space or surface, which means that the effect will be more dramatic than in the case of synthesis of nanostructured materials. Along with the above structure–property relationship issue, a proper understanding of this fact will provide better insight into the general synthetic mechanism for the synthesis of nanoparticles.

3. Surface Functionalization. The ultimate applications of nanoparticles will come in the form of nanodevices. This will require the controllable assembly of nanoparticles across desired and, in many cases, varied species of them into unified but functionalized contents. This is *controllable nanofabrication.* Details will be described in Chapter 13. In regard to advancements in nanoparticles, this issue challenges us to develop the concept of control of the intermolecular/colloidal forces between nanoparticles. Controlled functionalization of the surface of nanoparticles coupled with well-developed organics (ligands, capping agents, and amphiphiles) and proper control of their surfaces even without the organics will provide diverse routes to fulfilling this purpose. Also, the often-revealed "coupled properties" (noble properties of nanoparticles that are coupled with organics or that are induced by changes in the interparticle distances) will provide a wider spectrum for their application.

REFERENCES

Antonietti, M., Kuang, D., Smarsly, B., Zhou, Y. "Ionic Liquids for the Convenient Synthesis of Functional Nanoparticles and Other Inorganic Nanostructures," *Angew. Chem. Int. Ed.* **43**, 4988 (2004).

Bawendi, M. G., Steigerwald, M. L., Brus, L. E. "The Quantum Mechanics of Larger Semiconductor Clusters ('Quantum Dots')," *Annu. Rev. Phys. Chem.* **41**, 477 (1990).

Burda, C., Chen, X., Narayanan, R., El-Sayed, M. A. "Chemistry and Properties of Nanocrystals of Different Shapes," *Chem. Rev.* **105**, 1025 (2005).

Cushing, B. L., Kolesnichenko, V. L., O'Connor, C. J. "Recent Advances in the Liquid-Phase Synthesis of Inorganic Nanoparticles," *Chem. Rev.* **104**, 3893 (2004).

Daniel, M.-C., Astruc, D. "Gold Nanoparticles: Assembly, Supramolecular Chemistry, Quantum-Size-Related Properties, and Applications toward Biology, Catalysis, and Nanotechnology," *Chem. Rev.* **104**, 293 (2004).

Fendler, J. H., Meldrum, F. C. "The Colloidal Chemical Approach to Nanostructured Materials," *Adv. Mater.* **7**, 607 (1995).

Gaponenko, S. V. *Optical Properties of Semiconductor Nanocrystals* (Cambridge University Press: 1998).

Lisiecki, I. "Size, Shape, and Structural Control of Metallic Nanocrystals," *J. Phys. Chem. B* **109**, 12231 (2005).

Liz-Marzan, L. M. "Tailoring Surface Plasmons through the Morphology and Assembly of Metal Nanoparticles," *Langmuir*, **22**, 32 (2006).

Masala, O., Seshadri, R. "Synthesis Routes for Large Volumes of Nanoparticles," *Annu. Rev. Mater. Res.* **34**, 41 (2004).

Mie, G. "Contributions to the Optics of Turbid Media, Especially Colloidal Metal Solutions," *Ann. Phys.* **25**, 377 (1908).

Pileni, M. P. "Reverse Micelles as Microreactors," *J. Phys. Chem.* **97**, 6961 (1993).

Raimondi, F., Scherer, G. G., Kötz, R., Wokaun, A. "Nanoparticles in Energy Technology: Examples from Electrochemistry and Catalysis," *Angew. Chem. Int. Ed.* **44**, 2190 (2005).

Scott, R. W. J., Wilson, O. M., Crooks, R. M. "Synthesis, Characterization, and Applications of Dendrimer-Encapsulated Nanoparticles," *J. Phys. Chem. B* **109**, 692 (2005).

Shah, P. S., Hanrath, T., Johnston, K. P., Korgel, B. A. "Nanocrystal and Nanowire Synthesis and Dispersibility in Supercritical Fluids," *J. Phys. Chem. B* **108**, 9574 (2004).

Steytler, D. C., Gurgel, A., Ohly, R., Jung, M., Heenan, R. K. "Retention of Structure in Microemulsion Polymerization: Formation of Nanolattices," *Langmuir* **20**, 3509 (2004).

Uskoković, V., Drofenik, M. "Synthesis of Materials within Reverse Micelles," *Surf. Rev. Lett.* **12**, 239 (2005).

Vriezema, D. M., Aragonès, M. C., Elemans, J. A. A. W., Cornelissen, J. J. L. M., Rowan, A. E., Nolte, R. J. M. "Self-Assembled Nanoreactors," *Chem. Rev.* **105**, 1445 (2005).

Wasserscheid, P., Keim, W. "Ionic Liquids: New 'Solutions' for Transition Metal Catalysis," *Angew. Chem. Int. Ed.* **39**, 3772 (2000).

11

NANOSTRUCTURED FILMS

The self-assembly of various building units at different interfaces has been discussed in Chapter 6. This interfacial self-assembly phenomenon, with the more general name *surface science*, is one of the central concepts in the progression to nanostructured films (Van Hove, 2006). Definition, preparation, and properties of nanostructured films will be presented in this chapter. In relation to interfacial self-assembly, this chapter will particularly focus on how the interplay of the involved forces affects the formation of nanometer-scale structural features, what are the driving forces to make them into film forms, and what generates the unique properties of nanostructured films. The discussion will be extended to how this approach can be insightful for designing those films for further realization of various applications.

11.1. WHAT IS NANOSTRUCTURED FILM?

Generally, films can be defined as any types of matter including solid, liquid, and gas with only one dimension having definite size. Thus, nanostructured films can be defined as any matter with film-form that has structural features on the nanometer scale. These structural features can be nanoporous, nanoparticle, thickness that is on the nanometer scale, or surface roughness ranged at the nanometer scale.

Figure 11.1 shows the schematic representation of three typical nanostructured films that are commonly found in the field of nanotechnology. *Nanoporous* film has nanometer-scale porous structures either inside of it or on its surface. Its thickness is usually within the nanometer scale, too. Film with thickness above the nanometer scale should be also classified as nanoporous film as long as the structural feature of the pores belongs to this scale range. Most of the nanoporous films are prepared as composites or hybrid types of films. These films have nanometer-scale structural features inside, and the removal of porogen (or template) by various methods such as calcination or solvent extraction generates the desired nanopore. Thus, those films in the state of composites or hybrids also can be considered to belong to this classification. Chapter 9 (Figure 9.5) covered a good deal of this subject with the topic of nano- (or meso-) porous materials.

The second type is a *nanolayered* film. This type includes monolayer, bilayer, and multilayered films. The building unit of each layer can be atom, molecule, polymer, or colloidal object. Its composition can diversely range from organic, inorganic, and metallic to alloy. Like the nanoporous films, the total thickness of nanolayered films is usually within the nanometer scale. But, as long as the thickness of the individual layers is in the same range, films with total thickness greater than a nanometer also should be viewed as nanolayered films. Typical examples include self-assembled monolayer (SAM), Langmuir-Blodgett (LB) films, and polyelectrolyte multilayer films.

The third type is a *nanopatterned* film. This film has patterned structural features within the nanometer scale on its surface. It can be created either by direct patterned deposition (or growth) or by secondary fabrication on the surface of nanoporous or nanolayered films. Typical examples are superhydrophobic films, quantum dot–based devices, or other functional films.

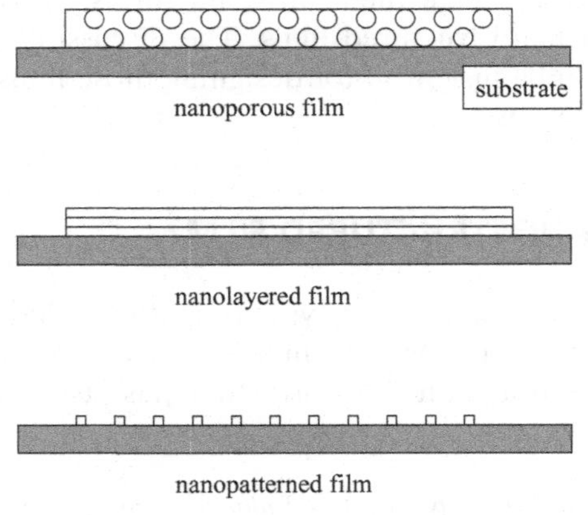

Figure 11.1. Various typical nanostructured films.

Of course, there can be different ways to classify nanostructured films. For example, by the properties of the films, they can be piezoelectric, conductive, reflective, magnetic, superhydrophobic, and so on. By the materials of components, they can be ceramic, semiconductic, polymeric, and so on. However, as will be shown later, the classification described here has a unique usefulness in clarifying them in relation to self-assembly processes at interfaces.

11.2. GENERAL SCHEME FOR NANOSTRUCTURED FILMS

Nanostructured films are two-dimensional objects (within the definition of the previous section); thus, they can be synthesized primarily with the support of substrates. There are some types of nanostructured films that are synthesized with the support of liquid surfaces (freestanding films). But most of them are synthesized on the surface of solid substrates. Preparation methods of the nanostructured films have a great deal of variation. A variety of seemingly different processes have been exploited (Khomutov, 2004; Petty, 2005). Figure 11.2 abstracts all these into simple general routes for the preparation of nanostructured films. Both flat and curved solid surfaces including spheres can be used as substrates as long as they have good affinity with the incoming building units or precursors. The way those building units or precursors interact with the solid surface primarily depends on each system. For LB films and layer-by-layer deposition, the physisorption of the building units on the solid surfaces is the main driving force for the film deposition. For SAMs and vapor-deposited films, chemisorption or possible subsequent surface reaction is the key driving force. This can include surface oxidation/reduction, surface dissociation, surface alloying, or surface metallization. When the sol–gel process is applied using precursor sols,

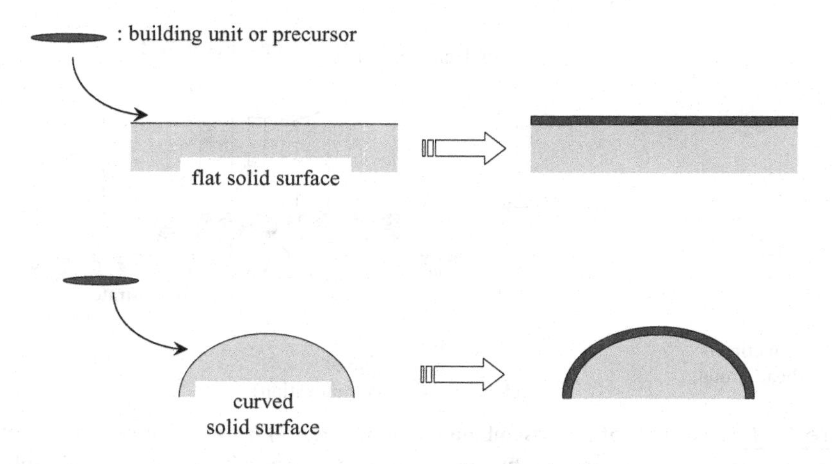

Figure 11.2. General scheme for the preparation of nanostructured films on solid surfaces.

wetting and spreading of the sols on the given solid surface becomes another key factor to be controlled. Nanostructured films often can be formed without any incoming building units or precursors. Surface reconstruction is a typical example (Section 6.6).

11.3. PREPARATION AND STRUCTURAL CONTROL OF NANOSTRUCTURED FILMS

11.3.1. Self-Assembled Monolayer (SAM)

A great deal of research has been amassed for the past several decades for SAMs, which has led to tremendous achievements in their preparations, characterizations, and applications. An excellent recent review (Love et al., 2005) is available for comprehensive detailed knowledge. This section, however, does not intend to cover these known facts again. The focus will be on the impact of the concept of force balance on the preparation, and the understanding and design of their characteristic structural features. The main driving forces to self-assemble SAMs are the strong chemical bond of the building units with the solid substrates and the intermolecular forces between the building units. Sequential interplay between these interactions determines the process, structure, and properties of SAMs. This will be discussed in relation to interfacial self-assembly as described in Chapter 6.

Figure 11.3 provides the schematic representation of the formation of SAMs. This self-assembly process can occur mainly at solid surfaces, from both liquid and gas phases. An exception is the SAMs that are formed on liquid-state metals at room temperature, such as mercury (Magnussen et al., 1996). The prerequisite for the formation of SAMs is the self-assembly building unit (almost exclusively

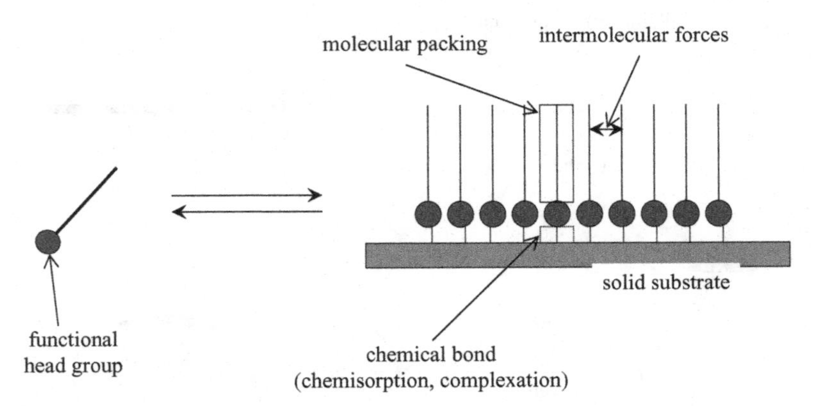

Figure 11.3. Formation of self-assembled monolayer (SAM) is the result of the interplay of intermolecular forces between the building units and a strong building unit–solid substrate bond.

molecular) with proper functional groups, mostly on its head group. These functional groups should have the capability to form a strong bond with the solid surfaces. When these types of building units approach the solid surfaces, the chemical bond formed with the solid surfaces (mostly by chemisorption or complexation) initially acts as the major driving force to self-assemble the building units on the solid surfaces. Since this is a much stronger force than intermolecular forces, those self-assembled building units can have quite limited ability to be rearranged. This is the key difference from interfacial self-assembly (Chapter 6), where the confinement was achieved by physisorption that is much weaker but comparable to that of intermolecular forces. The role of molecular packing and intermolecular forces between the building units is twofold. They contribute to determining the surface coverage (the average spacing between them) along with the surface structures. Also, they can be strong factors in accommodating the building units when there is a monomer exchange, which happens for many SAMs after the initial self-assembly.

The most common SAM system is the formation of long-chain thiol SAMs on noble metal surfaces such as gold and silver through thiolation. Also, popular systems include silylation of long-chain silanes on properly functionalized surfaces such as hydroxylated silicon and complexation of long-chain carboxylic acids on metal or metal oxide surfaces.

Formation of SAMs is a thermodynamic spontaneous self-assembly process with mainly molecular self-assembly building units. It mostly ends up at the primary self-assembly process. Chemisorption sites that are the crucial point for this are almost all consumed through the primary self-assembly process. The secondary self-assembly process can be possible only when the top region of the formed SAM is modified or functionalized to have the necessary functional groups.

Like those self-assembled systems in Part I and the nanostructured objects in the previous chapters in Part II, formation of SAMs can also be understood within the general scheme of the force balance. The difference here is that, with all the interplay between the intermolecular forces and colloidal forces, there is a strong involvement of strong bond formation (Ulman, 1996; 1999).

Figure 11.4 shows the schematic illustration of the concept of molecular packing on the molecular configuration of SAMs. Most of the SAMs have some degree of tilt that is ~10°—~30° from the surface normal and ~20°—~50° of twist from the plane of the chain and the surface normal. When the size of the head group of the building units is similar to the average spacing of the surface atoms (chemisorption site) of the solid substrates, the initial self-assembly occurs in such a way as to have the minimum possible room for the tail groups to be accommodated. Thus, the interplay of the intermolecular forces between the building units such as van der Waals and steric forces can be balanced without significant need of molecular rearrangement that can be presented as tilt and twist. Situation (b) is this case. A typical example is the SAMs of alkanethiol on a silver surface where the tilt and twist angles are the smallest among the same alkanethiol SAMs on different metal surfaces. When the additional attractive force is introduced

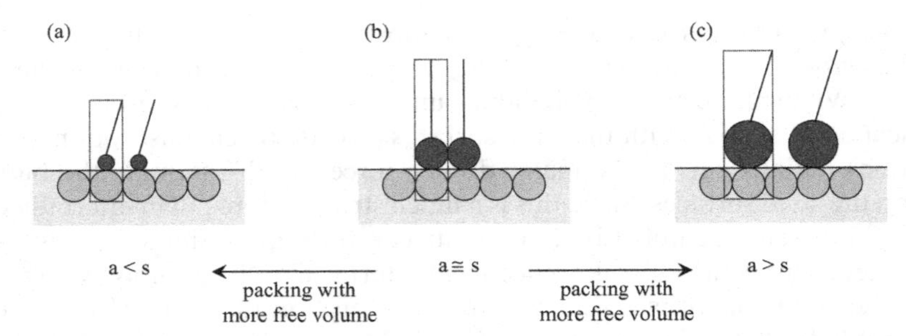

Figure 11.4. Effect of molecular packing on the configuration of self-assembled monolayer; a and s denote the diameter of the head group of the building unit and of the surface atom of the substrate (chemisorption site), respectively.

by, for example, the incorporation of a hydroxyl group on the other end of this building unit, its hydrogen-bonding capability shifts the force balance in such a way as to further restrict its mobility. This results in even smaller or no tilts and twists on the SAMs (Nemetz et al., 1993).

Situation (a) is the case where the diameter of head group is smaller than that of the surface atom. The adsorption site on the solid surface is fixed. And most of the actual adsorption point is in the middle of the solid atoms. So, the interplay of intermolecular forces between the building units is best balanced in such a way as to have more free space than case (b). This inevitably induces more conformational rearrangement, which is reflected in larger tilt and twist angles than case (b). A typical example is the alkanethiol SAMs on a gold surface compared with the same SAMs on a silver surface. The gold atom is larger than the silver atom, which means that the chemisorbed alkanethiols on gold have more spacing than on silver. This results in the larger tilt and twist angles of alkanethiol SAMs on gold than on silver. Conformational changes in building units on the same metal surface but with different lattices such as Au(111) and Au(100) can be reasonably tracked by this approach, too. Alkanethiol SAMs are formed on a liquid state of mercury surface with 0° of tilt (Magnussen et al., 1996). Mercury at its liquid state is fluidic but has enough capability to have a bond with alkanethiol. Therefore, the attractive van der Waals force between the alkyl chains can bring the mercury atoms to where the alkanethiol molecules are chemisorbed closely together, up to the point where the alkyl chains are also closely packed. The free space is now minimal; thus, the tilt and twist do not necessarily need to be compensated.

Figure 11.4c is the case where the head group diameter of the building units is bigger than the average spacing between the solid atoms. After the chemisorption of the first building unit, this condition induces the bonding of the next building unit not to be at the next nearest solid atom. There can be an atom that is not occupied by the building unit between the chemisorbed ones. This, of course, provides more room for the building units than case (b). Thus, the interplay of the intermolecular forces is balanced so as to have more tilt and twist. Its molecular packing occurs simply with more free volume. This condition also can be induced from case (b) by incorporation of a bulky group such as a phenyl ring in the middle of the alkyl chains. The steric factor from the phenyl ring should act as a repulsive force, but the π–π interaction between them is an additional attractive force. Generally, those SAMs with more tilt and twist configurations have greater tendency to have a less dense structure. As the geometrical matching between the size of functional groups (the possible distance between nearest head groups on SAMs) and the size of substrate atoms (the distance between the nearest chemisorption sites) is perfected (less mismatching or less difference), there can be a more effective van der Waals interaction, which results in better molecular packing.

SAMs have so much application potential for nanotechnology. Possibilities include biosensors, selective binding of functional entity, biocomparable substrate for cell engineering, circuit-boards for the creation of nanoelectronics and nanodevices, friction control such as lubricants, and many more. The key to the realization of all these potentials is the ability to control the structural and functional features of SAMs both on molecular and nanometer levels. This means the control of surface coverage, surface density, facile formation of proper SAMs with functional groups, control of tilt/twist angles, and so on. For example, there might be a need that comparable functional groups should be incorporated in the proper position of the SAM backbones. Also, the geometrical matching between the SAMs (the average spacing between the building units) and the incoming objects for the interaction (the average distance between the matching groups on that object) can be crucial to ensure the proper secondary functionalization. An excellent example can be found in the geometrical spacing matching to ensure the proper silanol group condensation of octadecyltrichlorosilane on top of the hydroxyl-terminated mercaptobiphenyl SAM formed at the gold surface (Stoycheva et al., 2006). Matching of hydroxyl group spacing with the spacing of the silanol group of siloxane trimer was the key to the formation of densely packed siloxane monolayer. A similar concept was examined in Chapter 9 (Figure 9.10) for epitaxial analysis of the formation of mesoporous silica on the surface of cationic surfactant micelles.

11.3.2. Layer-by-Layer Assembly

Layer-by-layer assembly is the process of preparing nanolayered films mainly through the control of the electrostatic interactions between the building units and the solid substrate, and between the building units (Biesalski et al., 2002; Decher and Schlenoff, 2003). Typical primary building units are polyelectrolytes with charged functional groups, which include poly(ethylene imine), poly(allylamine hydrochloride), poly(sodium styrene sulfonate), and so forth.

Figure 11.5 shows the process for the preparation of polyelectrolyte nanolayered films. Suppose that the solid surface is positively charged, so the anionic polyelectrolyte from the solution is easily layered on this surface via electrostatic attraction. This is truly the monolayer of molecular polyelectrolytes. Then, the

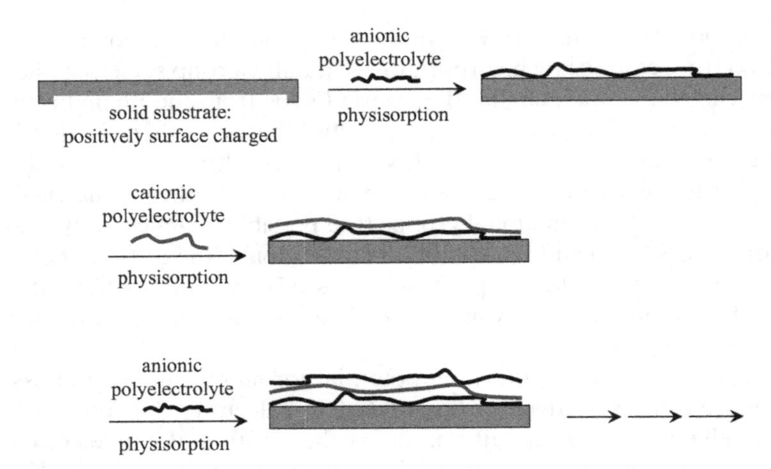

Figure 11.5. Preparation of polyelectrolyte nanolayered films via layer-by-layer assembly.

polyelectrolyte with cationic charge is employed on this surface, which will be layered on top of the first layer, also via electrostatic attraction. Successive layering can be achieved by alternatively employing cationic and anionic polyelectrolytes, which will produce multilayers of polyelectrolyte films.

Each layer of layer-by-layer films has molecular-scale thickness. A few tens of layers can be easily achieved and the total layered area can be extended above the centimeter scale. The force interplay for this process is the attractive electrostatic force that is balanced by the steric repulsion from the molecular backbone of the polyelectrolytes. The spacing between the layers is determined at the point where this balance is achieved. The charge density along the polyelectrolytes chain is usually very high, so the van der Waals force has quite limited influence on the attractive part of this picture.

This layer-by-layer assembly is one facile method to prepare nanolayered films. It does not require serious equipment. Also, the building units are not limited to polyelectrolytes. As long as the conditions (both building unit and solvent) for the electrostatic interaction are properly set, a wide variety of organic, inorganic, colloidal, and biological materials can be employed as its building units (Ai et al., 2003; Liu et al., 2003). They can be assembled as homogeneous nanolayered films and as heterogeneous composites. Curved solids such as spheres can be easily used as substrates, too (Caruso, 2001).

11.3.3. Vapor-Deposited Films

This is a method to prepare nanostructured films by direct exposure of the building units or decomposable building units (precursor) onto the solid surfaces. It thus has the capability to produce a much larger scale of films on the larger area of substrates.

When the building units are directly deposited without any significant decomposition, physisorption of those building units with solid surface becomes the main driving force. Therefore, the situation becomes the same as Figure 6.5. It is a molecular interfacial self-assembly, and the careful design of intermolecular forces, especially hydrogen bonds, between the building units is important to control the structural features of the films. This has led to the universal technique called *physical vapor deposition (PVD)* (Kippelen, 2004), which is very useful to produce well-developed organic thin films with a variety of building units.

For many other practical cases, the building units are in the form of molecules of high reactivity (precursor). They are usually inorganics such as alkoxides that have targeted elements or vaporized ions of targeted elements that are usually semiconductive or metallic. When they are collided or bombarded onto the solid surface, they are decomposed by the energy created by the collision, and the desired elements are deposited in film forms. A designed or preconstructed epitaxial match usually determines the final surface structures. Hence, it can be understood as an atomic self-assembly at a solid surface followed by a surface reaction such as chemisorption, surface decomposition, or surface dissociation. Most of the atomic shapes of atomic building units involved here can be assumed to be spherical or spherelike, and there are not many choices to vary them. This means that the packing geometry of the building units cannot have much of an impact for the formation of nanostructured films. The critical factor that determines the types and structures of these films is the force interplay between the building units and between the building units and the solid surfaces, which comes with the interplay between kinetics and thermodynamics. This has been described in Chapter 6 (Section 6.6). Thus, rather than molecular packing control, as for the above cases of SAMs, the control of external factors, such as the degree of vacuum, the energy of incident beam, precursor concentration, the degree of solid substrate to the incident beam, and the sputtering methods such as single-source or multiple-source, or pulsed laser ablation, has a much more important impact on the formation and structural features of the films. A variety of techniques have been developed based on these principles, which include chemical vapor deposition (CVD), vapor phase epitaxy (VPE), liquid phase epitaxy (LPE), molecular beam epitaxy (MBE), and metalorganic chemical vapor deposition (MOCVD).

This is a well-developed, practical method to prepare a wide spectrum of ultrathin films or multilayered films. Many structural features including number of layers, thickness, and surface roughness can be well controlled not only in nanolayered form but as a nanopatterned type as well. Also, the films prepared have high stability and good mechanical strength. They can be easily incorporated with a variety of functionalities, and high-speed mass production is possible. The semiconductor industry especially has a huge investment in this technique, which shows great promise to create many nanoscale devices in the near future. An excellent book is available to view the details (Francombe, 2001).

11.3.4. Sol–Gel Processed Films

The chemistry of the sol–gel process for the preparation of nanostructured films is the same as for the preparation of nanostructured bulk materials (Brinker, 2006). And its essential parts have been described in Chapter 9 (Section 9.3, Figures 9.3 and 9.12). The difference lies in the fact that the films are prepared to play the critical role of the substrate surface. As shown in Figure 11.6, this fact can help us classify the preparation of nanostructured films into two groups.

For the direct deposition, the precursor sol is directly deposited onto a given solid surface. This can be done either by direct dip-coating or spin-coating. Depending on the conditions of the sols, the proper choice of coating methods can be advantageous for the control of structural features such as thickness, roughness, and so on. The deposited sol then undergoes the sol-to-gel reaction to be formed into solid films. Environmental conditions including relative humidity and drying rate are the major controlling factors in this step. The films at this stage can be either nanocomposite or hybrid types. If necessary, the organic components inside the films, for instance, self-assembled micelles, can be removed, which will create films with nanoporous features (Brinker, 2004).

The pre-self-assembled precursor sol also can be deposited on prepatterned solid substrate. By controlling the structural or chemical characters of the patterned solid surface such as size/thickness of the patterned area or the types of surface micelles preassembled on a solid surface, the sol can be selectively introduced as the guide of the prepatterned solid. This can be applied as one of the

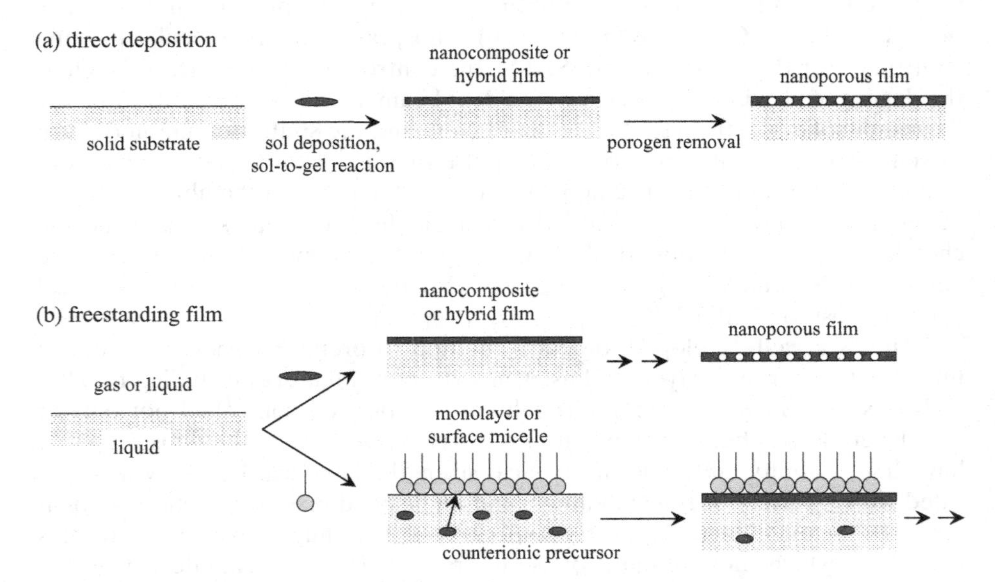

Figure 11.6. Different routes for the preparation of nanostructured films by sol–gel process.

facile nanofabrication methods for the secondary synthesis of nanoscale inorganic, metallic, and semiconductic films.

Nanostructured films also can be synthesized with the support of liquid surfaces. The films obtained are free on the liquid surfaces, which suggests the name *freestanding film*. The sols can be delivered either at the gas–liquid interface or liquid–liquid interface, and the subsequent sol-to-gel reaction provides the nanocomposite or hybrid types of films. They can be collected in as-synthesized forms for further treatment: additional postsynthetic treatment or the removal of organic component for the preparation of nanoporous films. For the other type of preparation approach, the monolayers or surface micelles can be preassembled on the gas–liquid or liquid–liquid interfaces. The precursor components that have counterionic capability can be preloaded in the subphase or delivered afterward. Then they can be attracted onto the surface of the organic component, and reacted. One typical advantage in this approach is that a mechanical factor (surface pressure) can be usefully introduced for a practical change in the films, which can be done by control of the preassembled layers. Also, an important factor to be examined is epitaxial matching (Section 9.6.3, Figure 9.10) between the preassembled layer and the structural parameters of synthesized films such as unit cell size of basic components or condensation length. This can sometimes have a great effect on the structural characteristics of the films. For example, when the reaction is restricted within the limited range of space set by intermolecular force conditions (surface well), the nanoparticles become the more likely products to be obtained, instead of the planar films (Section 10.3.2.5.).

11.3.5. Langmuir-Blodgett (LB) Films

LB films are obtained through a two-step process: (1) self-assembly on liquid surfaces as a Langmuir monolayer by the help of external mechanical force, and then (2) deposition on solid surfaces by dip-coating. Figure 11.7 shows the schematic representation of their preparation. This is for the case of upstroke-mode deposition, but the principle for downstroke mode is the same, except for the direction of the deposition. Depending on the given combination of upstroke and downstroke modes, LB films are commonly classified as X-, Y-, and Z-type of configuration. When a given Langmuir monolayer is best deposited by a given solid substrate with tilted or near-horizontal direction, it is commonly called Langmuir-Schafer (LS) film. This is an especially useful procedure when the physisorption between the building units and substrate is weak. LB films have been long studied, and much of the details can be found in excellent articles (Basu and Sanyal, 2002; Valli, 2003) and books (Roberts, 1990; Tredgold, 1994; Petty, 1996). As for the above SAMs, this section will not be spent on those known facts, but will be devoted to the issue of force balance for the formation and structural manipulation of LB films.

No chemisorption is involved during the formation of LB films. The primary driving force is the physisorption of the building units that are spread and self-assembled by an external mechanical force on the solid surfaces. Thus, consider-

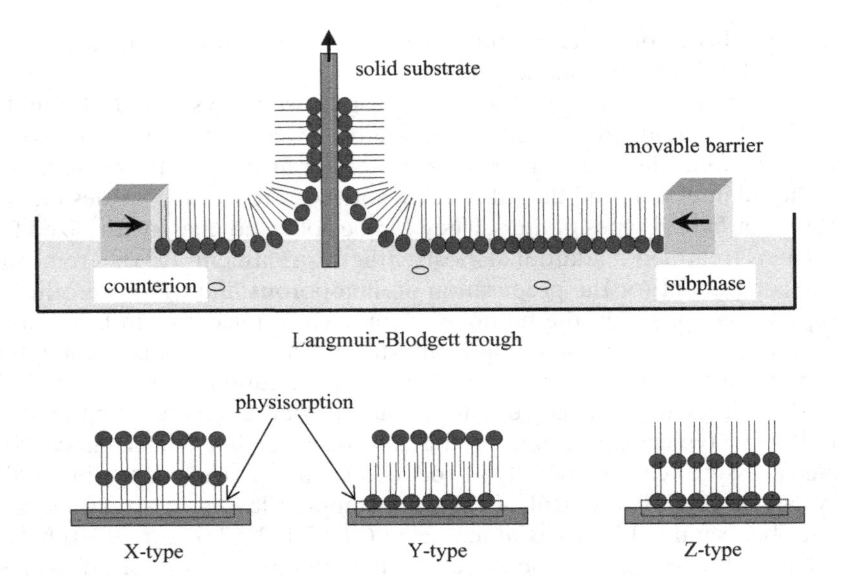

Figure 11.7. Schematic representation of the preparation of Langmuir-Blodgett film and its different types of multilayer deposition.

ing that the head groups of most of the molecular building units are ionic or hydrated (for nonionic), to ensure the proper force setting, the solid surfaces should be hydrophilic (if the subphase is water) or ionic with proper sign of surface charge. If a downstroke is necessary for the first layer deposition, since the common tail groups are hydrophobic, the solid surfaces should be hydrophobitized to ensure a proper initial interaction.

As shown in Figure 11.3, chemisorption determined the surface coverage of the building units for SAM: the chemisorbed site on the solid surface and its spacing relation with the size of head group of the building units. Then, the interplay of the intermolecular forces within the free space provided by this strong bond determined its structural features. On LB films, however, the external mechanical force acts as a primary factor to determine their surface coverage. This force is quantitatively defined as a surface pressure (π) that is the surface tension difference between the area without and with the building units. At low surface pressure of the gaseous phase of the Langmuir monolayer, the building units will be either dispersed or assembled as surface domain (or surface micelle) through the flat packing mode (Figure 6.4). With careful deposition techniques and conditions, they can be deposited on the solid surface without significant disruption of those structural features. This can be viewed as "patterned" LB films, which can provide layers of given building units that are fabricated at the nanometer scale.

As the surface pressure increases up to the liquid expand and solid phases, the building units now begin to be assembled with the upright packing mode

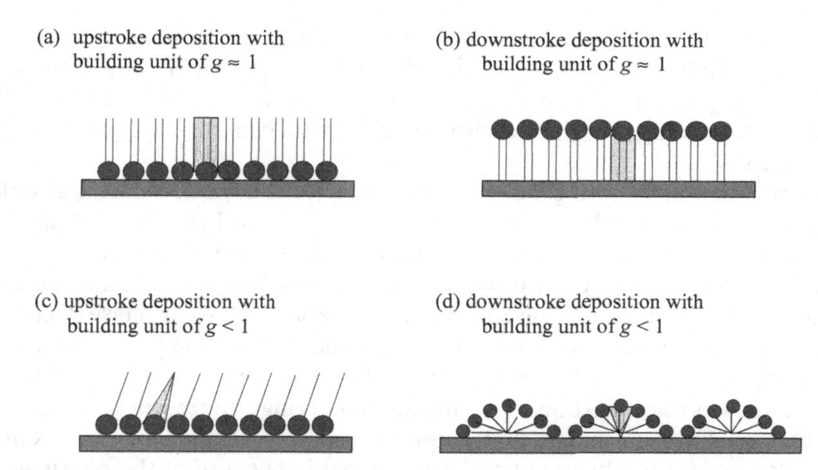

Figure 11.8. Different types of molecular packing on Langmuir-Blodgett films.

(Figure 6.4). This can be deposited on the solid surface as "fully covered" LB films from a simple monolayer to multilayers with structural complexity. Figure 11.8 shows the possible types of molecular packing at this condition (the condition of close packing by surface pressure), and its relation to the structural features of LB films assembled. This molecular packing is, of course, the result of the interplay of intermolecular forces between the building units. (a) is the case of upstroke deposition with molecular building units that have a packing geometry value close to 1. (b) is for the same building units but for downstroke deposition. Since this packing geometry ensures the lamellar type of assembly when they are packed closely, and that configuration can provide a minimized free space for the interplay of intermolecular forces, LB films assembled with this condition show a little configurational defect such as tilt, for both cases (a) and (b). This also becomes the likely case for all types of multilayered LB films. (c) is the case of upstroke-mode deposition with molecular building units that have a packing geometry less than 1. Building units like this mostly form spherical micelles (Chapter 3) when they are self-assembled in solution. Since the assembly for LB films is linear, and their head groups are closely tightened by external mechanical force, this assembly provides a free space where the interplay of intermolecular forces between mostly hydrocarbon chains has to occur to be balanced within. Like the SAM cases with mismatched head groups and substrate atoms (Figure 11.4a and c), this condition produces tilted configuration. A smaller g value results in a greater number of tilts. The difference from SAMs is that the interaction with the solid surface has little effect for LB films. (d) is for the same type of building units but for downstroke-mode deposition. Here, the primary driving force that maintains the integrity of the deposited monolayer is a physisorption between the hydrophobic tail group of the building units and the solid surface. This is even weaker than the interaction by the head groups. The interaction between the tail groups now can have enough strength to accommodate each building unit as its packing geometry to be expressed within the monolayer. (d) is for one possible extreme situation that is likely at relatively low surface pressure. Depending on the conditions, this also can be expressed as *tilted configuration*.

For building units with g value greater than 1, the expected packing configuration, in the other words, the structural features of the LB films prepared, can vary significantly. It mainly depends on their structural integrity. For example, if the building unit is a surfactant with three hydrocarbon chains attached on its head group, the surface pressure at the liquid expand or solid phase region will have enough strength to "compress" those flexible chains to be packed with linear configuration as in cases (a) or (b). If the building units have somewhat rigid structure such as fluorocarbon segments or phenyl rings, there will be a significant molecular packing mismatch that cannot be simply relaxed by, for example, tilting. Upward assembly, of course, is not possible because of the physisorption with solid surface unless the surface pressure applied was extremely high. Thus, often this condition induces a nonsymmetric packing of the building units, which can bring some directionality along its way. Superstructures such as spiral and ribbon can be understood as the result of this nonsymmetric force balance process (Mourran et al., 2005).

LB films can be prepared with well-controlled monolayers and multilayers. Precise control of the number of layers and thickness can be achieved with simple operation and choice of the building units. Secondary functionality can be easily incorporated. Building units should be always afloat on the surface of the liquids. But, as long as this condition is fulfilled, not only amphiphilic building units but nonamphiphilic ones such as nanoparticles, colloidal particles, polymers, and so on, can be practically assembled as nanostructured films as well.

The application possibilities of LB films for nanotechnology have been identified as very promising (Talham, 2004; Bertoncello, 2005; Diaz Martin and Cerro, 2005). For some, there is a requirement for LB films with a minimized number of defects such as pinholes and inhomogeneous packing (Takamoto et al., 2001). But for others, patterning possibly on a nanometer scale is more crucial, which might lead to new collective properties or usefulness for the structural template for secondary application possibilities.

Preparation of LB films is primarily the result of the delicate balance between all of the forces involved, which include van der Waals force, electrostatic force, hydration force, molecular geometry, and mechanical surface pressure. Thus, the correct identification and modification of these parameters and their relations can provide much better options for controlling the key factors to assemble the proper LB films for intended applications. Those factors are the degree of packing, its extension, and patterning. Also, since the LB method has mechanical factors to control the assembly of the building units, an additional degree of fabrication is possible by taking advantage of them. Controls of deposition degree, deposition rate, deposition angle, and so on, are among them. Fabrication of highly patterned stripe by simple control of deposition rate to the point of inducing contact angle inhomogeneity is a good example (Moraille and Badia, 2002; Kovalchuk et al., 2003).

11.4. PROPERTIES AND APPLICATIONS OF NANOSTRUCTURED FILMS

Many aspects of the properties of nanostructured films can be similar to those of nanostructured bulk materials. However, there are many properties and functions that can originate from the two-dimensional nanoscale structural features of nanostructured films. This section covers this along with their use in practical applications.

11.4.1. Nanoporous Films

The properties of nanoporous films are, in many ways, the same as those of meso- or nanostructured bulk materials shown in Chapter 9. The critical ones include high inner surface area, high pore volume, and tunable pore size. Thus, a variety of application possibilities for mesostructured materials (Section 9.9, Figure 9.16) can be directly shared with nanoporous films. Also, they can be found particularly useful in applications where the film structure is the essential part of them. Typical examples are in the areas of membrane, separation, or sensor. Robust films with nanometer-scale hexagonal pores that run through normal to their surface direction can be powerful novel substrates for membrane-related applications. The high inner surface area of nanoporous silica film can make it useful for the antihygroscopic protective film of organic light-emitting displays (OLEDs), with great potential to extend their lifetime substantially. Also, its low dielectric constant can provide critical base-coating for many film-based nanodevices.

11.4.2. Nanolayered Films

For nanolayered films, the best advantage comes from the relative ease of property control of individual layers on both structural and physicochemical aspects. This fact is greatly advantageous for creating nanolayered films with even multifunctionalities or multiproperties. Also, the composite and hybrid types of films can be fabricated with great diversity, which brings in unexpected synergistic novel properties. Mechanical strength such as tensile strength is often orders of degree higher than bulk materials. This is achieved by the layered structural features with different physical properties that can greatly reduce the propagation of mechanical defaults such as cracks or defects throughout a large or entire portion of the films.

Semiconductic nanolayered films with proper thickness and roughness are absolutely critical as the platform of potential nanosized circuits. Also, a variety of functional films such as magnetic, electric, piezoelectric, and so forth can find

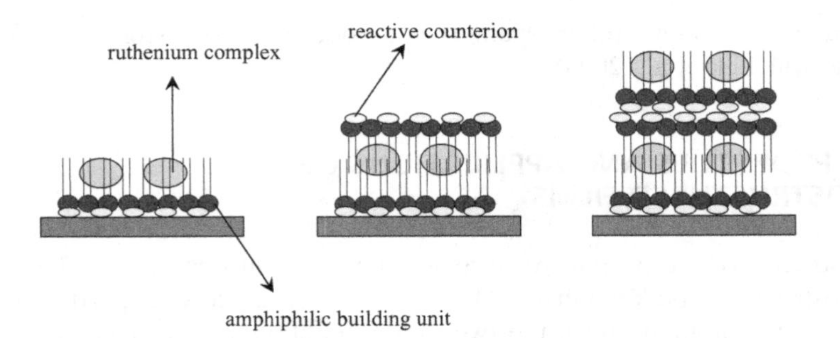

Figure 11.9. Schematic LB film designs of nanocomposite for oxygen sensors.

great usefulness for a wide range of applications, from the functional substrate preparation for spectroscopy to electrochemical cells.

A superhydrophobic surface (Section 5.1.2) combined with functional properties such as photocatalytic capability can find exciting novel applications. For example, a self-cleaning surface can be practically realized with this. One key lies in the proper nanolayering with hydrophobic building units on top of the nanopatterned substrate, possibly with multiscale hierarchy, which can greatly alter the surface energy so as to be superhydrophobic.

Figure 11.9 shows another example of the construction of functional nanocomposite based on nanolayered film. Ruthenium (II) complex is well known for its delicate sensing capability of oxygen gas. One of the issues is that it has to be well dispersed on its substrate so it can maximize the contact area. At the same time, the substrate has to have desirable stability and permeability for incoming oxygen gas. By constructing the mixed LB films of ruthenium complex with the proper choice of amphiphilic building units, the issues of uniform dispersion and permeability can be addressed. And, by the introduction of proper reactive counterions, the stability issue also can be addressed. This strategy can be applied to the preparation of a diverse LB films–based sensor (Valli, 2005) and also for an oxygen sensor with sol–gel-processed films (MacCraith et al., 1995).

11.4.3. Nanopatterned Films

Nanopatterned films can be fabricated by a variety of approaches. For example, SAMs can be patterned with the proper degree of physical or chemical modification such as mixed assembly or lithographic fabrication. Vapor deposition through epitaxial matching can create patterned nanoislands on solid surfaces. And a templating method using sacrificial masks or colloidal particles can provide intricate structural features on the nanopatterned surfaces.

Properties of the nanopatterned films are primarily determined by the properties of the original components. However, a more important aspect in the preparation of nanopatterned films is the creation of possible synergistic and even novel properties. This can be achieved by the manipulation of structural features, which includes the size/height of patterned domains, shape, and spacing between them. For example, arraying of patterned components with a correct spacing can maximize the collective spectroscopic or catalytic properties. Also, a proper spacing control between semiconductic or metallic quantum dots is one of the critical factors to ensure proper communication between them through electron transfer or light. This is possibly one of the most important perquisites for the realization of nanoelectronics.

11.4.4. Monolayer: Model Membrane

Chapter 6 (Section 6.3) introduced the formation of the Langmuir monolayer at the air–liquid interface, and the role of intermolecular force balance in its assembly. When its building units are lipids or lipid-based amphiphiles, their molecular structure, configuration, and the interplay of the intermolecular forces within are a lot like those of biological membranes. The difference is that this is a monolayer while biological membranes have a bilayer morphology consisting of mainly lipids and proteins. The key to the employment of the Langmuir monolayer as a model lipid membrane lies here. The structural, compositional, and configurational resemblances between the two provide the powerful insight that the Langmuir monolayer can mimic biological membranes with facile advantage (Brockman, 1999; Feng, 1999). Those valuable parameters including surface pressure, surface potential, surface density, and temperature are highly controllable, and the phase formation and transition on the surface of the monolayer can be tracked with nanoscale resolution *in situ* (Vollhardt and Fainerman, 2000; Brezesinski and Möhwald, 2003).

Figure 11.10 presents one possible design of a model lipid membrane. Lipid building units for the Langmuir monolayer can be selected based on the *representativeness* of the biological membrane to be studied. Membrane proteins can be employed, if necessary, but primarily based on the same criterion. For the

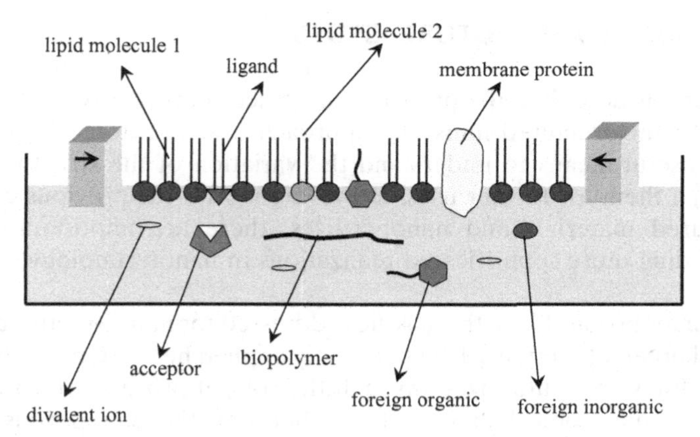

Figure 11.10. Design of a model lipid membrane to mimic a biological membrane.

study of ligand–acceptor interaction, either ligand or acceptor can be inserted as part of the Langmuir monolayer while the other component is introduced in subphase. Divalent cations such as calcium or magnesium ions, which are critical in many biological activities, polymeric or oligomeric biomolecules, and foreign organic and inorganic components to be studied also can be introduced in subphase. As these components interact with the monolayer, the force balance that was set for the given condition of the monolayer is subject to change. For example, divalent cations, inorganics, or biopolymers whose surfaces usually are highly charged will have a profound effect on the electrostatic interaction between the head groups of the lipid building units, while the hydrophobic moiety of biopolymers or organics will induce new van der Waals interaction with the alkyl groups of the monolayer. Acceptor that is complexed with ligand possibly causes steric repulsion. All these newly induced intermolecular forces will cause the whole system to be shifted into the new force balance. And this can be tracked both quantitatively and qualitatively by recording the change in the surface parameters mentioned above and in the surface phase. That in turn may allow extracting meaningful insights to understand the possible interaction mechanism of those additives with the biological membrane to be studied.

A model lipid membrane has been used for a variety of applications. This includes the physiological study of human lung surfactant (Zasadzinski et al., 2001), the development of drug discovery and delivery systems (Baksh et al., 2004), and the effect of foreign peptides on the biological membranes of human and other living organisms (Maget-Dana, 1999). One critical issue remains in the fact that most of the valid model membrane systems are assembled at the liquid–gas interface, which can raise the question of the validity of the results related to the interaction in the hydrocarbon chain regions. Further advances of this model membrane toward the area of *bionanotechnology* will depend on the reliable resolution of this correspondence issue between model and real systems.

11.5. SUMMARY AND FUTURE ISSUES

This chapter has described the preparation, structure, property, and application potential of nanostructured films. An approach based on force balance clearly provides a useful means to understand the variation of the structural features and to design them for further uses. However, as with the previous chapters on nanostructured materials and nanoparticles, there are important challenges ahead regarding more sophisticated realizations in nanotechnology:

1. For nanoporous films, those issues addressed for nanostructured materials in Chapter 9 (Section 9.10) are directly applied here. Many of the nanoporous films are synthesized through the sol–gel process, which brings in a very similar force balance scheme between the nanoporous films and nanoporous bulk materials. Thus, those challenges, including hierarchi-

cally ordered structure, thermal stability, polymorphism, and structure–functionality relationship, can be approached in the same manner.

2. For nanolayered films, there still is a good deal of limitation on the available building units for each process. SAMs and LB films are better fit for molecular building units, and vapor-deposition techniques work well for somewhat limited variations of atomic and molecular building units. The concept of these techniques is well developed and the assembly processes are generally facile. Also, each technique has its own uniqueness. Thus, the development of a richer spectrum of building units for each of the assembly methods and of the versatility between them will greatly help us expand the types and scopes of nanolayered films.
3. For nanopatterned films, more details will be presented in Chapter 13 on nanofabrication. However, on the physicochemical assembly side, two issues are worth mentioning here. First is the film assembly with multiple components such as binary, tertiary, and more. This should greatly expand the scope of possible heterogeneously patterned films both physically and chemically. Second is the development of multistep self-assembly processes for patterning. Multistep self-assembly, which is the case for most of the bulk self-assemblies, has great potential for the creation of a good degree of structural diversity. However, those multistep assembly processes are rare for the formation of films. Preparation of patterned films through multiple-step assembly should be helpful in increasing the possibility of morphological diversity of patterned surfaces.

REFERENCES

Ai, H., Jones, S. A., Lvov, Y. M. "Biomedical Applications of Electrostatic Layer-by-Layer Nano-Assembly of Polymers, Enzymes, and Nanoparticles," *Cell Biochem. Biophys.* **39**, 23 (2003).

Baksh, M. M., Jaros, M., Groves, J. T. "Detection of Molecular Interactions at Membrane Surfaces Through Colloid Phase Transitions," *Nature* **427**, 139 (2004).

Basu, J. K., Sanyal, M. K. "Ordering and Growth of Langmuir-Blodgett Films: X-ray Scattering Studies," *Phys. Rep.* **363**, 1 (2002).

Bertoncello, P. "Recent Advances in Langmuir-Blodgett Films," *Supramolecular Engineering of Conducting Materials*, Ram, M. K., ed., pp. 179–197 (Research Signpost: 2005).

Biesalski, M., Ruehe, J., Kuegler, R., Knoll, W. "Polyelectrolytes at Solid Surfaces: Multilayers and Brushes," *Handbook of Polyelectrolytes and Their Applications*, Vol. 1, Tripathy, S. K., Kumar, J., Nalwa, H. S., eds., pp. 39–63 (American Scientific Publishers: 2002).

Brezesinski, G., Möhwald, H. "Langmuir Monolayers to Study Interactions at Model Membrane Surfaces," *Adv. Coll. Inter. Sci.* **100–102**, 563 (2003).

Brinker, C. J. "Evaporation-Induced Self-Assembly: Functional Nanostructures made Easy," *MRS Bull.* **29**, 631 (2004).

Brinker, C. J. "Sol-Gel Processing of Silica," *Surfactant Science Series*, Vol. 131 (Colloidal Silica), pp. 615–636 (CRC Press: 2006).

Brockman, H. "Lipid Monolayers: Why Use Half a Membrane to Characterize Protein-Membrane Interactions," *Curr. Opin. Struct. Biol.* **9**, 438 (1999).

Caruso, F. "Generation of Complex Colloids by Polyelectrolytes-Assisted Electrostatic Self-Assembly," *Aust. J. Chem.* **54**, 349 (2001).

Decher, G., Schlenoff, J. B., eds. *Multilayer Thin Films: Sequential Assembly of Nanocomposite Materials* (Wiley-VCH: 2003).

Diaz Martin, M. E., Cerro, R. L. "Langmuir-Blodgett Films: A Window to Nanotechnology," *Chemical Engineering*, Galán, M. A., del Valle, E. M., eds., pp. 267–297 (Wiley: 2005).

Feng, S. "Interaction of Mechanochemical Properties of Lipid Bilayer Vesicles from the Equation of State or Pressure-Area Measurement of the Monolayer at the Air–Water or Oil–Water Interface," *Langmuir* **15**, 998 (1999).

Francombe, M. H., ed. *Thin Films: Frontiers of Thin Film Technology*, Vol. 28 (Academic Press: 2001).

Khomutov, G. B. "Interfacially Formed Organized Planar Inorganic, Polymeric and Composite Nanostructures," *Adv. Coll. Inter. Sci.* **111**, 79 (2004).

Kippelen, B. "Optical Materials: Self-Assembly Reaches New Heights," *Nature Mat.* **3**, 841 (2004).

Kovalchuk, V. I., Bondarenko, M. P., Zholkovskiy, E. K., Vollhardt, D. "Mechanism of Meniscus Oscillations and Stripe Pattern Formation in Langmuir-Blodgett Films," *J. Phys. Chem. B* **107**, 3486 (2003).

Liu, S., Volkmer, D., Kurth, D. G. "Functional Polyoxometalate Thin Films via Electrostatic Layer-by-Layer Self-Assembly," *J. Cluster Sci.* **14**, 405 (2003).

Love, J. C., Estroff, L. A., Kriebel, J. K., Nuzzo, R. G., Whitesides, G. M. "Self-Assembled Monolayers of Thiolates on Metals as a Form of Nanotechnology," *Chem. Rev.* **105**, 1103 (2005).

MacCraith, B. D., McDonagh, C. M., O'Keeffe, G., McEvoy, A. K., Butler, T., Sheridan, F. R. "Sol-Gel Coatings for Optical Chemical Sensors and Biosensors," *Sensors and Actuators, B: Chemical* **B29** (1–3), 51 (1995).

Maget-Dana, R. "The Monolayer Technique: A Potent Tool for Studying the Interfacial Properties of Antimicrobial and Membrane-lytic Peptides and Their Interactions with Lipid Membranes," *Biochim. Biophys. Acta* **1462**, 109 (1999).

Magnussen, O. M., Ocko, B. M., Duetsch, M., Regan, M. J., Pershan, P. S., Abernathy, D., Gurebel, G., Legrand, J.-F. "Self-Assembly of Organic Films on a Liquid Metal," *Nature* **384**, 250 (1996).

Moraille, P., Badia, A. "Highly Parallel, Nanoscale Stripe Morphology in Mixed Phospholipid Monolayers Formed by Langmuir-Blodgett Transfer," *Langmuir* **18**, 4414 (2002).

Mourran, A., Tartsch, B., Gallyamov, M., Magonov, S., Lambreva, D., Ostrovskii, B. I., Dolbnya, I. P., de Jeu, W. H., Moeller, M. "Self-Assembly of the Perfluoroalkyl-Alkane $F_{14}H_{20}$ in Ultrathin Films," *Langmuir* **21**, 2308 (2005).

Nemetz, A., Fischer, T., Ulman, A., Knoll, W. "Surface-Plasmon-Enhanced-Raman Spectroscopy with 21-hydroxyheneicosanethiol ($HS(CH_2)_{21}OH$) on Different Metals," *J. Chem. Phys.* **98**, 5912 (1993).

Petty, M. C. *Langmuir-Blodgett Films: An Introduction* (Cambridge University Press: 1996).

Petty, M. C. "Organic Thin Film Architectures: Fabrication and Properties," *Surfaces and Interfaces for Biomaterials*, Vadgama, P., ed., pp. 60–82 (Woodhead Publishing: 2005).

Roberts, G. G., ed. *Langmuir-Blodgett Films* (Plenum Press: 1990).

Stoycheva, S., Himmelhaus, M., Fick, J., Kornviakov, A., Grunze, M., Ulman, A. "Spectroscopic Characterization of ω-Substituted Biphenylthiolates on Gold and Their Use as Substrates for 'On-Top' Siloxane SAM Formation," *Langmuir* **22**, 4170 (2006).

Takamoto, D. Y., Aydil, E., Zasadzinski, J. A., Ivanova, A. T., Schwartz, D. K., Yang, T., Cremer, P. S. "Stable Ordering in Langmuir-Blodgett Films," *Science* **293**, 1292 (2001).

Talham, D. R. "Conducting and Magnetic Langmuir-Blodgett Films," *Chem. Rev.* **104**, 5479 (2004).

Tredgold, R. H. *Order in Thin Organic Films* (Cambridge University Press: 1994).

Ulman, A. "Formation and Structure of Self-Assembled Monolayers," *Chem. Rev.* **96**, 1533 (1996).

Ulman, A. "Self-Assembly of Surfactant Molecules on Solid Surfaces," *Organized Molecular Assemblies in the Solid State*, Whitesell, J. K., ed., Vol. **2**, pp. 1–38 (Wiley: 1999).

Valli, L. "Langmuir-Blodgett Films," *Thin Solid Films: Application, Preparation and Characterization*, Trusso, S., Mondio, G., Neri, F., eds., pp. 71–94 (Research Signpost: 2003).

Valli, L. "Phthalocyanine-based Langmuir-Blodgett Films as Chemical Sensors," *Adv. Coll. Inter. Sci.* **116**, 13 (2005).

Van Hove, M. A. "From Surface Science to Nanotechnology," *Catal. Today* **113**, 133 (2006).

Vollhardt, D., Fainerman, V. B. "Penetration of Dissolved Amphiphile into Two-dimensional Aggregating Lipid Monolayers," *Adv. Coll. Inter. Sci.* **86**, 103 (2000).

Zasadzinski, J. A., Ding, J., Warriner, H. E., Bringezu, F., Waring, A. J. "The Physics and Physiology of Lung Surfactants," *Curr. Opin. Coll. Inter. Sci.* **6**, 506 (2001).

12

NANOASSEMBLY BY EXTERNAL FORCES

As discussed in Chapter 5, many colloidal phenomena can induce forces that can have a degree of operating-length scale and strength similar to the intermolecular/colloidal forces. Thus, these colloidal phenomena–induced forces can have a great influence on the forces that act on individual colloidal particles and the balance scheme between them. The vectorial sum of these forces decides the direction and strength of the movement of the particles. When large numbers of colloidal particles are subject to the self-assembly process, these colloidal phenomena–induced forces can act as an attractive driving force and as a repulsive opposition force as well (Figure 5.11). However, as the strength of these forces is increased or as their interaction configuration with colloidal particles is adjusted, the influence of these forces can be strong enough to be dominant over the intermolecular/colloidal forces. This situation, thus, can lead to a new picture of the self-assembly of colloidal particles. Those forces can now take major control of the self-assembly process. For many cases, their action of direction is unilateral; this makes the self-assembly process directional. We will use the term *external force* to embrace this condition.

As discussed in Chapters 3, 4, and 5, both the molecular and colloidal self-assemblies are random by nature. The morphology of their self-assembled aggregates is mainly determined by the thermodynamics of individual systems. If we go further into the area of nanotechnology, this aspect of self-assembly provides a very unique advantage. As described in Chapter 8, this is the key point of bottom-up and hybrid approaches, which can make the assembly of nano–building units greatly efficient. However, what is deficient in this aspect is whether we can have the ability to manipulate the assembled structure against the rule of thermodynamics. Thus, there can be richer possibilities for constructing the assembled nano–building units with a "designed" structure. This can provide an additional level of means to control the nanoscale properties and even to discover new properties. For self-assembly, this is the issue of *directionality* of the self-assembly process. *Controllable* means to "direct" any axis of self-assembled aggregates; will provide a great level of fulfillment in this issue and therefore nanotechnology in general.

External forces can provide this directionality in self-assembly processes to a high degree. Many colloidal self-assemblies are strongly affected by the existence of external forces. This also is the case for some molecular self-assembies at given conditions. Typical examples of external force include capillary force, electric force, magnetic field, flow, spatial confinement, optical force, ultrasound, and centrifugal force. Proper introduction and control of these forces can help significantly enhance the efficiency of nanofabrication (details in Chapter 13; Löwen, 2001; van Blaaderen, 2004). It can also provide efficient ways to control the structural factors for the preparation of colloidal crystal. Colloidal crystal is colloidal particles assembled into a crystalline array that can show unusual or novel optical and thermodynamic properties. Structural factors including precise position, interparticle distance, and number of layers are keys to manipulating those properties.

Though all categories of self-assemblies depicted in Figure 1.1 are subject to the influence of external forces, the focus in this chapter will be primarily on colloidal self-assembly and molecular self-assembly. The picture of force balance for these self-assemblies under external forces will be discussed first, followed by their implications for nanotechnology, such as the photonic band gap.

12.1. FORCE BALANCE AND THE GENERAL SCHEME OF SELF-ASSEMBLY UNDER EXTERNAL FORCES

The general scheme of force balance for self-assembly under external forces has already been introduced in Chapter 5 (Section 5.2). Table 5.1 described the classification of three types of forces involved in this process. This scheme mainly works for self-assembly of colloidal particles. But it is valid for some molecular self-assembly processes, too.

The general scheme of self-assembly under external forces is difficult to establish. It certainly can be derived from the scheme without external forces (Figure 5.12). But, since external forces are now the dominant forces over the entire self-assembly process, the general picture itself can be greatly manipulated by them, which actually is the purpose of applying the external forces. The definition of the critical concentration can now be a much less significant parameter. Instead, the critical physical quantity of the applied forces becomes an important factor. In most cases, this factor determines the point (more likely, narrow region) where self-assembly is initiated, which means the point where the force balance for the entire system is set. It also determines the point where the existing self-assembled aggregates begin to be manipulated under given external forces. Examples include evaporation rate, the voltage or frequency of AC current, the strength or frequency of magnetic field, the stress or frequency of flow, the physical dimension of confined space, and so on. Also, an ambiguity between primary, secondary, and higher-order self-assembly processes becomes common under external forces. However, the sequential type of self-assembly by the sequential or multiple application of external forces can now be a new option.

12.2. COLLOIDAL SELF-ASSEMBLY UNDER EXTERNAL FORCES

This section will be devoted to colloidal self-assembly under external forces. In most cases, hard colloidal particles with various sizes and shapes are considered a primary building unit for this self-assembly. Molecular self-assembly under external forces will be examined in the next section.

12.2.1. Capillary Force

Whenever colloidal particles make contact with liquid-based interfaces (liquid–gas interface, liquid–liquid interface, wetting liquid film, foam film, etc.), some degree of interaction is always induced between them. More specifically, a process toward a balance between the forces that act on the colloidal particles is executed. This force balance process always results in the apparent interaction between the particles. This is *capillary force*. Their interface with the liquid is critical for the determination of the degree and direction of this force. Thus, the surface properties of colloidal particles play a major role during the capillary action. Surface wetting and surface roughness are among important factors. Gravity, which can act on the colloidal particles, is often comparable to this capillary force. This means the size and density of colloidal particles becomes another important factor, especially when they interact on a free liquid surface. Capillary force can be attractive or repulsive. Also, it operates between the colloidal particles and between the colloidal particles and solid substrate as well. Details on these aspects of capillary force can be found in an excellent review (Kralchevsky and Denkov, 2001) and book (Kralchevsky and Nagayama, 2001).

Figure 12.1 shows the schematic representation for three major types of capillary forces: immersion, flotation, and bridged. For immersion and flotation capillary forces, the direction of the capillary force is lateral with respect to the liquid–particle contact line (or wetting line). This is why these types are often

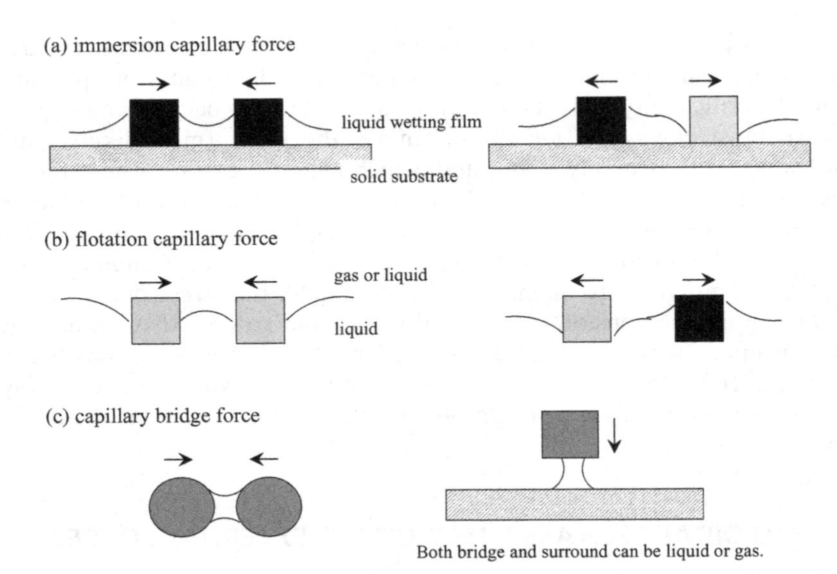

Figure 12.1. Capillary force–induced self-assembly of colloidal particles.

called *lateral capillary force*. For the bridged type, its direction is normal to the contact line. This is the only type of *normal capillary force*.

Immersion capillary force appears when the colloidal particles are partially immersed in the liquid films. This liquid film can be a wetting film on solid surface or a freestanding film such as the films in a bicontinuous microemulsion system or foam films. Wetting of liquid on the surface of the particles generates the menisci around them. Depending on the wettability of the liquid, the direction and magnitude of the force exerted by the formed menisci are determined. The sign of the contact angle (at the particle–liquid–gas contact line) is determined by the hydrophobicity or hydrophilicity of the particles (when the liquid is water). When the menisci from the neighboring particles that are deformed by the wetting are interfered with in such a way as to minimize the deformation by getting closer, an attractive capillary force is induced between them. The opposite situation creates a repulsive force. If the particle does not induce any deformation on the liquid interface, no interaction can arise between the particles.

Flotation capillary force appears between colloidal particles when they are afloat on the surface of liquid (liquid–gas or liquid–liquid). When the force exerted by the menisci causes them to interact with each other (meaning when they begin to overlap), they also get to interact with the gravity and buoyant forces acting on the particles (these forces are vertical in respect to the liquid interface, and thus interact with the vertical component of the force exerted by the menisci). When the balance between these forces requires the decrease of the interparticle distance (meaning the meniscus deforms in such a way), an attractive force is generated between the particles (usually between similar particles). The increase of the distance results in a repulsive force (usually between dissimilar particles). Gravity on the particle is one of the major players for this force, so the weight of the particles (or relative density to liquid) becomes one primary factor.

Because the immersion capillary attractive energy is much larger than the thermal energy (kT) at this range, the immersion capillary force becomes significant up to the particle size with a few nanometer radii. But flotation capillary force becomes negligible for particles with radii less than a few microns. Below this range, thermal energy becomes comparable to capillary attractive energy.

Bridged capillary force is induced when a liquid or gas phase connects the third solid phases: particle–particle or particle–substrate. The situation can vary depending on the surface property. When the solid surface is hydrophilic, the water or aqueous phase can be the bridged phase. A hydrophobic solid surface can manifest only gas or oil as the bridged phase. In the former case, the surrounding phase (the phase that surrounds the bridged phase) can be gas or oil. The water or aqueous phase is the most common surrounding medium in the latter case. Capillary condensation is the situation when this bridged phase is water. Capillary cavitation is for gas (or vapor) as the bridged phase. As discussed in Section 5.1, a curved interface always generates the pressure difference across it (Laplace pressure). The bridged phases between solid phases are always curved toward the bridged phase from the surrounding medium; thus, the interaction is induced so as to decrease the distance between the solid phases by the action of Laplace pressure. Also, the surface (interfacial) tension that is exerted around the interface acts in the same direction. Bridged capillary force is the result of this interaction, and is always attractive.

In most cases, these capillary forces are orders of magnitude stronger than van der Waals force, electrostatic force, and other colloidal forces. This aspect of capillary force, in many cases, can help achieve self-assembly with those colloidal particles whose force balance among colloidal forces does not favor their assembly. Much stronger attractive capillary force can overcome those repulsive colloidal forces. Examples include two-dimensional self-assembly with lateral capillary force, three-dimensional self-assembly with normal capillary force, colloidal crystallization by capillary evaporation, and self-assembly of colloidal objects on emulsion surface film. Another excellent example is *mesoscale self-assembly*. A designed application of capillary force created a variety of superstructures of colloidal objects of mm or sub-mm size with great precision and selectivity (Bowden et al., 2001).

12.2.2. Electric Force

Electric field–induced colloidal self-assembly can be accomplished under both DC and AC electric currents. But mainly the work has been done under an AC electric field. Since the flow around a colloidal particle critically affects the attractive force between the particles, the existence of a wall (electrode) is critical for this assembly. Self-assembly occurs mainly on this liquid–solid (electrode) interface (Gong et al., 2002; Zhang and Liu, 2004; Negi et al., 2005). Self-assembly in bulk solution under the influence of an electric field can be more complicated because of this flow-induced force. Without the interaction with the solid surface, this force can be more often random than directional, which can interrupt the self-assembly process.

Figure 12.2 shows the schematic representation of self-assembly of charged colloidal particles under an AC electric field. Randomly dispersed colloidal particles begin to assemble on the solid (electrode) surface at the critical magnitude of voltage applied and/or the critical value of frequency. These critical values can vary, depending primarily on colloidal factors such as charge type and surface density, particle size, and weight. Solution factors including pH, dielectric constant, and salt type and concentration can also greatly affect those critical values. The exact mechanism of the origin of the *dielectrophoretic force* in this picture is still under debate. But the electrohydrodynamic fluidic convective current that flows around each particle with the same direction near the surface of the solid acts as the main attractive driving force for this assembly. This is balanced with the electrostatic repulsive opposition force. Both the original repulsion by the surface charge and the electric field–induced repulsion contribute to this repulsive force. The existence of the critical voltage and frequency for the colloidal particle at a given concentration can be viewed as an analogy of the existence of the critical micellar concentration (*cmc*) for the surfactant self-assembly (Chapter 3). The difference lies in the length scale of surfactant for molecular self-assembly and this is colloidal-scale self-assembly. But, in both cases, the driving

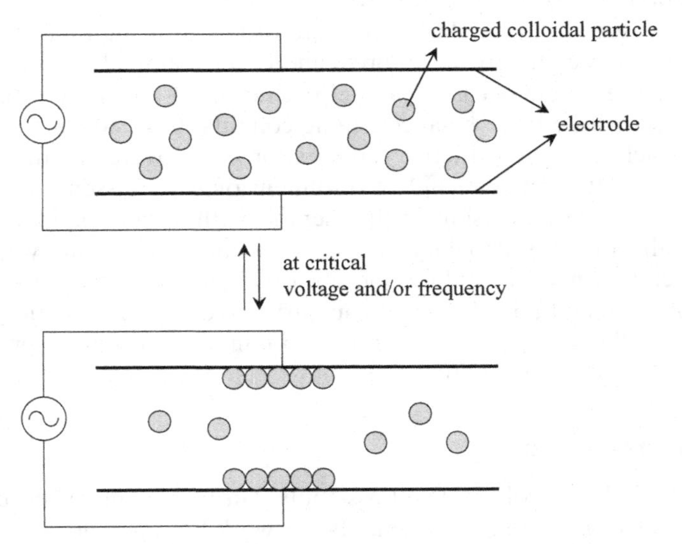

Figure 12.2. Self-assembly of colloidal particles under AC electric current.

and opposition forces are balanced at the critical value. For surfactant, the entropy-origin attractive force is intrinsic. For colloidal particles, the attractive force is generated by the external force. But, in both cases, as the value approaches to the critical level, the attractive force gradually accumulates with the increase in the concentration (for surfactant) or with the change in the voltage and/or frequency (colloidal particle). When these forces begin to assemble the self-assembly building units, the repulsive forces begin to get involved and balanced. The morphology of self-assembled aggregates is determined by this balance.

The morphology of self-assembled aggregates can be controlled by the change in the voltage and frequency. Also, when asymmetric colloidal particles such as rod or tube are subject to AC electric field, the direction of the aligned particles is controlled by the same direction. If the electrode is patterned (e.g., lithographically with alternating areas within the nm–μm range of electric and nonelectric areas), this electrohydrodynamic force can be repulsive at the boundary of the electric–nonelectric area.

12.2.3. Magnetic Force

Figure 12.3 shows the self-assembly process of magnetic particles under the external magnetic field. Stable magnetic colloidal particles (particles that have enough repulsive force between them to keep them apart without external magnetic field) will be magnetized once the magnetic field begins to impose. Two different types of magnetic fields can be applied: homogeneous and inhomogeneous. This is quite an analog of the AC and DC electric fields in the previous subsection. External magnetic field can induce the magnetization on the colloidal particles in a way so as to induce the more repulsive force and also to induce the attractive force between them. The relative direction of the magnetization and of the hydrodynamic flow that is generated around the particles cooperatively

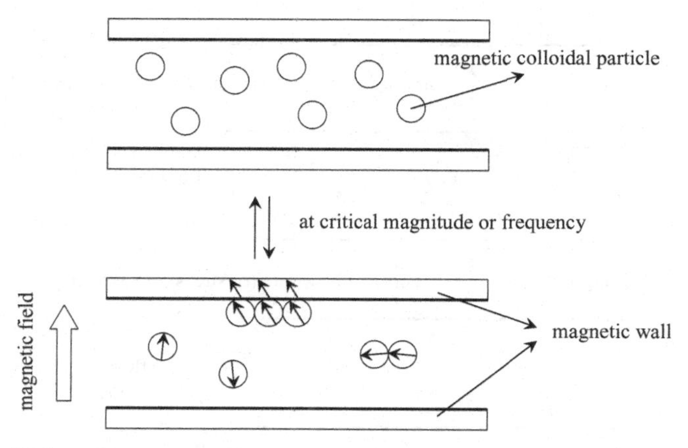

Figure 12.3. Self-assembly of magnetic colloidal particles under magnetic field.

determine the degree and direction of this force. By the control of the direction and strength of homogeneous field, and of the frequency and strength of inhomogeneous field, these repulsive and attractive forces between the particles can be controlled. Thus, depending on the system, there can exist the critical value of the strength and frequency of the magnetic field, where the particles can start to assemble together. The critical number of the particles that are assembled is determined by the force balance that acts on the particles. Chainlike aggregates that are often found in ferrofluids are a good example (Satoh et al., 1996; Zubarev and Iskakova, 2003).

When the walls are magnetized or they have permanent magnetism, the assembly of the particles is strongly affected by the strength and direction of the magnetism. Proper manipulation and selection of the direction and strength can induce the assembly of the particles on the surface of the wall or can induce their repulsive dissipation from the wall. This provides one of the important principles for patterned nanofabrication in the direct (Helseth et al., 2004; Yellen and Friedman, 2004; Helseth et al., 2005) or indirect way (Yellen et al., 2005).

12.2.4. Flow

Figure 12.4 shows the schematic representation of the assembly process under the influence of external flow force. This picture can be applied to both the shear-type and elongational-type of flow profiles. As discussed in Section 5.1, the forces generated by the hydrodynamics around individual colloidal particles are mostly directional. It is usually toward the direction of the flow, but there are many complex flow modes that generate somewhat random forces around the particles. These forces compete and balance with other forces that exist (Vermant and

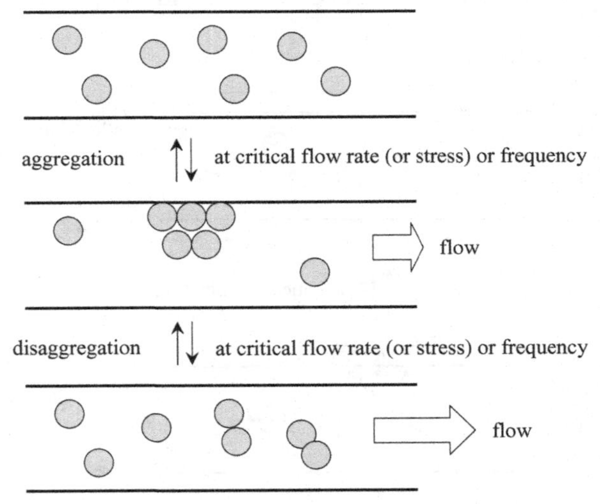

Figure 12.4. Self-assembly of colloidal particles under flow.

Solomon, 2005). Two different types of flow modes can be applied to both types of flow profiles: constant flow or oscillation flow. When the force balance is achieved, the colloidal particles begin to self-assemble into self-assembled aggregates. Since the system starts with the repulsive force dominant condition, this happens when the hydrodynamic force has generated enough attractive forces between the particles. For most systems, there are critical values of the flow force that are needed to achieve this: critical shear flow for constant flow or critical frequency for oscillation flow. Like the situation for the electric field–induced self-assembly that is induced by the electrohydrodynamic force, this usually happens with cooperative action with the walls; thus, the self-assembled aggregates should form on the surface of the walls.

This self-assembled aggregate is subject to the influence of additional flow force. When the shear force is increased beyond the point where the aggregates can maintain their structural integrity, they begin to disaggregate. This also happens at (or near) the critical flow rate or frequency. Additionally generated hydrodynamic force now acts as additional repulsive force. This situation suggests an important application for the construction of colloidal crystals. With preconstructed colloidal crystal (e.g., by evaporation or by electric field–induced assembly, which usually has a significant amount of defect sites and crack sites), when this colloidal crystal is placed under the flow force and the flow applied is at or just below the point where it begins to disaggregate, the flow-induced repulsive force can act to minimize those defects and crack sites by reassembling the particles on those sites. This can help us obtain well-developed larger-area colloidal crystals (Amos et al., 2000).

12.2.5. Mechanical Force

As a part of surface self-assembly, the processes for the formation of Langmuir monolayer and Langmuir-Blodgett (LB) film have been discussed in Chapter 6. The key for this process was the mechanical force that was directly imposed on the individual building unit by using the movable barrier. This picture of self-assembly can be directly applied to the formation of a large-scale colloidal monolayer and, in some cases, multilayer films of nanoparticles or colloidal particles. Figure 12.5 shows the process. As long as the colloidal particles stay afloat on the surface of the liquid, this process should be possible for particles with a wide range of size, surface properties such as charge, and physical properties. However, unlike in cases of amphiphilic molecules, since the weight of the particles can induce the gravity force, modulation of the interaction force of the particles with the surface of the substrate, which can ensure the stability of the monolayer formed, is critical for this process. For light particles, the capillary force should be enough to hold the particles on the substrate. For heavier particles, different chemical strategies can be adopted. For example, either the substrate or the particle surface or both can be prefunctionalized with functional groups to ensure the proper linkage between them. If necessary, the Langmuir-Schafer method (Chapter 6) can be used to minimize the effect of gravity.

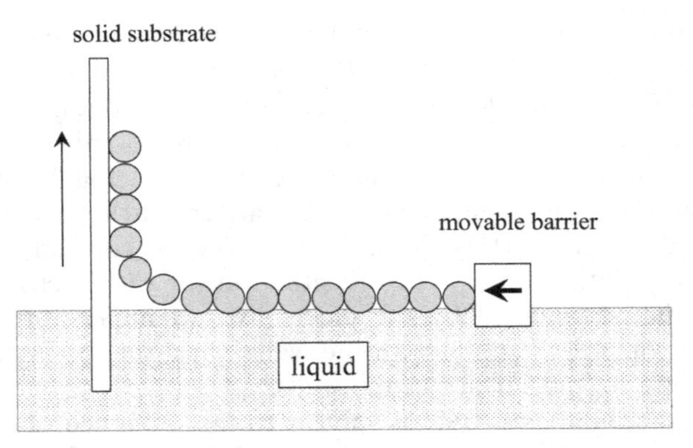

Figure 12.5. Self-assembly of colloidal particles using mechanical force.

The first unique benefit of this approach is that the mechanical force can be controllably manipulated at will with very high precision. Careful manipulation allows preparing a highly controllable monolayer and multilayer. Also, it makes us control the interparticle distance, and provides control of the surface phase structure by the simple control of surface pressure. This is possible even for systems of colloidal particles with strongly charged surfaces, since this mechanical force can be adjusted to easily overcome the electrostatic repulsive force between the particles.

12.2.6. Force by Spatial Confinement

As many of the physical and chemical properties of matter are unexpectedly and dramatically changed by spatial confinement, the self-assembly process and aggregate structures of colloid particles can be also greatly altered when the process is confined within limited geometry (Wang and Möhwald, 2004). When at least two of its three dimensions limit the direction and propagation of the self-assembly physically, chemically, or spatially, it can be considered as a confined or directed self-assembly. When only one dimension is limited, it belongs to the category of surface or interfacial self-assembly, which was discussed in Chapter 6.

Figure 12.6 shows typical examples of colloidal self-assembly within different geometries. The confined space can be a nm–μm size (diameter) channel, the same range of pores, and grooves with different geometries such as V shaped, rectangular shaped, or hemispherical shaped. These can be categorized as two-dimensional confinement. A typical three-dimensional confinement is emulsion droplets, bubbles, or liquid droplets on a solid surface. When colloidal particles are introduced into confined space mainly by capillary force (in some cases of narrow space, electrocapillary force is needed to push the colloidal sols into the space) or by physical mixing such as in emulsion, the movement of the particles toward walls is greatly restricted. The degree of freedom for cases of channel or

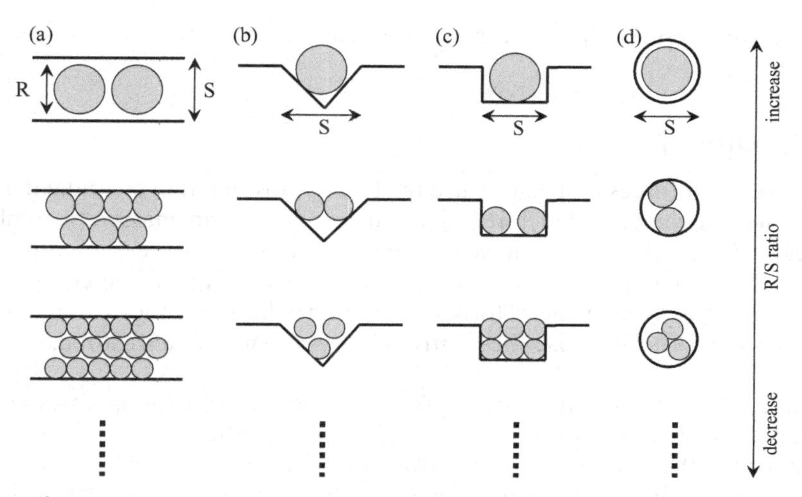

Figure 12.6. Self-assembly of colloidal particles within confined spaces.

pore is only one; while it is zero for emulsion or bubble. This physical confinement primarily makes this self-assembly process directional or confined (directed). Also, the interaction of the particles with the inside of the space becomes significant, and becomes comparable to the interaction between the particles. This interaction is usually unsymmetrical, while the interaction between the particles is symmetrical and acts as additional driving force for this directional or confined self-assembly. In this sense, the geometrical confinement itself is driving and directional forces. For example, colloidal particles with strong electric repulsion force, which have a strongly restricted self-assembly tendency, can be easily self-assembled once they are successfully introduced inside the confined space.

Self-assembly of colloidal particles inside the confined space often provides the facile opportunity for symmetry breaking or even symmetry manipulation. For most self-assembled aggregates of colloidal particles, whether they were obtained without external forces or with external forces, the most stable, in other words, the most common form of colloidal self-assembled aggregates is the face-centered cubic (*fcc*) structure. However, the aggregates obtained inside the confined space often show a variety of different stable geometries. This includes not only close-packed *fcc* structures, but helical, that is, chiral, structures (Yin and Xia, 2003; Li et al., 2005), and many other exotic structures (Manoharan et al., 2003; Schaak et al., 2004) as well. This is one of the most promising advantages of self-assembly inside confined space. Those structures and also the handedness of the helical structure can be easily controlled by the design of the ratio of the particle size (radius for spherical ones, radius/length ratio for rod-shaped ones) (*R* in Figure 12.6) to the size of confined space (pore size, channel radius, size of emulsion droplet, etc.) (*S* in Figure 12.6) and simple control of colloidal forces. This aspect can be understood as packing geometry without the control of the geometry of individual building units (with the design of geometry, see Section

5.4). Relatively simple physical restriction can easily force a different packing geometry.

12.2.7. Other Forces

There are other forces that can induce the forced movement of colloidal particles. While the abovementioned forces have direct control in massive assembly of particles, these other forces have somewhat limited ability to move particles. However, they have important roles and meanings as part of the colloidal self-assembly process, since these forces can be an additional tool to increase the efficiency and/or to increase the controllability of particle fabrication.

12.2.7.1. Laser-Optical Force. When dielectric colloidal particles with the size range of ~10 nm–~10 μm have enough contrast difference to media such as solvent or air, they can be trapped, moved, and even guided by a laser beam. This is the direct result of the radiation pressure that acts on the particles. This force has magnitude roughly comparable to gravity force. By designing the direction, strength, and intensity gradient of the laser beam, this concept can be rendered more sophisticated, as in *optical tweezers*. Small atoms and molecules and biological objects such as cells, proteins, and viruses can also be manipulated with this technique (Ashkin, 1997).

Laser-optical force–based methods have not evolved to the degree of larger-scale assembly of particles. But, with proper combination with other abovementioned methods, this approach can be helpful to increase the quality of self-assembled aggregates. It can also be a useful concept to manipulate colloidal particles one by one.

12.2.7.2. Ultrasound. Ultrasound itself does not provide any degree of directional force on self-assembly building units; rather it can help agitate the entire system with force comparable to the abovementioned forces. But this can be a useful force to increase the overall quality of preassembled aggregates. For example, the crystallinity of colloidal crystal can be significantly improved by proper imposition of ultrasound agitation. This is done by increasing the possibility of more homogeneous interparticle interaction and by decreasing the possible defective or mismatched sites. By agitation of the force balance, thus decreasing the defaults and/or finding and correcting the hindered points on force balance, this concept can find useful applications in the area of nanofabrication of colloidal particles (Jung and Livermore, 2005).

12.2.7.3. Gravity and Centrifugal Forces. Sedimentation induced by gravity can sometimes be the best way to obtain well-developed colloidal crystal, especially of medium-charged, somewhat heavy particles. Not imposing any other external forces or minimizing their interferences can often yield a minimum number of defect sites. An example can be found in the construction of chromatography columns with colloidal packing materials.

Mass inertia force can be generated by centrifugal force, which can be comparable to or often overcome colloidal and intermolecular forces. This can provide an opportunity to assemble those building units whose repulsive forces between them are dominant. It can also be applied, as in the case of ultrasound, to increase the quality of self-assembled aggregates by reducing the defect sites.

12.3. MOLECULAR SELF-ASSEMBLY UNDER EXTERNAL FORCES

This section will describe molecular self-assembly under the influence of external forces, mainly surfactant micellization under external forces and the alignment of liquid crystals by those forces. For the colloidal self-assembly mentioned in the previous section, external forces directly act as major driving and directional forces, and in some cases as opposition forces. This is largely because the strength of those forces on the particle is comparable to intermolecular and colloidal forces. However, for the molecular level of self-assembly building units, the magnitude of the external forces acting on individual molecules is orders of magnitude smaller than those of intermolecular forces. Thus, for the typical molecular self-assembly such as micellization, these external forces do not induce significant influence on the primary self-assembly process of the surfactant molecules. But, once the micelles have formed, the size of this primary self-aggregate is, in most cases, large enough so that the external forces become comparable to the intermolecular forces. These forces now can significantly alter the size (length of rod- or wormlike micelles) and orientation of the micelles. This fact can be understood as external force–induced secondary self-assembly of surfactant molecules.

For liquid crystals and certain types of polymers, the primary building units themselves are big enough to be affected by external forces. While their sizes are not changed by these forces, their orientation can be controlled to some degree with these forces. The external force is the directional force for the primary self-assembly of these building units.

External force–induced processes are largely reversible. Thus, once the force is removed, the assembled systems tend to go back to their original state. Also, it usually is accompanied by critical values. But often these critical values are not well-defined. Also, unlike colloidal self-assembly, the electric field does not have much influence on molecular self-assembly or its aggregate systems.

12.3.1. Flow

Both types of flows (Figure 5.10a) and both of their modes (constant or oscillation) can have significant impact on most molecular self-assembled aggregates of surfactant, polymers, and lipids. This includes spherical micelles, rodlike and wormlike micelles, and even liquid crystals. Generally, this phenomenon is called *flow-induced phase transition* of micellar solution. In the sense of its primary building units, this is molecular self-assembly. But, given the fact that it is the colloidal-size micelles that are subject to change, it can be considered as colloidal self-assembly, too. No flow-induced phase transition occurs below the *cmc* of any of those primary building units.

Figure 12.7 shows the schematic representation. This can be the alignment of micelles, such as rodlike micelles, along the direction of the flow. It can be the deformation of micelles, such as stretching of wormlike micelles, and their alignment along the direction of the flow. It also can be the case for some degree of structural change such as the growth of spherical micelles into rodlike micelles and alignment. When the liquid crystals are placed under the flow, deformation and/or rearrangement of their structures often occur. For example, the hexagonal structure can be better aligned through the better alignment of its building units (rod- or wormlike micelles). This can eventually offer a hexagonal liquid crystal with better crystallinity. The alignment is always along the direction of the flow. Also, the lamellar structure can be deformed into a better-aligned one, and the deformation of cubic phase into hexagonal or lamellar structure is common. For most cases, these changes occur at the critical values (or narrow range) of the flow. For the constant flow, this value can be a critical shear rate or shear stress. For the oscillation type of flow, this is a critical frequency. In some conditions, the tumbling, not the alignment, also can happen along the direction of the flow, with the long axis rolling along the flow but the short axis perpendicular to the flow direction.

An application example can be found for synthesis of mesoporous materials through flow-induced better-aligned surfactant micelles. Better-aligned micelles mean a better overall crystallinity of the micelles, which can provide the chance for a better-structured mesoporous material. Also, the change in optical, magnetic, and electric properties of liquid crystals under flow has many useful potentials applications. Another important application is mass fluid transport with less energy. For example, the transport of crude oil through long pipelines inevitably requires the massive buildup of internal pressure to ensure the proper flow. This requires regular pumping throughout the long pipeline. A small amount of "drag reduction agents" (which increase the fluidity of the fluid by their alignment along the direction of flow) can reduce this pressure dramatically, which makes this process economically sound. Often polymers with chainlike structures are used. But the problem is that these polymers are cut when they flow through the propeller of pumping site. This requires constant re-addition of polymers through-

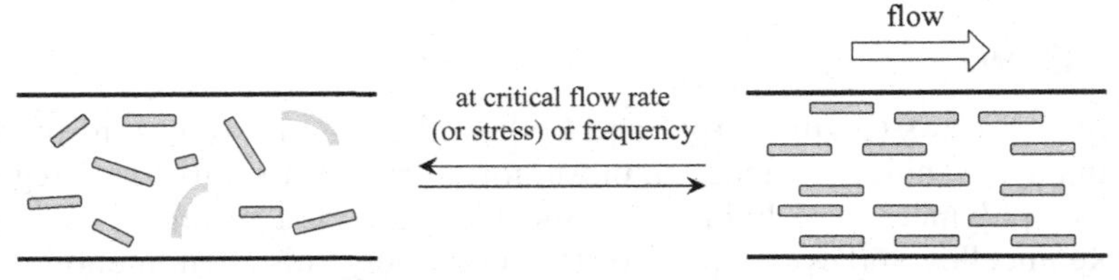

Figure 12.7. Flow-induced phase (or structural) transition of micellar solutions.

out the whole pipeline. If surfactant micelles (rodlike or wormlike) were replaced with polymers, the micelles would be simply reformed (re-self-assembled) just after they are broken down by the propeller; thus, the surfactants need not be re-added. This is why surfactant drag reduction agents are often called *living polymers*. The flow systems of heating and cooling components can be equipped in the same way to save energy (Zakin et al., 1998).

12.3.2. Magnetic Field

The origin of magnetic force–induced changes in self-assembled systems is the anisotropy of the magnetic susceptibility ($\Delta\chi$) that can appear in many liquid crystalline states of surfactants, lipids, and polymers. Many bilayer types of self-assembled systems also show this property. When the value of $\Delta\chi$ is negative, the states have a tendency to be oriented along the long axes (or phase director) perpendicular to the direction of the magnetic field (2 in Figure 12.8). The states with positive value of $\Delta\chi$ have a tendency to be oriented parallel to the direction of the magnetic field (1 in Figure 12.8) (Jansson et al., 1990; Cho et al., 2001). As with the electric field effect, the force induced by the external magnetic field is not strong enough compared with the flow-induced force. Thus, its effect is somewhat limited to the change of orientation of these liquid crystals or liquid crystal-like structures. Sometimes, it can have an effect on the structural deformation of primary self-assembled aggregates such as liposomes to a somewhat limited extent.

This magnetic field–induced force can be understood as the directional force that is applied after the primary force balance is set; that is, the self-assembled system is formed. This force does not operate at any significant level during the self-assembly process itself. But it does have the ability to perturb the force balance and help induce a better or new alignment of self-assembled aggregates.

12.3.3. Concentration Gradient

When micellar solutions make contact with solid substrates, the edge of the three-phase contact line always has a gradual concentration gradient with increased concentration toward the front of the edge. This front line can be

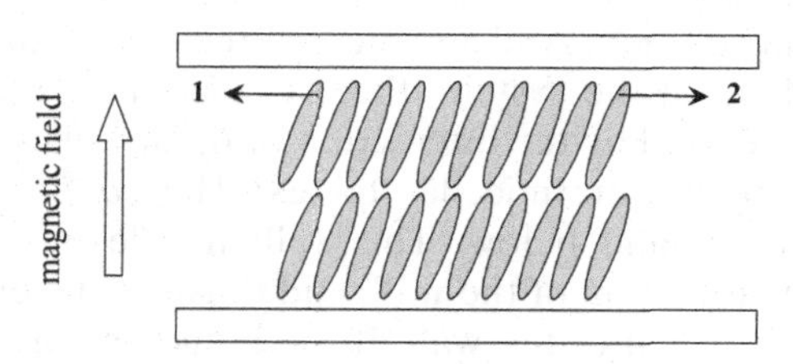

Figure 12.8. Effect of external magnetic field on self-assembled liquid crystal systems.

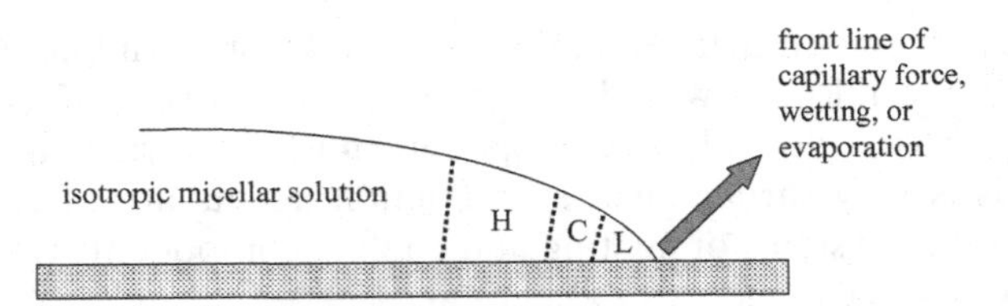

Figure 12.9. Concentration gradient–induced molecular self-assembly induced by capillary force, wetting, or evaporation. H, C, and L represent hexagonal, cubic, and lamellar structure, respectively.

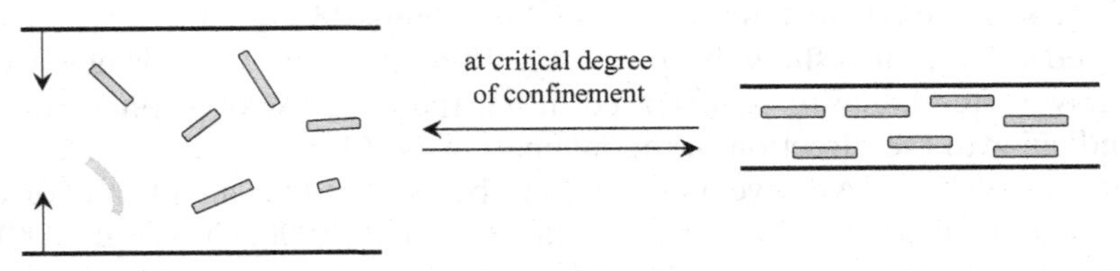

Figure 12.10. Molecular self-assembly within confined space.

induced either by capillary force, by wetting, or by selective local evaporation. Therefore, in getting closer toward the edge, the micelle solution experiences a gradual increase in the intermicellar interaction because of the increased concentration. This should induce the change of the micellar solution into liquid crystal phases, most probably as a sequence of hexagonal, cubic, and lamellar structures (Figure 12.9). Details of the formation of liquid crystal phases in surfactant self-assembly have been presented in Chapter 4. A manipulated but well-controlled concentration gradient can induce a change in the force balance of micelles. For the hexagonal, its structure tends to be aligned along its long axis parallel to the direction of the front line, and for the lamellar, its layers tends to be aligned along the solution–solid interface. The induced three-phase front line is a directional force for this type of self-assembly.

12.3.4. Confinement

As shown in Figure 12.10, let us assume that there is a given space that contains the isotropic micellar solution. And assume that the size of that space is gradually decreased. Depending on the system, there can be a critical size of the space where the forces generated by the interaction of the micelles with the wall begin to be comparable to the intermicellar forces. The *confinement effect* can be defined from this point and below. The wall–micelle interaction is naturally directional along the direction of the wall, and thus has the effect of aligning the micelles along the direction of the wall. If the liquid crystals are confined, both their structures and lattice parameters will be significantly affected.

12.3.5. Gravity and Centrifugal Forces

Some biological self-assembly systems, such as for assembly of microtubules, show a very interesting dependency on gravity. It seems that the presence of gravity is crucial for this self-assembly process and the direction of gravity is critical for the determination of its morphology. In the sense of force balance, gravity here can be considered as both the driving force and directional force (Tabony et al., 2002). It will be reasonable to anticipate that some molecular self-assembly systems of this type can show dependency on centrifugal force in such a way that it contributes to the force balance and possibly directs the self-assembled systems.

12.4. APPLICATIONS OF COLLOIDAL AGGREGATES

As did Chapters 9, 10, and 11, this chapter will address *lower-order applications* only. Higher-order ones will be introduced in Chapter 14. One main focus will be on how the force balance concept can help improve applications. In practice, many factors (external forces) are employed in a cooperative way, which can maximize the properties of the aggregates that are required for the given or intended applications.

12.4.1. Optical Band Gap

This is the most-studied system among the self-assembled aggregates of colloidal particles, and has the greatest potential future application (Xia et al., 2000). The size of colloidal particles is comparable to the wavelength of light. This provides the very important fact that light can strongly interact with them when they are dielectric. This is like the case where the current within well-organized semiconductor materials can be controlled through their interaction with electrons. Spatially organized or structured colloidal aggregates (colloidal crystals) of dielectric particles can confine and control light through their influence on photons. Thus, there exists a band gap that blocks light of a given wavelength. Figure 12.11 shows the schematic explanation. The terms *photonic band gap* or *photonic crystal* are

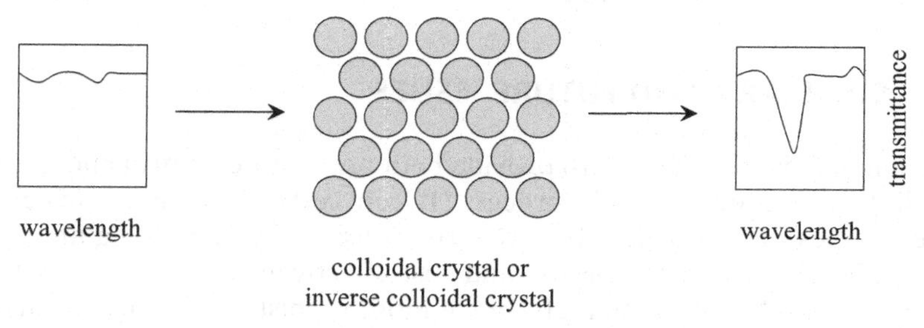

Figure 12.11. Photonic band gap.

derived from this phenomenon. This is a totally novel way to express "color" compared with traditional dye molecules. While dye molecules express their color by *absorbing* a certain wavelength of light, photonic band-gap crystals do it through the *blocking* of light. The change in the color of dye molecules is possible by manipulation of the functional group or by the change in the molecular structure itself. For photonic band-gap crystals, this can be accomplished by the change in their structural factors. This includes not only particle factors such as the size and shape of particles, but more importantly organization factors, including their entire morphology, size of lattice constant, and interparticle distance as well. This is the key fact in the realization of microphotonics through the control of light, just like the currently booming area of microelectronics using the control of current.

The importance of self-assembly and the introduction of external forces to achieve the better control of colloidal self-assembly lies here. It provides better routes, compared with the top-down-like carving procedure, to create colloidal crystals, and a facile way to assemble particles into larger-scale crystals with a high degree of control. Controlled inverse colloidal crystals also show the photonic band gap, with some additional advantages such as concurrent use of their large pore and large surface area. Their synthesis and relation with molecular self-assembly were introduced in Chapter 9.

12.4.2. Nanostructured Materials

As described in Chapter 9, the synthesis of nanostructured materials largely starts from the pre-self-assembled "template" or cooperative self-assembly of the precursor with the molecular self-assembly building units such as surfactants or polymers. External forces that can provide additional degrees of control over their structure, crystallinity, and even self-assembly process can ensure a better outcome in the quality of the nanostructured materials. Examples include the synthesis of mesoporous silica under magnetic field (Tolbert et al., 1997), the control of morphology of mesoporous silica by flow, and evaporation-induced self-assembly for the synthesis of functional mesostructured silica. Preparation of aligned nanostructured films between two narrowly sandwiched solid surfaces or at liquid–liquid interfaces is another excellent example. The effect of this spatial confinement can propagate up to thousands of layers.

12.5. SUMMARY AND FUTURE ISSUES

External field is a useful tool to provide better and more control and good directionality in the self-assembly process. It can certainly help us achieve some current issues such as mass fabrication (mass assembly of building units on macroscopic scale), structural control, and cost-effectiveness.

However, since we ultimately aim at more sophisticated applications such as microelectronics, whether it will be with a purely bottom-up approach or hybrid approaches, there remain great challenges ahead (Parviz et al., 2003; Arsenault et al., 2004). The key issue, of course, is better or more precise controllability of the process. For example, structural defects need to be eliminated or at least kept to a minimum for some applications so that they can be tolerated by the system operation. But other applications need to be created with a high degree of controllability. Also, there is need for more control of the entire morphology, the symmetry, and more precise control of structural parameters such as interparticle distance. The following are some future issues:

1. Diversity of Building Units. The primary source of the novel properties of colloidal self-assembled aggregates is their primary building units. A variety of colloidal building units with diverse shapes, components, and compositions have been developed. However, research efforts have been largely focused on spherical-shaped, single-component particles. Though knowledge from this certainly will provide the foundation of self-assembly of those other building units, systematic research on this issue would greatly help uncover more interesting new properties, which might provide additional usefulness in existing applications or even lead to novel applications.

2. Cooperative or Sequential Processes. Each of the external forces introduced in this chapter has its own advantages and disadvantages. This suggests that the cooperative and/or sequential way of applying external force is necessary for some self-assembly systems. It will, of course, have to be designed to maximize its efficiency. For example, ultrasound agitation during self-assembly under magnetic or electric field can help us minimize default points, and thus minimize defects. This approach might make the fabrication setup unnecessarily complicated, but a simple engineering solution will help overcome this in many cases.

3. External Field–Property Relationship. The importance of the structure–property relationship of self-assembled systems in the development of their practical applications has been repeatedly described in previous chapters. For colloidal self-assembly, this becomes the relationship of the physical parameters of the external fields with the properties of colloidal self-assembled aggregates. A well-set-up relationship should not only help clarify the underlying physical mechanism but streamline the application trials as well.

4. Heterogeneous Self-Assembly. This implies colloidal self-assembly not only between particles with different structures and compositions but between different properties. Examples include metal–polymer, inorganic–organic, metal–biological, and so on. This approach aims, of course, at the preparation of "hybrid" colloidal self-assembled aggregates, which could hybrid the useful properties as well. This might be critical to expanding the reality of applications using colloidal aggregates and eventually the reality of functional devices. It will certainly help create more diverse pools of structural varieties for these applications.

REFERENCES

Amos, R. M., Rarity, J. G., Tapster, P. R., Shepherd, T. J., Kitson, S. C. "Fabrication of Large-Area Face-Centered-Cubic Hard-Sphere Colloidal Crystals by Shear Alignment," *Phys. Rev. E* **61**, 2929 (2000).

Arsenault, A., Fournier-Bidoz, S., Hatton, B., Míguez, H., Tétreault, N., Vekris, E., Wong, S., Yang, S. M., Kitaev, V., Ozin, G. A. "Towards the Synthetic All-Optical Computer: Science Fiction or Reality?" *J. Mater. Chem.* **14**, 781 (2004).

Ashkin, A. "Optical Trapping and Manipulation of Neutral Particles Using Lasers," *Proc. Natl. Acad. Sci. USA* **94**, 4853 (1997).

Bowden, N. B., Weck, M., Choi, I. S., Whitesides, G. M. "Molecule-Mimetic Chemistry and Mesoscale Self-Assembly," *Acc. Chem. Res.* **34**, 231 (2001).

Cho, G., Fung, B. M., Reddy, V. B. "Phospholipid Bicelles with Positive Anisotropy of the Magnetic Susceptibility," *J. Am. Chem. Soc.* **123**, 1537 (2001).

Gong, T., Wu, D. T., Marr, D. W. M. "Two-Dimensional Electrohydrodynamically Induced Colloidal Phases," *Langmuir* **18**, 10064 (2002).

Helseth, L. E., Backus, T., Johansen, T. H., Fischer, T. M. "Colloidal Crystallization and Transport in Stripes and Mazes," *Langmuir* **21**, 7518 (2005).

Helseth, L. E., Wen, H. Z., Hansen, R. W., Johansen, T. H., Heinig, P., Fischer, T. M. "Assembling and Manipulating Two-Dimensional Colloidal Crystals with Movable Nanomagnets," *Langmuir* **20**, 7323 (2004).

Jansson, M., Thurmond, R. L., Trouard, T. P., Brown, M. F. "Magnetic Alignment and Orientational Order of Dipalmitoyl Phosphatidyl Choline Bilayers Containing Palmitoyl Lysophosphatidyl Choline," *Chem. Phys. Lipids* **54**, 157 (1990).

Jung, S., Livermore, C. "Achieving Selective Assembly with Template Topography and Ultrasonically Induced Fluid Forces," *Nano Lett.* **5**, 2188 (2005).

Kralchevsky, P. A., Denkov, N. D. "Capillary Forces and Structuring in Layers of Colloid Particles," *Curr. Opin. Coll. Inter. Sci.* **6**, 383 (2001).

Kralchevsky, P. A., Nagayama, K. *Particles At Fluid Interfaces and Membranes: Attachment of Colloid Particles and Proteins to Interfaces and Formation of Two-Dimensional Arrays* (Elsevier: 2001).

Li, F., Badel, X., Linnros, J., Wiley, J. B. "Fabrication of Colloidal Crystals with Tubular-like Packings," *J. Am. Chem. Soc.* **127**, 3268 (2005).

Löwen, H. "Colloidal Soft Matter Under External Control," *J. Phys.: Condens. Matter* **13**, R415 (2001).

Manoharan, V. N., Elsesser, M. T., Pine, D. J. "Dense Packing and Symmetry in Small Clusters of Microspheres," *Science* **301**, 483 (2003).

Negi, A. S., Sengupta, K., Sood, A. K. "Frequency-Dependent Shape Changes of Colloidal Clusters under Transverse Electric Field," *Langmuir* **21**, 11623 (2005).

Parviz, B. A., Member IEEE, Ryan, D., Whitesides, G. M. "Using Self-Assembly for the Fabrication of Nano-Scale Electronic and Photonic Devices," *IEEE Trans. Adv. Pack.* **26**, 233 (2003).

Satoh, A., Chantrell, R. W., Kamiyama, S.-I., Coverdale, G. N. "Three-Dimensional Monte Carlo Simulations of Thick Chainlike Clusters Composed of Ferromagnetic Fine Particles," *J. Colloid Interface Sci.* **181**, 422 (1996).

Schaak, R. E., Cable, R. E., Leonard, B. M., Norris, B. C. "Colloidal Crystal Microarrays and Two-Dimensional Superstructures: A Versatile Approach for Patterned Surface Assembly," *Langmuir* **20**, 7293 (2004).

Tabony, J., Glade, N., Papaseit, C., Demongeot, J. "Microtubule Self-organization and its Gravity Dependence," *Advances in Space Biology and Medicine*, 8 (Cell Biology and Biotechnology in Space), Cogoli, A., eds., pp. 19–58 (Elsevier: 2002).

Tolbert, S. H., Firouzi, A., Stucky, G. D., Chmelka, B. F. "Magnetic Field Alignment of Ordered Silicate-Surfactant Composites and Mesoporous Silica," *Science* **278**, 264 (1997).

Van Blaaderen, A. "Colloids Under External Control," *MRS Bull.* **29**, 85 (2004).

Vermant, J., Solomon, M. J. "Flow-Induced Structure in Colloidal Suspensions," *J. Phys.: Condens. Matter* **17**, R187 (2005).

Wang, D., Möhwald, H. "Template-Directed Colloidal Self-Assembly, The Route to 'Top-down' Nanochemical Engineering," *J. Mater. Chem.* **14**, 459 (2004).

Xia, Y., Gates, B., Yin, Y., Lu, Y. "Monodispersed Colloidal Spheres: Old Materials with New Applications," *Adv. Mater.* **12**, 693 (2000).

Yellen, B. B., Friedman, G. "Programmable Assembly of Colloidal Particles Using Magnetic Microwell Templates," *Langmuir* **20**, 2553 (2004).

Yellen, B. B., Hovorka, O., Friedman, G. "Arranging Matter by Magnetic Nanoparticle Assemblers," *Proc. Natl. Acad. Sci. USA* **102**, 8860 (2005).

Yin, Y., Xia, Y. "Self-Assembly of Spherical Colloids into Helical Chains with Well-Controlled Handedness," *J. Am. Chem. Soc.* **125**, 2048 (2003).

Zakin, J. L., Lu, B., Bewersdorff, H.-W. "Surfactant Drag Reduction," *Rev. Chem. Engin.* **14**, 253 (1998).

Zhang, K. Q., Liu, X. Y. "In-situ Observation of Colloidal Monolayer Nucleation Driven by an Alternating Electric Field," *Nature* **429**, 739 (2004).

Zubarev, A. Y., Iskakova, L. Y. "Structural Transformations in Ferrofluids," *Phys. Rev. E* **68**, 061203 (2003).

13

NANOFABRICATION

There is an old Asian proverb, "Even though you have 3 *mals* (*mal* is one of the traditional units that is used to measure the volume mainly of grains, and equals approximately 4.8 *gal*) of gemstones, they are not jewelry unless they are fabricated." With a simple analogy, we could say, "All those marvelous nanostructured objects will not become valuable until they are fabricated into something useful to us." This is the topic of this chapter.

Nanofabrication can be defined as "a process that is necessary to fabricate nanostructured objects." It should also include a process that is necessary to fabricate bigger-than-nanoscale objects as long as it performs within the nanometer operating range. This is a parallel term with the well-established *microfabrication*, which represents a process to fabricate microscale objects or fabrication within the micrometer operating range. For the past few decades, microfabrication has well been developed as a major technique, especially in the semiconductor industry. This is a *top-down* approach. However, as this development keeps pushing down its scale limit into the nanometer region, the top-down techniques inevitably face their sacrifice of *throughout* efficiency; the *throughout-resolution dilemma* (Marrian and Tennant, 2003). A radical new bottom-up approach has been envisioned to overcome this hurdle, and it is rapidly

Self-Assembly and Nanotechnology: A Force Balance Approach, by Yoon S. Lee

positioning itself as a key trend (Whitesides et al., 1991; Robinson, 2003; Whitesides, 2003).

The *bottom-up* approach is based on the self-assembly of nanostructured objects. Instead of carving out the nanoscale features from the bulk, the fabricated system is designed to be assembled through the force interactions between its building units. The preparation of nanostructured objects has been described so far in this part of the text, which includes nanoparticles, nanostructured materials, and nanostructured films. They make up the essential parts of the nanofabrication building units. Nanofabrication with the aspect of self-assembly is all about the assembling of these varieties of building units with a satisfactory degree of control. They could be assembled homogeneously to bring out the physical/chemical properties we desire. There could be a requirement that they be assembled through heterogeneous processes, between the different sizes and shapes of the building units, or through hybrid processes, between the different natures of building units. These issues will be the main focus of this chapter. Also, continuing through the next chapter, some aspects of nanofabrication will be correlated with the issue of nanodevices.

The top-down side of nanofabrication, which includes different types of lithography, has its own hurdles, such as high cost, expensive equipment needs, and multicomplex process. But it has its own unique advantages, too; precise operation and high resolution. The details of this topic will not be discussed here. An excellent book is available for further study (Cui, 2005). However, its cooperative or synergistic relationship with the self-assembly approach will be discussed throughout the chapter.

13.1. SELF-ASSEMBLY AND NANOFABRICATION

Figure 13.1 represents the schematic view of the relationship of self-assembly with nanofabrication. It also encompasses its relationship with nanodevices and nanomachines.

Building units for nanofabrication (designated here as "fabrication building unit") are those objects that can be embraced by the definition of nanostructured materials (see Chapter 8). In some cases, those self-assembly building units that are discussed in Part I of this book become the fabrication building units without any treatment. However, in many cases, it is necessary to modify those self-assembly building units to be able to apply them as fabrication building units. It is mainly to provide them with a proper processing ability through the change of interaction between them, or to expect or control the final properties once they are fabricated. This is achieved mainly by self-assembly. Some typical examples include metal nanoclusters whose surfaces are modified with self-assembled monolayers, carbon nanotubes whose surfaces are equipped with specific ligands, solid substrates whose surfaces are decorated with surface micelles, and lipid vesicles decorated with poly(ethylene glycol) chains.

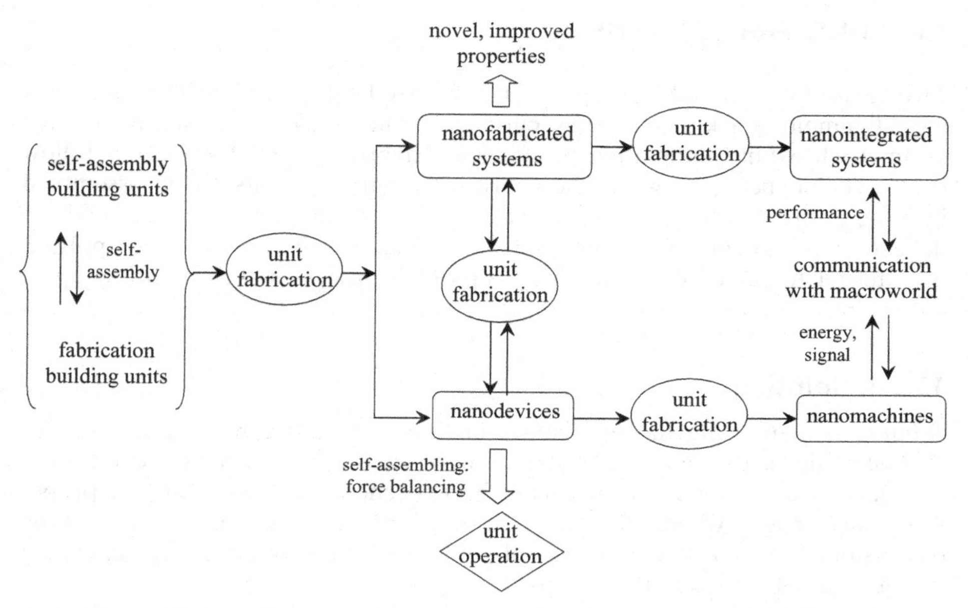

Figure 13.1. Self-assembly, nanofabrication, nanodevices, and nanomachines.

Fabrication building units (also self-assembly building units) are fabricated through a variety of fabrication processes. When the fabricated systems show novel phenomena and/or improved properties that are different from their bulk counterparts of individual components, they can be designated as *useful nanofabricated systems*. When they show a capability for unit operations (through force balancing; details in Chapter 14), they can be designated as *nanodevice*.

Fabrication for nanofabricated systems and nanodevices can be tricky and complicated. It might be acquired with just one step of the process. It also might take a series of multisteps. However, I propose here that the majority of those fabrication processes can be broken into a small number of fundamental processes that commonly appear throughout different nanofabrication processes. These will be designated as *unit fabrications* (details in the next section). Nanofabrication is performed through a unit fabrication or a series or combination of unit fabrications. In some cases, nanofabricated systems can be transformed into nanodevices, and vice versa. Unit fabrication is also crucial for these processes. To be able to express their potentialities into actual application systems, nanofabricated systems, for the most part, need to be fabricated further. This is also performed by unit fabrication. These fabricated systems will be designated as *nanointegrated systems* here. They are the ones that have the ability to communicate with the macroworld through their physical, chemical, or mechanical performances. The detailed scheme for the fabrication of nanomachines will be presented in Chapter 14.

13.2. UNIT FABRICATIONS

Twelve different unit fabrications will be discussed in this section. They are jointing, alignment, stacking, crossing, curving, reconstruction, deposition, coating, symmetry breaking, templating, masking, and hybridization. Most of them follow the force balancing between the fabrication building units to be assembled into the desired structures without external forces. But some of them need the delicate external control of force balancing. This usually requires the applying of external forces or pre- (or post-) top-down processes such as lithographic patterning.

13.2.1. Jointing

Jointing is connecting the fabrication building units. It can be homogeneous (between the same building units) or heterogeneous (between the different units). This unit fabrication requires the existence (or attachment) of proper functional groups where the actual connection occurs. It can be linear (one dimensional), planar (two dimensional), or bulky (three dimensional), depending on the number of functional groups.

Three basic types of jointing are possible: direct jointing, jointing through connector, and jointing through specific recognition. Figure 13.2 shows the schematic representation. A typical example of direct jointing is jointing through hydrogen bonding or electrostatic attraction between the functional groups. When the fabrication building units have functional groups whose force status is similar (e.g., both groups are hydrogen bonding donor, both are hydrogen bonding acceptor, or both have the same charge), the jointing can be induced by using a proper connector that could mediate the force status in between the fabrication building units. The third type is a connection through functional groups with recognition capability. The lock-and-key type of self-assembly mode that is abundant in bio-mimetic systems can be useful for this particular fabrication action (Mann et al., 2000; McNally et al., 2003). DNA complementarities and biotin-avidin complexation are excellent examples. The geometrical recognition that can be obtained from some types of self-assembly building units such

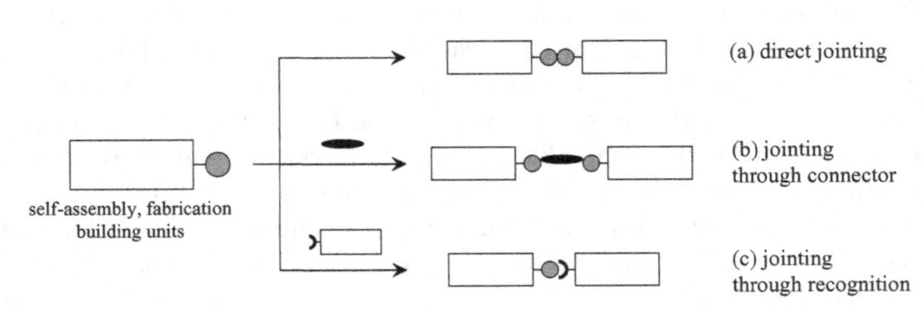

Figure 13.2. Unit fabrication: jointing.

as calixarene- and cyclodextrin-based derivatives can serve as an efficient recognitive functional group as well (Asfari et al., 2001).

Jointing is one of the most frequently appearing unit fabrications. Its proper design can be key for the nanofabrication processes such as linear-connecting, surface-patterning, mass-assembling, and networking.

13.2.2. Crossing and Curving

Crossing and curving are used to create junctions between the fabrication building units. Usually, these unit fabrications are with building units that have high length-to-diameter ratio, such as nanowire, nanotube, or nanorod. Most of the cases are two-dimensional surface fabrication, but three-dimensional fabrication is also possible through combination with other unit fabrications. Figure 13.3 shows the schematic representation.

The most representative need for crossing will be found in the area of nanoelectronics, where the development of precise networking of those high length-to-diameter ratio components is important. Directional assembling of the fabrication building units is inevitably critical. This often requires physically guided assembly with external forces such as flow or capillary force (Messer et al., 2000). It also requires precise control of the direction of the guided assembly itself to ensure the right position, angle, and number of junctions. Often, this can be practically performed through merging with top-down techniques (Whang et al., 2003).

Curving becomes important when there is a need for nonlinear networking or to have a junction between nearby crossed fabrication units such as edges. This fabrication also can be practically performed through directional

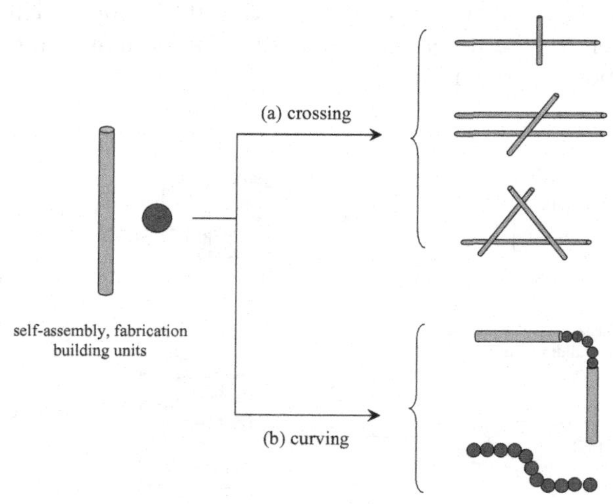

Figure 13.3. Unit fabrication: crossing and curving.

assembling, which might need successive operations with continuous variation of assembling direction. Direct writing using dip-pen lithography (Salaita et al., 2006) will be another effective method for this.

13.2.3. Alignment and Stacking

Alignment and stacking are used to assemble large amounts of fabrication building units with controllable direction and morphology. They are to have all dimensions of fabrications. Alignment is for assembling along the direction of the short axis of fabrication units; thus, it is a side-by-side mode. Stacking usually occurs along the larger area of the units; thus, it is a face-by-face mode. Figure 13.4 shows the schematic representation.

The symmetric packing characteristic of molecular self-assembly is very useful for this mass-assembling. For example, the fabrication of multilayer films of polyelectrolytes can be done with the precision of a single layer through layer-by-layer self-assembly (Chapter 11). This fact is critical not only for the strength and stability of the nanofabricated systems, but for their property control as well. It can provide useful means to control the electron or charge transfer, charge density, and diffusion.

When alignment or stacking are required for the bigger sizes of fabrication building units, such as nanoparticle, nanorod, and nanotube, the principles of colloidal self-assembly (Chapter 5) can be applied. It may be necessary to perform these unit fabrications using building units that have repulsive force-dominant force status. External force can be properly applied as an attractive force to overcome this repulsive one. Alignment or stacking itself will be mainly determined by the geometric factors of the building units. But this external force can also serve as the directional force to control the whole direction of the alignment or stacking. Assembly of nanorods by external mechanical force (Langmuir-Blodgett technique) is a typical example (Kim et al., 2001). This concept also should work well for the mass-assembling of common nanocomponents such as carbon nanotubes and fullerenes.

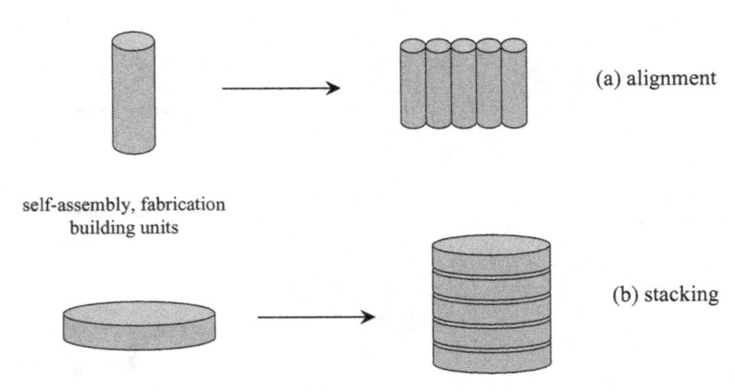

Figure 13.4. Unit fabrication: alignment and stacking.

Alignment and stacking are useful unit fabrications for the proper manipulation and/or expression of the collective properties of the fabrication building units within nanofabricated systems. For example, the surface plasmon can be greatly changed by the mode and direction of the assembling of the building units. Alignment and stacking are also useful to obtain mass-assembled fabrication building units, which could well serve as the initial foundation for the next unit fabrications. Well-assembled, large-area semiconductive nanostructures could become an excellent "bread-board" to set up the circuits for nanoelectronics.

13.2.4. Reconstruction, Deposition, and Coating

Reconstruction, *deposition*, and *coating* are the three main unit fabrications for patterning at solid surfaces. They are rarely applied for three-dimensional fabrication except in some cases of reconstruction. Figure 13.5 shows the schematic representation.

Reconstruction is done to create patterned solid surfaces by perturbation–reorganization process. For certain solid surfaces, an input of external energy can induce the shifting of their force balance at the surface or near the subsurface. This can trigger the rebalancing of the force balance, which can result in the creation of patterning. Annealing through thermal energy is the most common example. But any other energy input can be applied as long as it does not cause physical or chemical damage on the surface. Reconstruction usually works well with multicomponent solid surfaces such as semiconductors and alloys.

Deposition is for creating a regular pattern or film on a solid surface using decomposable precursor. Some details were described in Chapter 11. Unlike

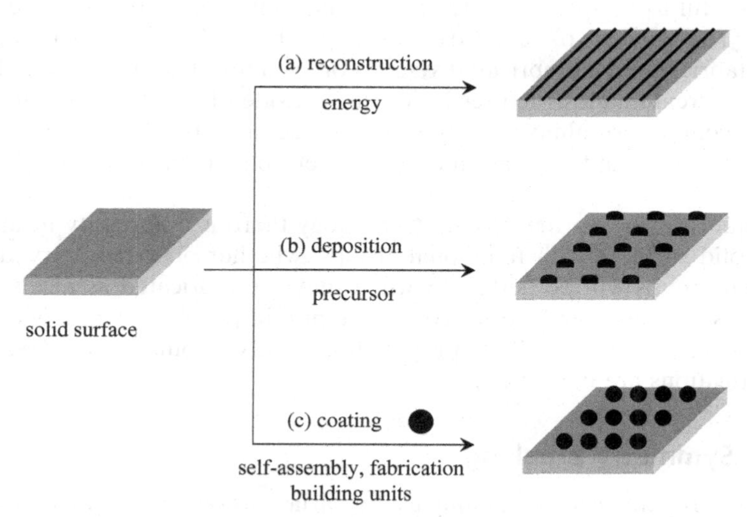

Figure 13.5. Unit fabrication: reconstruction, deposition, and coating.

reconstruction, it is the precursor that is to be patterned or filmed; thus, in many cases it needs prepatterning of the target surface through top-down techniques. This unit fabrication has the most diverse choices of patterning materials and the highest resolution. It also have the highest requirement to be merged with top-down techniques. Many lithographic techniques can be viewed as hybrid approaches of fabrication, and not as purely top-down. For example, a self-assembled monolayer might need to be preformed for the lithographic patterning that follows, or a self-assembled sacrificial layer such as colloidal crystal might be needed for multiscale patterning.

Coating is also done to create a regular pattern or film on solid surfaces. But the difference from deposition is that coating is performed through fabrication (or self-assembly) building units that are already preassembled. Unlike the precursor, these preassembled building units are not subject to decomposition. Their delivery to the surface can be carried out in diverse ways. They can be adsorbed from the solution, exposed from gas phase, or even delivered from another solid surface. Once on the surface, they undergo force balancing between them, and the patterns or films are formed as the result of this force balancing. Interaction with the surface will be determined by their nature, but usually it is limited on the first layer. A good example is the formation of amphiphile surface micelles or of the regular nanostructure of organic molecules adsorbed on solid surfaces. Fabrication of film-based devices through Langmuir and Langmuir-Blodgett techniques (Chapter 11) are another typical example.

These surface-patterning unit fabrications are extremely useful for mass-patterning or mass-layering on solid surfaces. So, the nanofabricated systems themselves often become nanointegrated systems without any additional unit fabrications. They can be bases for other unit fabrications such as for bread-boards for nanoelectronics and many types of sensoring devices. They can also be useful as templates for templating unit fabrication (details later in this section). The concept of reconstruction might be useful for the fabrication of already-fabricated nanofabricated systems or even for bulk materials that can be induced to create nanostructures inside and outside of the surfaces. But deposition and coating are almost solely solid surface actions. Deposition is also an effective approach in the creation of one-dimensional nanostructures (Xia et al., 2003).

In almost every corner of nanotechnology there is a necessity to have patterned solid surfaces with functional groups, whether planar or curved. These functional groups will be better off when they are fabricated as nicely defined geometries such as monolayer or controlled multilayered films, also with a satisfactory degree of stability. Thus, proper choice and combination of these surface unit fabrications are important.

13.2.5. Symmetry Breaking

Symmetry breaking is for obtaining useful variations on nanofabricated systems. When a perturbative force is imposed on fabricated systems, its force rebalancing can induce the reorganization of the systems locally or even entirely. This can help control the given properties of the systems or even create novel properties. It can also provide useful means to modify the given pattern or morphology of the fabricated systems.

Symmetry breaking can be performed via three basic modes: along the long axis of the fabrication building units, along their short axis, and through multiple directions. Figure 13.6 shows the scheme. Long-axis breaking is useful when the fabrication building units are mainly coupled in end-to-tail mode. Even a slight change in the coupling can result in the significant shifting of their collective properties such as surface plasmon. Short-axis breaking is good for the side-by-side fabrication mode. It can provide a means to control the degree of contact area between the building units. Multidirection breaking is useful for the control of the packing symmetry. Since these unit fabrications require the perturbation of the force status throughout the whole system, the fabricated systems are usually subject to external forces. For certain systems, the principles of bio-mimetic self-assembly can be another effective method for symmetry breaking. Using fabrication units that have asymmetric packing capability and employing the impurities that can induce reorganization via the asymmetric packing mode are good examples. For some nanofabricated systems that in contact with a given surface, a simple change of the surface properties can often trigger an abrupt change in the force balance, and that can propagate throughout the entire system. This is another good example of multiple-direction breaking.

Symmetry breaking is a less-developed unit fabrication process than the others. It may require careful analysis to follow the correct mechanism as the symmetry is changing. But it can offer rich fabrication diversities for a wide range

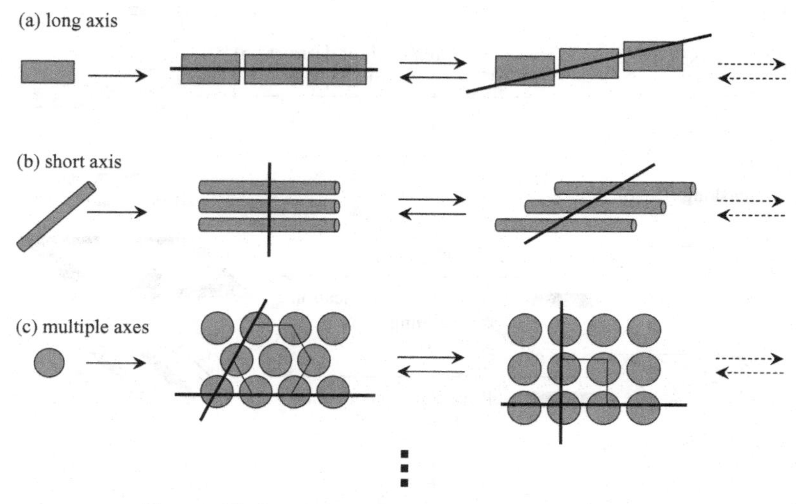

Figure 13.6. Unit fabrication: symmetry breaking.

of building units. It also provides good implications for the operation principle of nanodevices (details in Chapter 14).

13.2.6. Templating and Masking

Templating and *masking* are done to create structures or patterns that are copied or determined by the already-existing and prefabricated nanostructures (template and mask, respectively). Figure 13.7 shows the scheme.

Templating always begins with nanostructures that can serve as nanoscale "molds" for the intended nanofabricated systems. A variety of materials can be used as templates as long as they are not damaged by the interaction with precursors. This includes amphiphilic micelles, liquid crystals, colloidal crystals, surface structures, and biological entities (Yodh et al., 2001; Wang and Möhwald, 2004). The precursors interact with the surfaces of the templates and react into the hard structure. Removal of the template leaves the copies of the nanostructures originated from the template. The most typical example of this unit fabrication is the formation of the mesostructured materials in Chapter 9.

Masking means to use a mask (or shield) to create mainly surface patterns. This mask is not destroyed during the process. Rather, it sometimes becomes a part of the nanofabricated systems (*soft masking*). Using photomasks for photolithography and using prefabricated colloidal crystals for other lithographic patterning are good examples of *hard masking*. The most abundant example of soft masking is found in the fabrication of patterned self-assembled monolayers

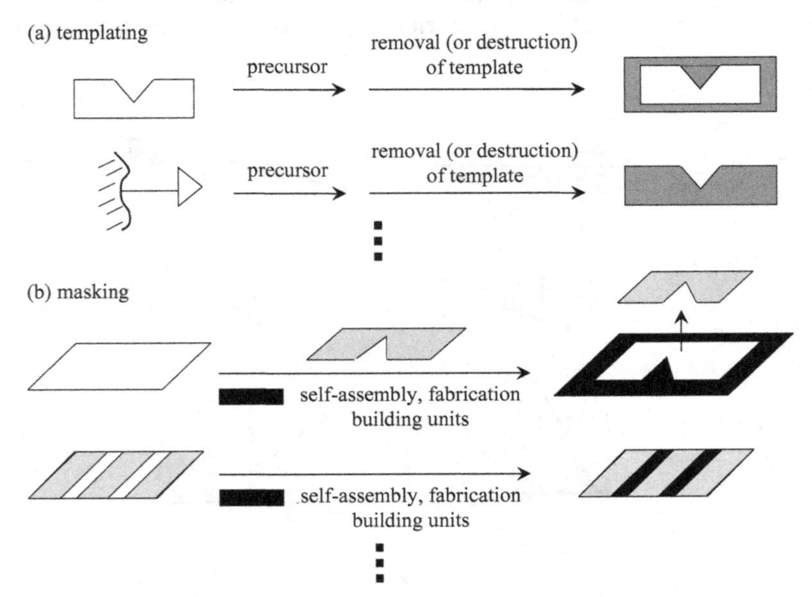

Figure 13.7. Unit fabrication: templating and masking.

(SAMs). There are many possible noninvasive means that can be employed to create physical barriers on the substrate. This includes surface phase transition, surface self-assembly, wetting, and spreading. By proper choice of fabrication units, the first set of SAMs can be fabricated with the patterns determined by this barrier. The physical barriers then can be simply removed through, for example, evaporation or washing. The second set of SAMs can now be fabricated in this empty area. By minimizing the exchange reaction with the first set of SAMs, this second set of SAMs is patterned by following the first set of SAMs (soft mask). Usually, both components become the final part of the nanofabricated systems. The well-known microcontact printing method uses a micropatterned stamp (usually made of PDMS) to direct prints on solid surfaces using SAM-forming building units as "ink" (Whitesides et al., 2001). This "stamping" fabrication can be also viewed as a masking unit fabrication in the sense that the prepatterned structure guides the structures of fabrication systems.

Templating and masking are suitable for mass-patterning of solid surfaces, mass-fabricating of highly regular structures, and even for mass-synthesizing of hybrid materials.

13.2.7. Hybridization

Hybridization means to perform the nanofabrication with more than two different natures of building units. Two types of processes are common: intermixing and interfacial reaction. Figure 13.8 shows the scheme.

Intermixing is physically mixing the fabrication building units. This can be performed on the surface and also inside the materials. The phase separation usually serves as the main driving force to create nanoscale structures. Surface alloying, which can provide a variety of nanopatterned solid surfaces (often called nanoislands), and bulk alloying, which usually is done to create nanofabricated systems with improved physical properties, are good examples. *Interfacial reaction* obtains nanofabricated systems in the form of nanocomposites or nanohybrids (some details in Chapter 9). Preformed or preassembled structures are

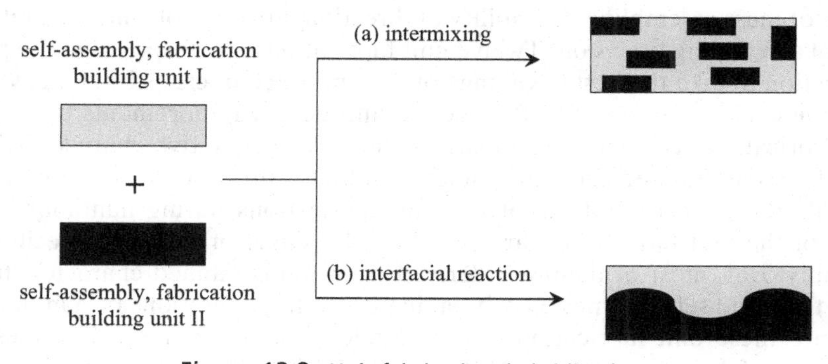

Figure 13.8. Unit fabrication: hybridization.

usually the direct source of nanostructures. But often the desired nanostructures are formed through the fluctuation on the interface during the reaction.

A typical example of hybridization is *biomineralization* (Mann, 2001; Tirrell et al., 2002). Since many building units for bio-mimetic self-assembly are the biological-origins or derived from them (Chapter 7), many of the bio-mimetic self-assembled aggregates possess geometrical matching characteristics that can guide the growth of biominerals. This can create biominerals with a great degree of structural diversity. Also, the *hybridized* nature of nanofabricated systems (such as biological-inorganic and biological-semiconductor) often yields a superior synergistic effect on their physical properties. The extreme fracture-resistance of many biostructures such as the shells of shellfish is a good example from nature. The superior strength of spider webs to stainless steel wire with the same size (diameter) and length is another example. This property is believed to come from the hybridization of its building unit (collagen), which is hierarchically self-assembled with the second building unit (hydroxyapatite).

Hybridization can be a useful approach for mass-assembling or mass-patterning of either or both of the fabrication building units. Mass-assembling of semiconductor nanoparticles using bacteriophage that has specific affinity to that semiconductor (Lee et al., 2002) and mass-assembling of integrated nanoparticle-biomolecule systems (Katz and Willner, 2004) are excellent examples.

13.3. NANOINTEGRATED SYSTEMS

Ultimately, what we are aiming at is *nanointegrated systems* that can have intended or novel performances in the macroworld. Nanointegrated systems are fabricated with nanofabricated systems through unit fabrications. And nanofabricated systems are fabricated from fabrication (and/or self-assembly) building units also through unit fabrications. The performance of the nanointegrated systems can be primarily expected from the properties of the fabrication building units. However, the success of the multistep fabrication process can practically ensure the intended performance of nanointegrated systems.

The success of nanofabrication requires three key elements. They are the ability of mass assembling, the ability of directionality control, and the ability to control assembling precision. Twelve unit fabrications are proposed in the previous section to take the best advantage of the unique characteristics of the variety of self-assemblies, so we can effectively fulfill the key requirements throughout nanofabrication. Table 13.1 summarizes those representative characteristics of the self-assemblies and each unit fabrication that is proposed from them. It also presents the expected effects of the unit fabrications during nanofabrication. Some of the unit fabrications are based on the single principle of specific self-assembly. But, most of them are derived from the combined characteristics of more than one self-assembly. By breaking down the operations of nanofabrication into these unit fabrications and by understanding their specific roles, the whole nanofabrication can be better performed: by logical planning, not by

TABLE 13.1. Relationship of self-assembly with unit fabrication and nanofabrication.

Self-Assembly	Characteristics	Unit Fabrication	Nanofabrication
Atomic	Site specificity: surface well	Deposition	Surface patterning, jointing
	Kinetics/ thermodynamics	Reconstruction, deposition	Surface patterning, filming
Molecular	Symmetric packing	Coating, hybridization, templating	Mass assembling, filming
	Distinct aggregation number	Coating, templating	Surface patterning, clustering
Colloidal	Symmetric packing	Coating, hybridization, templating	Mass-assembling, filming
	Larger length scale	Masking	Patterning
	Colloidal force	Symmetric breaking, alignment, stacking	Directional-assembling, mass-assembling
Bio-mimetic	Asymmetric packing	Symmetric breaking, alignment, stacking	Directional-assembling, mass-assembling
	Lock-and-key	Jointing, crossing, curving	Selective assembling
	Directionality	Alignment, stacking	Directional-assembling
	Strong bonding	Hybridization, jointing	Mass-assembling, stability
	Hierarchy	—	Successive unit fabrication
Surface	Surface confinement	Symmetric breaking	Directional-assembling
	Surface directional	Alignment, stacking	Directional-assembling

trial-and-error. This is important because the proper coupling of individual fabrication building units and the control/maximization of their collective properties are almost solely determined by how they are fabricated each other.

Figure 13.9 shows the schematic process toward a nanointegrated system. Typically, three different routes can be explored: successive, sequential, and hierarchical. When the first group of fabrication building units is fabricated through a certain unit fabrication, it can be defined as a primary nanofabricated

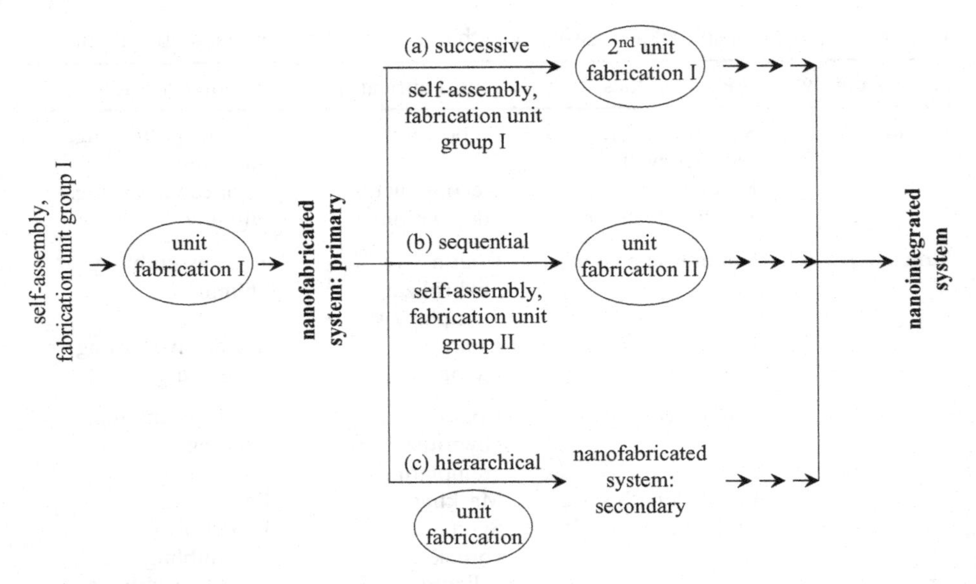

Figure 13.9. Toward nanointegrated systems via successive, sequential, or hierarchical unit fabrication.

system. The successive route is to fabricate nanointegrated systems through repetition of the same unit fabrication with the same (or similar) group of building units. This can be an ideal approach when the regular structures of building units are needed throughout entire nanointegrated systems. The sequential route is to keep fabricating the different groups of building units onto the primary nanofabricated system possibly through the different unit fabrications. The fabrication of electron transfer and charge storage devices through different unit fabrications with polyelectrolytes, nanoparticles, and nanoplatelets (Cassagneau and Fendler, 2001), and other examples (Crespo-Biel et al., 2006), are representative ones for this route. By analogy of the general scheme of self-assembly (Figure 1.3), the primary nanofabricated system itself often can be employed as the building unit for the next step of the fabrication. This will generate the secondary nanofabricated system, and higher, if necessary, until the desired nanointegrated system is reached. This is the *hierarchical route.*

Of course, there can be other routes toward nanointegrated systems. However, whichever route is chosen for a specific system, the choice of the unit fabrication usually is the essential key that can ensure the proper modification of important parameters, which include interparticle distances, packing geometry/density, and the degree of coupling. For instance, symmetry breaking might be a better unit fabrication choice to ensure the expression of some delicate optical properties over alignment or stacking, which are symmetric.

Often, we need to design nanointegrated systems with a specific application in mind, rather than pursuing the potential of the newly discovered properties of the individual fabrication building units. It can be helpful, in this case, to perform the nanofabrication backward. First, the required performances of the nanointegrated system should be identified. This includes types of applications, such as sensor, optical device, friction control, wettability control, and so on. This should drive the choices of forms such as three-dimensional, film, porous, and so forth. Next, the environmental conditions they will be facing, such as gaseous, solution, temperature, pressure, and so on, must be identified. Second, nanofabricated system(s) that can fulfill those requirements should be designed. This then should lead us to identify the proper unit fabrication(s) and the proper fabrication building units to be employed.

For example, one possible scheme for the fabrication of nanointegrated systems is shown in Figure 13.10. This is for the preparation of regularly patterned nanoparticles on solid surfaces through a sequential route of jointing–coating–jointing unit fabrications. Biotinylated amphiphiles, lipids, and streptavidin are the self-assembly building units that participate in this nanofabrication, and biotin-modified nanoparticles are the fabrication building units. The first unit fabrication is the jointing (through recognition) of streptavidins onto the Langmuir monolayer of biotinylated amphiphiles, whose patterns can be preformed in various ways (Chapter 11). This patterned streptavidin/amphiphile monolayer is then transferred and stabilized onto a given solid surface as Langmuir-Blodgett (LB) film via a second unit fabrication of coating. The third step is to fabricate the nanoparticles through another jointing unit fabrication—between the biotin groups on the surface of nanoparticles and streptavidins. This example representatively shows that the proper choice/combination of unit fabrications can be an effective route toward nanointegrated systems. Mass-

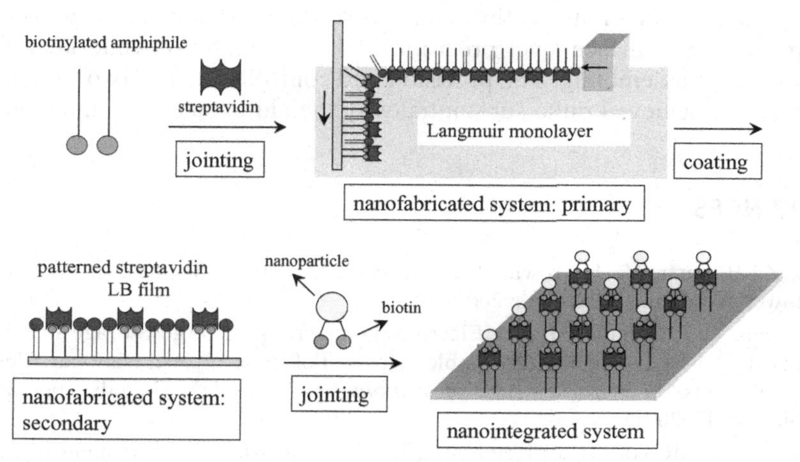

Figure 13.10. Fabrication of a nanointegrated system of regularly patterned nanoparticles via the jointing–coating–jointing sequential route.

assembling was achieved through coating unit fabrication; jointing contributed to the directionality of the fabrication; and assembling precision, which is the ability to control the interparticle distance, could be achieved in various ways throughout the fabrication.

13.4. SUMMARY AND FUTURE ISSUES

Nanofabrication is the assembling of a variety of nanostructured fabrication units into functional systems. The bottom-up approach toward this goal is to borrow useful principles of different self-assemblies. I have proposed that a majority of the nanofabrication processes can be broken down into 12 different self-assembly-based unit fabrications. Most of the nanofabrication processes can be performed (and planned) by the proper combination of these 12 unit fabrications. They are also directly applicable for the fabrication of nanodevices, for interconversion between nanofabricated systems and nanodevices, and ultimately for fabrication of nanointegrated systems and nanomachines. The idea of unit fabrication will help clarify the ever-more complex nanofabrication processes more systematically and will be useful in developing further nanofabrication processes more efficiently.

As the nanofabrication processes keep going forward, especially as ever-increasing numbers of newly developed nanostructures keep entering as potential candidates for nanofabrication, it will be inevitably necessary to come up with more unit fabrications in the near future. Novel behaviors of the new fabrication building units might require us to expand the scope of self-assembly or even redefine it. Also, current nanofabrication processes are mostly based on surface operation. It is obvious that in the future the dimensions of nanofabrication should be expanded beyond this surface confinement. The biggest challenge for these unit fabrications for three-dimensional nanofabricated systems will be how to maintain surface-level precision and controllability while keeping the ability of mass-assembling. The precision and controllability of two-dimensional systems were achieved quite substantially by the aid of surface confinement.

REFERENCES

Asfari, Z., Böhmer, V., Harrowfield, J., Vicens, J., Saadioui, M., eds. *Calixarenes 2001* (Kluwer Academic Publishers: 2001).

Cassagneau, T. P., Fendler, J. H. "Electron Transfer and Charge Storage in Ultrathin Films Layer-by-Layer Self-Assembled from Polyelectrolytes, Nanoparticles, and Nanoplatelets," *Electrochemistry of Nanomaterials*, Hodes, G., eds., pp. 247–286 (Wiley-VCH: 2001).

Crespo-Biel, O., Ravoo, B. J., Reinhoudt, D. N., Huskens, J. "Noncovalent Nanoarchitectures on Surfaces: From 2D to 3D Nanostructures," *J. Mater. Chem.* **16**, 3997 (2006).

Cui, Z. *Micro-Nanofabrication: Technologies and Applications* (Springer: 2005).

Katz, E., Willner, I. "Integrated Nanoparticle-Biomolecule Hybrid Systems: Synthesis, Properties, and Applications," *Angew. Chem. Int. Ed.* **43**, 6042 (2004).

Kim, F., Kwan, S., Akana, J., Yang, P. "Langmuir-Blodgett Nanorod Assembly" *J. Am. Chem. Soc.* **123**, 4360 (2001).

Lee, S.-W., Mao, C., Flynn, C. E., Belcher, A. M. "Ordering of Quantum Dots Using Genetically Engineered Viruses," *Science* **296**, 892 (2002).

Mann, S. *Biomineralization: Principles and Concepts in Bioinorganic Materials Chemistry* (Oxford University Press: 2001).

Mann, S., Shenton, W., Li, M., Connolly, S., Fitzmaurice, D. "Biologically Programmed Nanoparticles Assembly," *Adv. Mater.* **12**, 147 (2000).

Marrian, C. R. K., Tennant, D. M. "Nanofabrication," *J. Vac. Sci. Technol. A* **21**, S207 (2003).

McNally, H., Pingle, M., Lee, S. W., Guo, D., Bergstrom, D. E., Bashir, R. "Self-Assembly of Micro- and Nano-Scale Particles Using Bio-inspired Events," *Appl. Surf. Sci.* **214**, 109 (2003).

Messer, B., Song, J. H., Yang, P. "Microchannel Networks for Nanowire Patterning," *J. Am. Chem. Soc.* **122**, 10232 (2000).

Robinson, B. H., ed. *Self-Assembly* (IOS Press: 2003).

Salaita, K., Wang, Y., Fragala, J., Vega, R. A., Liu, C., Mirkin, C. A. "Massively Parallel Dip-Pen Nanolithography with 55000-Pen Two-Dimensional Arrays," *Angew. Chem. Int. Ed.* **45**, 7220 (2006).

Tirrell, M., Kokkoli, E., Biesalski, M. "The Role of Surface Science in Bioengineered Materials," *Surf. Sci.* **500**, 61 (2002).

Wang, D., Möhwald, H. "Template-Directed Colloidal Self-Assembly: The Route to 'Top-Down' Nanochemical Engineering," *J. Mater. Chem.* **14**, 459 (2004).

Whang, D., Jin, S., Wu, Y., Lieber, C. M. "Large-Scale Hierarchical Organization of Nanowire Arrays for Integrated Nanosystems," *Nano Lett.* **3**, 1255 (2003).

Whitesides, G. M. "The 'Right' Size in Nanobiotechnology," *Nature Biotech.* **21**, 1161 (2003).

Whitesides, G. M., Mathias, J. P., Seto, C. T. "Molecular Self-Assembly and Nanochemistry: A Chemical Strategy for the Synthesis of Nanostructures," *Science* **254**, 1312 (1991).

Whitesides, G. M., Ostuni, E., Takayama, S., Jiang, X., Ingber, D. E. "Soft Lithography in Biology and Biochemistry," *Annu. Rev. Biomed. Eng.* **3**, 335 (2001).

Xia, Y., Yang, P., Sun, Y., Wu, Y., Mayers, B., Gates, B., Yin, Y., Kim, F., Yan, H. "One-Dimensional Nanostructures: Synthesis, Characterization, and Applications," *Adv. Mater.* **15**, 353 (2003).

Yodh, A. G., Lin, K.-H., Crocker, J. C., Dinsmore, A. D., Verma, R., Kaplan, P. D. "Entropically Driven Self-Assembly and Interaction in Suspension," *Phil. Trans. R. Soc. Lond. A* **359**, 921 (2001).

14

NANODEVICES AND NANOMACHINES

Devices and even machines that function at nanometer scale, once a scientific fantasy, are now becoming a reality. Fundamental types of nanometer-scale devices (*nanodevices*) are rapidly being developed, the principles of their operation are being established, the performance of the prototypes of some nanoscale machines (*nanomachines*) is being demonstrated, and more sophisticated nanoscale machines with more complex structures and/or networks are now envisioned (Browne and Feringa, 2006).

There has not been much general consensus on clear definitions of *nanodevice* and *nanomachine*. It has been tempting to deduce them from the definitions of macroscale devices and machines. *Webster's Collegiate Dictionary* (2001) defines *device* as "an invention or contrivance, esp. a mechanical one, for some specific purpose" and *machine* as "a structure consisting of a framework and various fixed and moving parts, for doing some kind of work." Thus, whether "work" has been performed serves as a distinction between devices and machines at the macroscale. And this definition does not provide any relationship between devices and machines.

As will be shown throughout this chapter, the working principles of nanodevices and nanomachines are, in most cases, fundamentally different from their

Self-Assembly and Nanotechnology: A Force Balance Approach, by Yoon S. Lee

macroscale counterparts. Nanodevices and nanomachines are correlated with a significant degree of hierarchy. And some nanodevices (especially biological ones) actually show the capability of work to be done. In the previous chapter (Figure 13.1), I defined *nanodevice* as "a fabricated system of fabrication and/or self-assembly building units that can show a capability for unit operation," and defined *nanomachine* as "an integrated nanodevice that can show a communication capability with the macroworld." The uniqueness of nanodevices and nanomachines can be better addressed this way, and their relationship with nanofabricated and nanointegrated systems, which in some cases is not easy to distinguish, can be clarified, too. Most importantly, this definition can provide a clear scheme for understanding the role of self-assembly in their fabrication and operation.

A general view of nanodevices and nanomachines can be provided with their typical operating ranges. They operate at up to yocto (10^{-24}) Joule of energy with usually nano (10^{-9}) to pico (10^{-12}), but sometimes up to atto (10^{-18}), Newton of force. Detection up to a single small molecule is not uncommon. Movement of their parts is usually in nm ranges with frequency of a few nm to nanosecond, which makes their moving speed ~nm—~μm per second. Voltage is also applied in the range of nV. For example, the biological motor kinesin generates ~5×10^{-18} W per molecule. Also, a variety of nanoscale spaces that can be used for nanoscale reactions or as nanochambers can have capacities down to femtoliter (10^{-15}) and sometimes to yoctoliter.

This chapter is organized to provide a systematic understanding of nanodevices and nanomachines. First, I address the issue of what are the physical, chemical, and mechanical criteria of fundamental building units of nanodevices (fabrication and self-assembly building units: *nanocomponents*) to be a useful part of nanodevices, and how they can make nanodevices while others cannot. Second, I propose three "element motions" whose balance and correlation can be easily applied to clarify working principles of nanodevices that are seemingly different from and independent of each other. As in the case of nanofabrication in the previous chapter, where a small number of unit fabrications can represent the majority of nanofabrication processes, the operation of the majority of the nanodevices developed so far can be represented with a relatively small number of fundamental operation modes. This issue will be proposed under the name *unit operation*. Some representative examples of nanodevices and some thoughts about the fabrication and future challenges of nanomachines will follow.

14.1. GENERAL SCHEME OF NANODEVICES

When fabricated systems of nanocomponents show capability of unit operation(s), they can be defined as nanodevices. Their operation principles are mainly established at the nanometer scale and their operating components are in the range

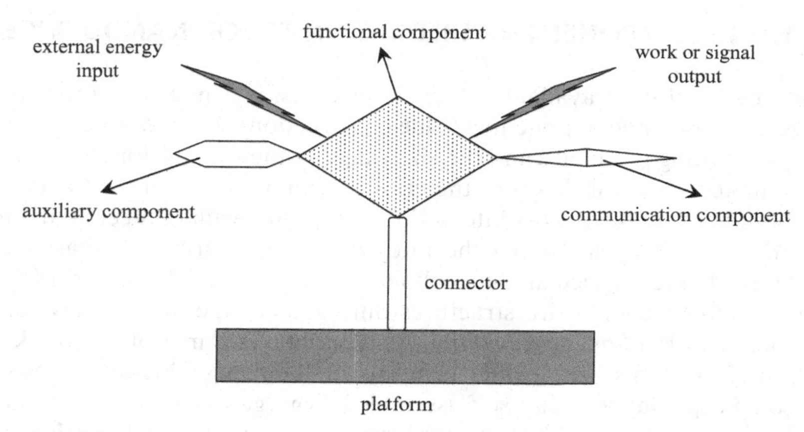

Figure 14.1. Conceptual schematic representation of nanodevices.

of nanometers. But sometimes their operation is at the nanoscale even though the whole device is above the nanometer scale.

Figure 14.1 shows the schematic representation of nanodevices. This is a conceptual one to provide the general view; their actual structural features can vary significantly. The functional component is to generate the *push* and *pull* element motions (details in the third section) whose balance and correlation induce the desired unit operation(s). This is mainly done by the input of external energy, but it is not uncommon for the energy source to be incorporated as a part of the nanodevices. The results of the unit operation(s) are reflected in the work done or signal output. Usually it is required for the nanodevices to have the component that can generate the *guide* element motion, whose role is to make the unit operation(s) controllable, such as directional rotation or directional movement. The auxiliary component is for this purpose, and the incorporated power supply can be a part of this component, too. Networking of the individual nanodevices is critical to create nanomachines that can eventually communicate with the macroworld. The communication component is used to make effective connections with adjacent nanodevices. In many cases, nanodevices have to be anchored on the surfaces. When this is necessary, the nanodevices can be immobilized on the surfaces (platform) through the proper connector, such as a shaft component for molecular motor.

Unit fabrications based on self-assembly principles (Chapter 13) can be the key to fabricating nanodevices. They will also ultimately facilitate massive parallel production of nanodevices and their networking. For example, the multistep self-assembly principles (Figure 1.3) can have a significant impact, since properly working nanodevices will have to have multiple functions, multiple compartments, and multiple heterogeneous components that are assembled together. A top-down approach alone is simply too ineffective to perform this task.

14.2. NANOCOMPONENTS: BUILDING UNITS FOR NANODEVICES

It would be ideal if all available atoms, molecules, and nanostructured objects could be used as nanocomponents for the construction of nanodevices. However, the major working principles and the size range of their operations are mainly at the nanometer scale; this leads to the critical requirement that the nanocomponents should ensure the proper interaction of intermolecular and colloidal forces among them, while maintaining the integrity of their structural features. The interaction of intermolecular and colloidal forces should be controllable but strong enough to hold entire structures during their operation. These are the requirements of *rigidity* and *flexibility*. As described in Part I of this book, most self-assembly systems are flexible in terms of their assembly-and-disassembly processes. Surprisingly, many self-assembled aggregates are strong enough to endure unit operations and even perform mechanical work under the right conditions. For example, self-assembled lipid bilayers have mechanical strength comparable to stainless steel of the same thickness, yet they are extremely flexible (Seifert, 1997; Karlsson et al., 2006). The next set of requirements is *diversity* and *hierarchicality*. Diversity of nanocomponents is important to expand the scope of possible nanodevices. And the hierarchicality is critical to paving the way toward nanomachines. Self-assembly building units fulfill this set of requirements, too.

External force–induced self-assembly systems are especially important for proper operation of functional components (Figure 14.1) under the control of external energy. For electrochemical energy, ferrocene-based and sexithiophene-based derivatives are typical examples. For temperature-sensitive systems, thermosensitive polymers that can form thermosensitive self-assembled gels are among the most often applied. Azo group–based derivatives, ring open-closure systems, carbon–carbon double-bond systems, and other *cis–trans* isomerization systems are good for photo-induced operations. Systems with chirality are also good for this, since enantiomers respond oppositely in polarized light, which can be converted into light-induced opposite operations.

Three representative groups of examples for nanocomponents that satisfy these requirements will be presented with details in the rest of this section. They are *interlocked* and *interwinded molecules, DNA*, and *carbon nanotubes* and *fullerenes*.

14.2.1. Interlocked and Interwinded Molecules

Rotaxane, catenane, and pseudorotaxane are those for the former cases (interlocked) (Balzani et al., 2000) and knotane (Lukin and Vögtle, 2005) is for the latter case (interwinded). Rotaxane is a molecule that is composed of a macrocyclic component and a dumbbell-shaped component that are interlocked with each other. Catenane is a molecule composed of two interlocked macrocyclic components, not covalently but geometrically. Pseudorotaxane is a molecule composed of a macrocyclic and a linear-shaped component. These *pre-self-assembled* molecules have a unique nanometer range of space within their dynamic structures, and at the same time their molecular backbones are rigid. They are also flexible in the sense of incorporation of functional groups. These features give them a strong potential for a diverse range of unit operations, which makes them quite universal nanocomponents. Also, the interlocked or inter-winded (i.e., strongly paired) parts of these molecules make them highly stable against the possibility of losing a part during the unit operation under strong external forces. This may provide them with good durability (a long lifetime for nanodevices) and stability, too.

A variety of unit operations have been demonstrated with nanodevices based on these molecules. These include molecular shuttle and molecular elevator, both with the force of ~ a few hundred pN, molecular gate, molecular switching, and molecular motor (Balzani et al., 2000; Lukin and Vögtle, 2005). As will be shown later in this chapter, the fundamental principle of their operation is about the control of "push" and "pull" element motions with the help of the "guide" element motion. For these nanodevices, the guide motion is provided intrinsically by their structural designs. Another key advantage of these molecules is that by proper incorporation of functional groups, diverse types of external forces including chemical, electrochemical, and photochemical can be employed to control these element motions.

14.2.2. DNA

DNA has perfect site-specific, programmable self-assembly capability through base-pairing. It also shows incredible molecular recognition, and can self-assemble with chiral and helical structures. This is important to generate the force for element motions by releasing the energy stored within these conformational features. They are rigid but conductive, come with a variety of structural diversities, and can be conformationally responsive to a variety of external forces. They are easy to synthesize, easy to functionalize, and easy to manipulate (easy to cut-and-paste with a variety of enzymes). They have enough stiffness but also flexibility to be a nanodevice.

Forces from DNA-based nanodevices are mainly generated through push or pull element motions that are strongly involved with hydrogen bonding during hybridization and dehybridization. This may be the reason that DNA-based nanodevices are strongly anticipated to generate a force comparable to the protein-based molecular motors with high controllability of directionality. A variety of DNA-based nanodevices have been demonstrated (Simmel and Dittmer, 2005). Their unit operations include switching and gating, tweezering that generates a few pN to a few tens of pN of force, walking, shuttling, directional motor, and information processing performance.

14.2.3. Carbon Nanotubes and Fullerenes

Carbon nanotubes have remarkable mechanical toughness and electrical conductivity along with many other useful properties up to a diameter as small as 1 nm.

So do fullerenes. They also come with structural diversity and easiness of diverse surface functionalization.

Diverse ranges of nanodevices have been demonstrated with these. A nanoscale rotational actuator whose metal plate rotor is operated by an external voltage is a good example. This takes advantage of the mechanical toughness and electrical conductivity of the multiwalled carbon nanotube, which works here as the support shaft and as the source of rotational freedom (Fennimore et al., 2003).

14.3. THREE ELEMENT MOTIONS: FORCE BALANCE AT WORK

So far in this text, a variety of self-assembly processes, and preparation and fabrication of nanostructured objects and patterns, have been successfully described based on the processes of force balancing among intermolecular and colloidal forces. These forces have been classified into three groups: attractive, repulsive, and functional/directional.

Here, I propose that a variety of seemingly different unit operations of nanodevices can be systematically described, once we introduce three element motions of their nanocomponents. These are *push, pull*, and *guide motions*. Regardless of the types of nanodevices, their unit operations can be understood as the balancing processes of these three element motions, which are generated through the force balancing of intermolecular and colloidal forces between nanocomponents of the nanodevices. Figure 14.2 shows the schematic representation. Push motion is generated when there is a repulsive interaction between the nanocomponents of the nanodevice. The moving component is being pushed away from the other relatively or permanently static component. Pull motion is the opposite case. This motion is generated when the moving component is being pulled by the attractive force or attractive functional interactions. The driving forces for the unit operations are the result of either push or pull motion, or the

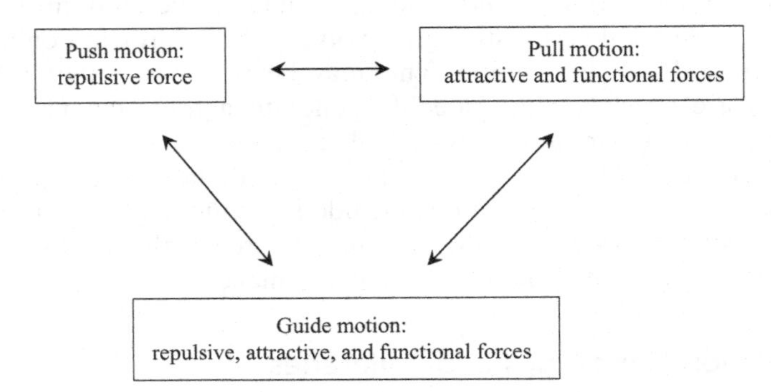

Figure 14.2. Balance among three element motions for unit operations; each element motion is generated through the force balance of intermolecular and colloidal forces.

vector sum of both motions. When each motion is successively induced, the resultant unit operation can be controlled to be a successive operation. When both motions are alternatively induced, the unit operation can be controlled to be oscillatory. What fundamentally makes a nanodevice a nanodevice is its controllable directionality at nanometer scale, which should be clearly differentiated from Brownian motion. Guide motion provides this controllability. This motion is generated when either one of, or the balancing of, repulsive, attractive, and functional forces can guide/control the direction/degree of either or both of the push and pull motions during the operation of the nanodevices.

There have been numerous biological studies over the past few decades to clarify directional motion in biological systems, and also synthetic efforts to mimic this directional motion in synthetic systems (Kinbara and Aida, 2005). Their operations can be fairly easily and systematically understood by following the idea of the balancing of these three element motions. It can be also applicable to controlling the degree of the operations. For example, *cis–trans* isomerization of stilbene induced by light irradiation is of the order of ps (Kwok et al., 2003), but the process of *trans*-to-*cis* is ~100 times slower than that of *cis*-to-*trans*, which is obviously caused by steric barrier. This shows that a properly designed steric repulsion could be very effective to guide and control the operation of the nanodevice.

14.4. UNIT OPERATIONS

Twelve different unit operations will be proposed in this section, and details of their operating principles based on the balancing of the three element motions and some representative examples will be discussed.

The operation of a variety of different nanodevices can be complicated to understand. However, their apparent operating behaviors can be broken down into a number of fundamentally simple modes of operation. Any single one of the fundamental modes or the combination of different modes results in their apparent operation. This fundamental mode will be referred to here as *unit operation*. Criteria for the definition of unit operation can vary. But since it is our ultimate goal to control the nanodevices at our will, it will be reasonable to define it as "a representative mode of operation that can be directional and controllable that repeatedly appears in different types of nanodevices." This way, it can also encompass both the operations without actual work to be done and those that actually generate work in response to external or internal energy input such as biological motors.

At the core of the balancing of element motions for unit operations are the intermolecular and colloidal forces. They are advantageous since they are easier to control than covalent bonds (where there should be constant cut-and-bond processes). They can be reversible without much limitation, and can be operated in solution, in air, and in vacuum with one-, two-, and three-dimensional geometries. They can also provide a number of advantages for the fabrication of nanodevices into nanomachines, which can include fast communication between nanocomponents, fast and low-cost operation, self–error correcting, self-repairing, and self-reinforcing.

14.4.1. Gating and Switching

Unit operation of gating or switching performs open-and-close (for gate) and on-and-off (for switch) modes of operation. As will be shown in some examples later in this subsection, there can be a variety of different forms of gates and switches. However, for most cases, gate and switch share the same or very similar operation principles. Figure 14.3 shows the representative scheme for this. Either one or both of the nanocomponents that are composed of the gate or switch can perform push and/or pull element motions with the direction of guide element motion in a way that they get closer and make contact. The gate or switch will be closed (for gate) or on (for switch) in this case. Balancing of element motions that can reverse this process will result in the gate-open or switch-off.

Gate operation with pseudorotaxanes with external reducing agent as external chemical stimulus has been nicely demonstrated (Hernandez et al., 2004). And molecular gate has been realized by manipulating the assembly/disassembly processes of molecular self-assembly building units through the environmental condition (pH)–induced control of attractive and repulsive forces (Casasus et al., 2004). A typical biological example can be found in Chaperonin. It performs open-and-close operations through the conformational change of its components that are asymmetrical packed. ATP is an energy source for this, and the binding of ATP triggers this conformational change (Saibil and Ranson, 2002).

Many of the switch operations occur at surfaces. Whatever the external stimulus is and whatever properties are being changed (such as wettability, friction, or conductance), the switch operation is performed through the control of attractive and repulsive forces that can be understood as the balancing of the element motions (Liu et al., 2005). Effective switching operation has been performed using azo-based (Banerjee et al., 2003) and ferrocene-based (Chen et al., 2004) nanocomponents under the input of light and electric field, respectively,

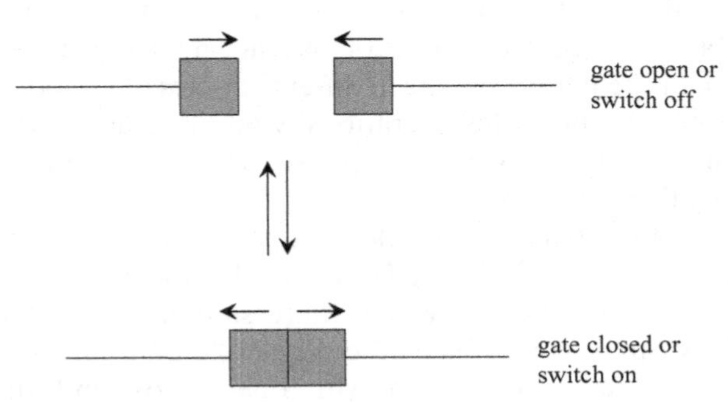

Figure 14.3. Unit operation: gating and switching.

as external energy. Switching unit operation also has been demonstrated with catenane, which shows impressive two- and three-way positional switching with its guide motion provided by intrinsic molecular frame (Kay and Leigh, 2005). DNA-based switching operation that takes advantage of metal ion–induced balancing of the element motions is a good example. The systematic change of DNA handedness was key for this (Simmel and Dittmer, 2005).

14.4.2. Directional Rotation and Oscillation

Directional rotation and oscillation are unit operations that perform controllable rotation and oscillation of one nanocomponent (rotor) on another one (stator) with controllable directionality that overcomes the random motion caused by Brownian motion.

Figure 14.4 shows the schematic representation. Any type of possible intermolecular and colloidal force can be applied as the driving force for this unit operation. It can be designed to be operated only with push element motion that acts on the rotor. It can be designed to be operated only with pull element motion that is exerted on the rotor from the surrounding parts. It can also be designed to be operated by successive and/or alternative action of push and pull element motions. When it is guided by the guide element motion to be a rotation with a controllable degree of directionality, it is directional rotation. When it is guided to be oscillatory, it is, of course, an oscillation operation. For example, electrochemical external stimulus has been successfully employed to induce the push mode through electrostatic repulsive force between rotor and stator parts for the push-back types of rotation of ferrocene-based molecular motors. The key to the unidirectional rotation of this motor is the subsequent successive electrostatic attractive force that is designed to ensure the directionality of that rotation (Rapenne, 2005). Another example is the unidirectional molecular motor that is driven by photo-induced *cis–trans* isomerization where steric repulsive force generates both push and guide modes (Van Delden et al., 2005). Some of the literature uses the term *molecular motor* to encompass diverse kinds of biological entities, as long as they show the ability to convert the chemical energy to mechanical or some kinds of motion. But, with the definition of *directionality* here, we will use the term *motor* for only the molecular devices or nanodevices that show directional rotational motions.

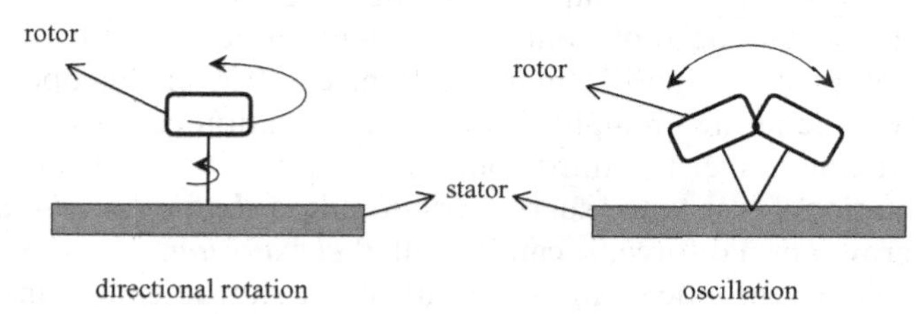

Figure 14.4. Unit operation: directional rotation and oscillation.

Numerous systems have been developed that show the unique rotational motion and thus have potential to be developed as molecular or nanoscale motors (Kottas et al., 2005). One typical synthetic example is the nice demonstration of unidirectional rotation of a catenane system. One macrocyclic ring provides the guide element motion, while the external stimulus–driven change in intermolecular forces generates push and/or pull element motions that drive the other ring to rotate (Kay and Leigh, 2005). A well-designed steric shielding system showed that it can guide the rotation of a phenylene-based rotor (molecular gyroscope) around a carbon–carbon single bond in air whose speed can reach up to GHz range even at ambient temperature (Karlen and Garcia-Garibay, 2005).

One well-known biological example is the bacterial flagellar motor, which provides many species of bacteria such as *E. coli* the power to swim through the directional rotary motion. It is generated by the successive protonation and deprotonation processes that induce the unbalance of the force balance that was set before the initial step. This force unbalance is rebalanced through the conformation changes, which generates the torque for the directional rotation and thus the power to swim. The asymmetrical packing of proteins seems to be critical for this torque generation and also to provide the guide motion. Another sophisticated example from biological systems is ATP synthase. This highly energy-efficient (nearly 100%) biological motor uses ATP (adenosine triphosphate) or protons as energy sources. As ATP binds and is hydrolyzed on the specific sites of its surface or as proton fluxes, the asymmetrically packed (or assembled) rotary part of proteins changes its conformational geometry as the force balance is being shifted to a new balance. This induces the rotation, and the guide motion by the other proteins around it makes it directional. It generates up to ~100 pN/nm of torque. The Ni-based nanopropellers attached on the γ-subunits of the F_1-ATPase demonstrate the seemingly great potential of this molecular motor for the development of working nanodevices (Soong and Montemagno, 2005). Two unit fabrications of "hybridization" (between biological and inorganic entities) and "jointing through recognition" (biotin–streptavidin linking) were the key to fabricating this nanodevice.

14.4.3. Shafting, Shuttling, and Elevatoring

Shafting, shuttling, and elevatoring are the unit operations that perform the linear movement of one nanocomponent along the guide of the other one. Figure 14.5 shows the schematic representation. Again, either one of, or the proper combination of, push and pull element motions can trigger this operation. With the capability to generate enough "work," the operation can become a "cargo" operation that can deliver the third component. When the movement is against any nongravitational resistance such as friction, it is *shafting* or *shuttling*. When it is against gravitational force, it can be called *elevatoring*.

One excellent synthetic example of molecular shuttle is demonstrated with a rotaxane-based system that shows controllable reversible linear movement of

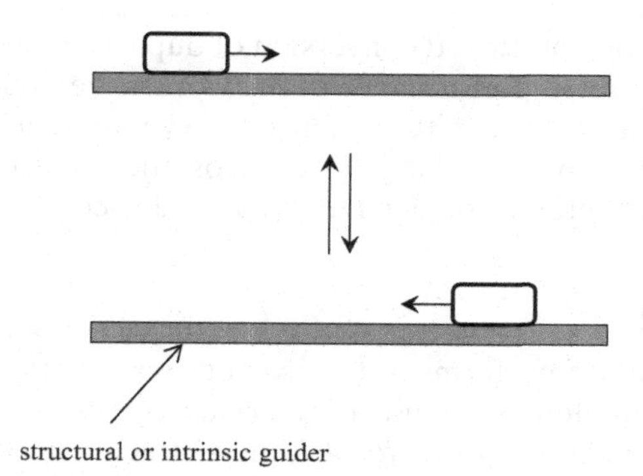

Figure 14.5. Unit operation: shafting, shuttling, and elevatoring.

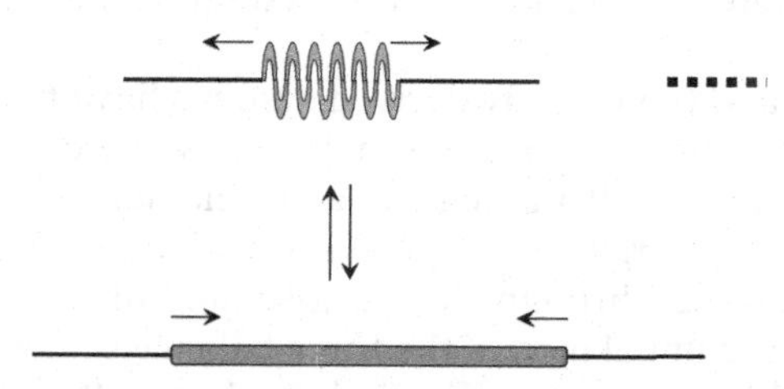

Figure 14.6. Unit operation: contraction-and-extension.

the macrocyclic part with the guide of the dumbbell-shaped part (Credi, 2006). Various external stimuli, including redox, acid–base, and photochemical, were able to generate this movement. An example of elevator operation was also found in a rotaxane-based system. Energy supplied by acid–base reaction triggered the controllable reversible change of hydrogen bonding between the two parts, which generated the linear directional elevator motion of the macrocyclic part with the guide of the other part. The movement was ~0.7 nm with force generated up to 200 pN (Badji et al., 2004).

14.4.4. Contraction-and-Extension

Contraction-and-extension unit operation performs a repeatable linear operation within a single nanocomponent or assembled ones that work like a single component. Figure 14.6 shows the schematic illustration. For most cases, this operation is generated through intramolecular-like self-assembly. Extension is mainly triggered by push element motion within the chain (intra-push), and contraction is by pull element motion (intra-pull). Guide element motion is usually intrinsic.

The cycle of contraction-and-extension operation has been demonstrated by employing an electrochemical signal to ferrocenyl-based polymer chains, which has generated mechanical work (Butt, 2006). A DNA chain–based system that takes advantage of proton-fueled conversion of duplex to i-motif conformational change has demonstrated performance of the work to be done (Shu et al., 2005). It is also likely that the key is the change of asymmetrically assembled DNA chain that is triggered by the change of electrostatic repulsive forces; ~11 pN/m of force has been generated for this DNA-based device.

14.4.5. Walking

Walking unit operation performs a step-by-step type of reversible movement of one nanocomponent along the guide of the other component with a high degree of directionality. As shown in Figure 14.7, the characteristic design of two "legs" that are contacted with the guide part and the "stem" part that connects the leg parts is critical for this unit operation. Alternative movement of each leg by push-and-pull element motion moves the system along the direction of the bottom guide part.

Only biological and biologically derived systems have been demonstrated to perform this unit operation so far. Myosin, which is one of the cytoplasmic proteins, is known to move (walk) along the actin filament unidirectionally at the expense of ATP as an energy source. The successive binding and hydrolysis of ATP on the functional site (through bio-self-assembly-like process) in the myosin-actin system trigger the unbalance of the force balance that was set initially. This unbalance is relaxed in such a way as to induce the "walking" motion of myosin along the direction of the actin filament through the reversible change of its helical conformation. Usually, it creates a walking motion of a few nm to a few tens of nm each cycle, depending on the types of myosins. Another biological example can be found in the other molecular motor protein, kinesin, which walks along the direction of microtubules. The energy source is APT, as with myosin. It is also highly involved with the force rebalancing of hierarchically self-assembled nanocomponents, which induces the pull element motion on kinesin. Guide element motion is induced by its interaction with microtubules.

Interesting nanodevices that actually show their ability to perform work have been developed using these myosin-actin (Månsson et al., 2005) and kinesin-microtubule systems (Hess and Vogel, 2001). This demonstrates that it is possible to develop a molecular shuttle that can actually load/unload nanoscale cargo. As

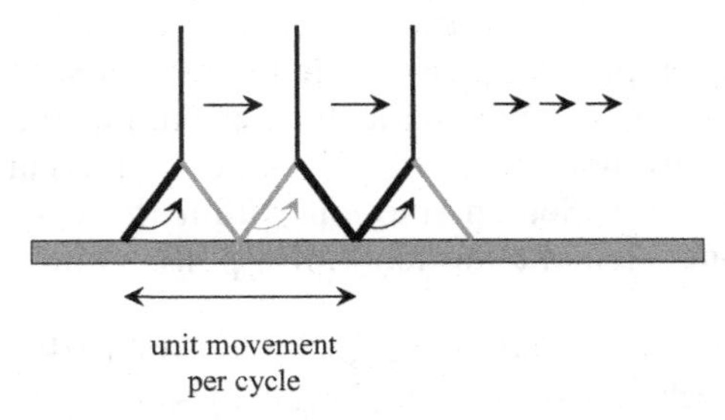

Figure 14.7. Unit operation: walking.

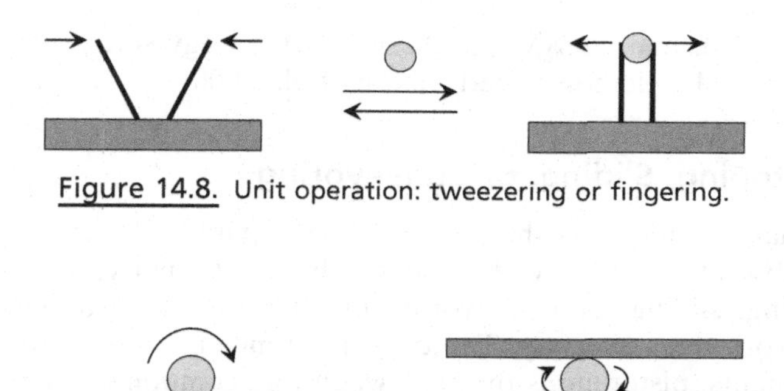

Figure 14.8. Unit operation: tweezering or fingering.

rolling bearing

Figure 14.9. Unit operation: rolling and bearing.

in the case of ATPase-based nanodevices, two unit fabrications were the key to the fabrication of these nanodevices: biological-inorganic hybridization and jointing through recognition.

14.4.6. Tweezering or Fingering

This unit operation performs an open-and-close motion of paired nanocomponents that have the capability to grab and move nanoscale object(s). Figure 14.8 is the schematic illustration. Pull element motion between the two paired nanocomponents mainly works to grab the object, while push element motion works to release it. Usually, guide element motion is not necessary.

An excellent nanotweezer has been fabricated with carbon nanotubes through deposition and direct jointing unit fabrications. The operation to grab and move nanoclusters was achieved through the precise control of the force balance between the external electric field–induced electrostatic repulsion [between the two nanotubes (hand)] and the elastic force of the nanotube (Kim and Lieber, 1999). For a biological example, it has been shown that by selective self-assembly of DNA base-pair, the open-and-close motion can be generated through the force rebalancing of the asymmetrically packed DNA pair (Simmel and Dittmer, 2005).

14.4.7. Rolling and Bearing

Rolling and bearing are unit operations that perform the directional rotation of spherical-shaped nanocomponents on the surface of other components (rolling) or between the surfaces of two components (bearing) (Figure 14.9).

Guided rolling of the fullerene-based *nanocar* has been demonstrated. A well-designed molecular structure makes the rolling motion possible on a solid surface in air. Thermal energy was the primary energy source, and the use of light energy was also demonstrated (Shirai et al., 2006).

14.4.8. Pistoning, Sliding, or Conveyoring

When two nanocomponents that are positioned parallel to each other move a relative or absolute distance with controllable directionality, this operation can be a pistoning, sliding, or conveyoring unit operation. Pistoning and sliding operations work the same way. The only difference is the structural feature of their components; pistoning is the case when one component is moving in and out of the other component. Figure 14.10 is the scheme: part (a) is when both of the components move relatively to each other, while part (b) is when one is static, so the other one moves absolutely. Both *pistoning* and *sliding* are mainly back-and-forth types of reversible operations. When one component is moving absolutely along the other one that is static in a successive way, it becomes a *conveyoring* operation.

Directional transport of indium along the surface of multiwalled carbon nanotubes by the control of the external current can be viewed as a nanoscale conveyor (Regan et al., 2004).

14.4.9. Self-Directional Movement

As shown in Figure 14.11, self-directional movement is for when a certain nano-component acquires the force to move directionally, which overcomes the Brown-

(a) Pistoning or sliding with moving platform

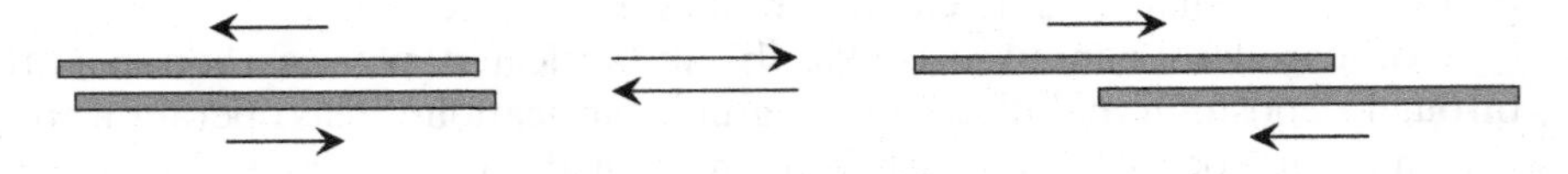

(b) Pistoning or sliding with static platform

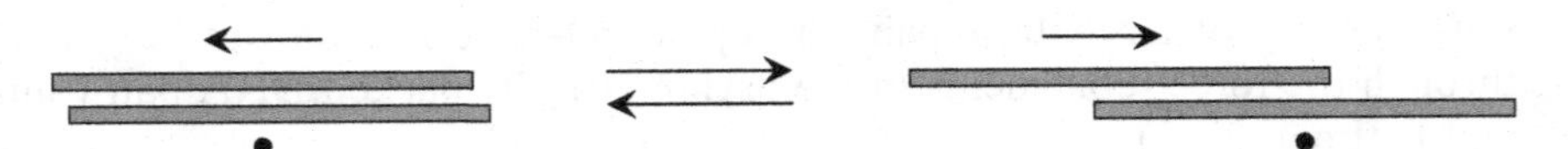

(c) Conveyoring

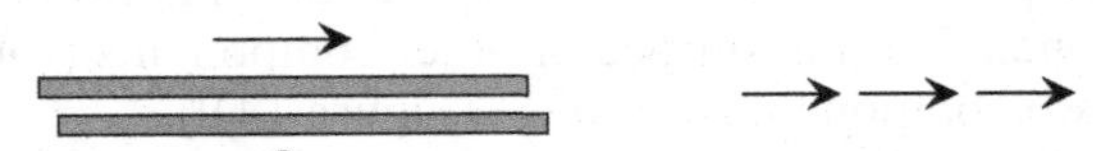

Figure 14.10. Unit operation: pistoning, sliding, or conveyoring.

Figure 14.11. Unit operation: self-directional movement.

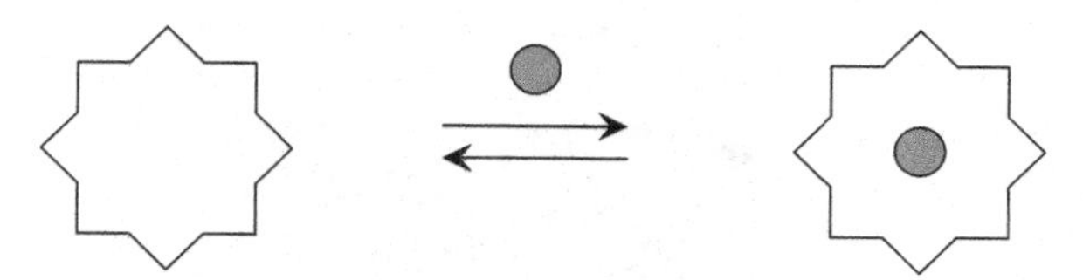

Figure 14.12. Unit operation: capture-and-release.

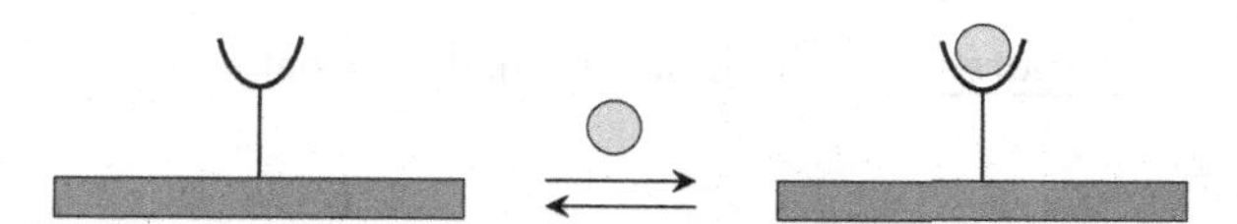

Figure 14.13. Unit operation: sensoring.

ian motion. It has been demonstrated that a catalytically generated concentration gradient can induce self-generated flow in the specific area of the nanoscale object, so the object moves unidirectionally through the liquid (Paxton et al., 2005).

14.4.10. Capture-and-Release

This unit operation captures and releases one nanoscale object from the other nanocomponent in a controllable way. Figure 14.12 is the scheme. It can be mainly performed by systems with adsorption-and-desorption capability. The difference from the sensoring operation in the next subsection is its selectivity. A capture-and-release operation should be able to work with a number of various nanoscale objects, so it can control their numbers that are captured and released. It also should be able to control their patterns as all-at-once or successive types of operation.

14.4.11. Sensoring

A sensoring unit operation detects a specific object (or signal) for the operation of nanodevices. It does not have to be an all-at-once or successive type, but has to work at the specific conditions (concentration, solution, air, etc.) required for a given system. Figure 14.13 is the scheme. It is usually reversible, and able to work through specific recognition, size complementarity, or a shape-selective process, depending on the requirements.

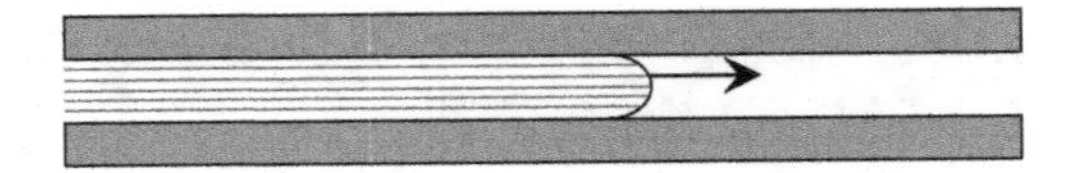

directional flow inside nanochannel

directional flow on nanogroove

Figure 14.14. Unit operation: directional flow.

14.4.12. Directional Flow

This unit operation performs the controllable flow not only of fluids, but of gas, particles, and even single molecules as well. As shown in Figure 14.14, it can occur inside the nanoscale channel or on the surface of a nanopatterned groove. The key parameters, including its directionality, amount, and speed, should be controllable. This unit operation is critical not only for the construction of nanofluidic types of devices such as lab-on-a-chip, but for nanoreactors and nanoscale communication devices as well.

Well-designed construction of lipid-based vesicles networked with a series of nanotubes has been demonstrated. This system shows the directional and controllable transport of desired liquids or particles in desired amounts through its nanotube components. It is also shown that its flow can be guided through the manipulation of push or pull element motion (Karlsson et al., 2006).

14.5. NANODEVICES: FABRICATED NANOCOMPONENTS TO OPERATE

The definition of *nanodevices* was given earlier in this chapter. For real applications, some of them can have a single operation, while some others may have to have multiple unit operations that are coupled to each other in one device.

It may not be wise to attempt a definite classification of nanodevices at this stage, since this is an area that is rapidly advancing. However, it will be helpful to see how the concepts of nanofabrication and nanodevices work for the actual nanoscale systems at work. Two representative examples will be presented here: a delivery system whose applications are widely in use, and nano- (or molecular) electronics, whose development has been advancing over the past decade.

14.5.1. Delivery Systems

Delivery vehicles are a variety of usually nanometer (often micrometer)-scale multicomponent-assembled systems that have the capability to deliver materials of interest on the specific target or in the way we planned. They do not necessarily have to be limited to the long-studied drug-delivery systems. They should include systems to deliver genes, cosmetic materials, and functional objects such as imaging contrast nanoparticles, as well. Even though there are many notably different aspects, depending on the specific systems, including the morphology and formulation process, their fundamental principles of operation remain very similar. This is in reality a huge topic whose entire scope can hardly be covered in a single section. Issues of delivery systems related to self-assembly, nanofabrication, and nanodevices will be mainly addressed here.

Figure 14.15 presents the general scheme of different types of delivery systems. O/W microemulsion- or normal micelle–based systems possess the capability to retain hydrophobic substances within their interior through micellar solubilization (Chapters 3 and 4). This makes them good candidates to deliver water-insoluble or less-soluble substances. Considering ~60% of potential drug molecules are water insoluble or less soluble, which can cause delivery problems, these systems can have a great impact in addressing this issue. W/O microemulsion- or reverse micelle–based systems can be, on the other hand, useful for hydrophilic substances.

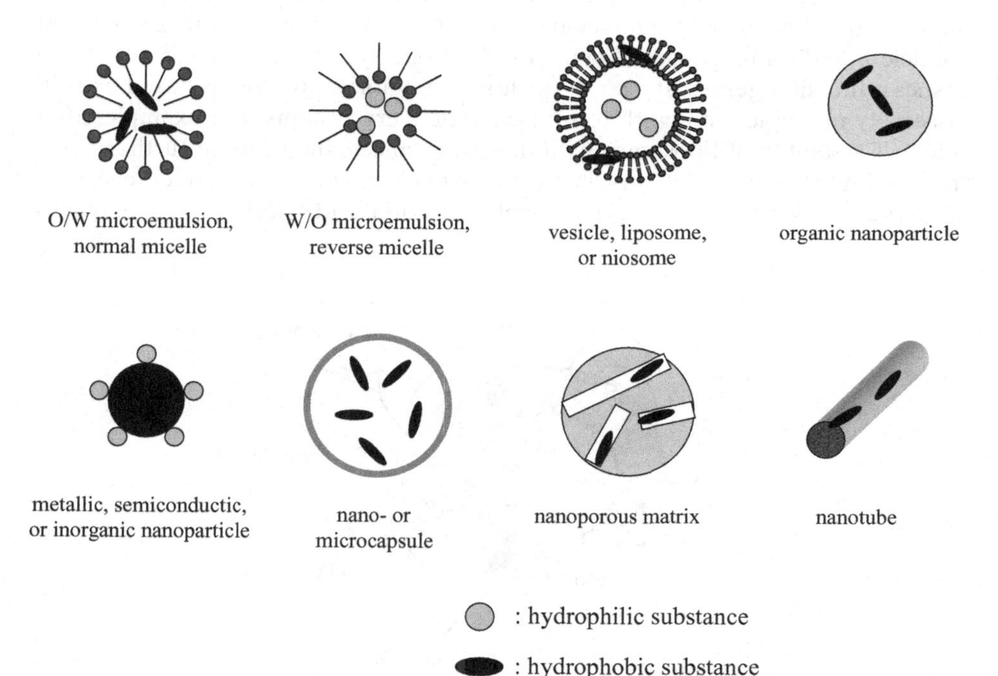

Figure 14.15. Schematic view of different types of delivery systems.

By taking advantage of their structural characteristics, vesicles, liposomes, or noisomes can retain both hydrophilic and hydrophobic substances within their water–core region and bilayer area, respectively. This feature makes them not only quite versatile candidates for a delivery system, but a platform for the development of highly functional systems such as target delivery vehicles (one conceptual scheme is shown in Figure 14.16).

Organic nanoparticles that can include gellified organic, macromolecular, or lipid molecules can retain hydrophobic substances. And the surfaces of solid nanoparticles can be functionalized to be able to deliver hydrophilic substances. Nano- or microcapsules such as aerogel, nanoporous materials, and nanotubes also can be assembled as a system to deliver hydrophobic substances.

Regardless of the types of substances that are being delivered, all systems presented here have inevitable ties with self-assembly principles based on the interplay of intermolecular and colloidal forces. The core processes of their formulation, transport, and substance release are much better off when they are progressed without any significant formation or breaking of strong bonds such as covalent, which can impose a great risk of delaying the processes and changing the properties of the substances involved.

The mode of delivering a drug substance can have a tremendous impact on its efficacy. Key challenges for better clinical practice include maximizing the drug loading, stability in the bloodstream, specific targeting to deseased cells, circulation lifetime against the immune system, controlled release of drug molecules, and diffusion of drug molecules through cell membranes. A target delivery vehicle based on liposomes (Figure 14.16) is one system that can address these issues quite all together. It also shows how nanofabrication that is based on self-assembly principles can work to formulate delivery systems. For example, after the self-assembly of liposomes with drug substances, they can be further decorated with targeting agent, attached with channel protein for drug release, coated with polyethylene glycol (PEG) for stealth function, and stabilized with polymer-

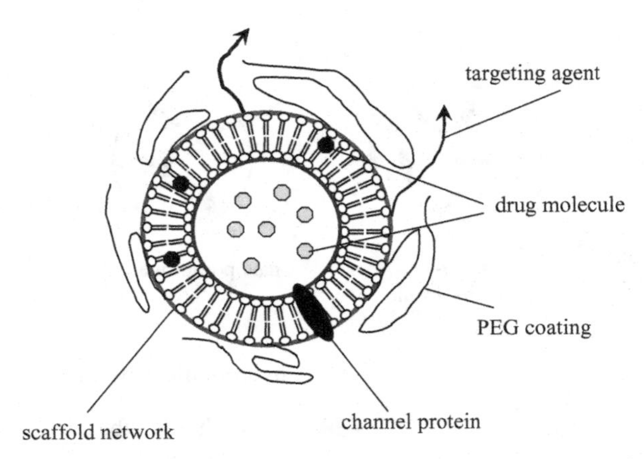

Figure 14.16. Schematic illustration of liposome-based target drug delivery systems.

ized scaffold network for longer circulation lifetime in the bloodstream. Jointing unit fabrication is useful for agent attachment and protein embedment. Coating is for PEG coating and polymerized scaffold.

Different unit operations for nanodevices are critical for the end-point operation of this delivery system. For example, the actual targeting of the system on the specific site of the targeted cells usually occurs with one of the bio-mimetic self-assembly principles such as recognition or ligand-acceptor binding. Upon binding of the system on the targeted cells, there should be a change in its force balance, and its rebalancing should be able to disassemble the system and thus release the drug substances. Depending on the detailed system and conditions, gating or switching, contraction-and-extension, capture-and-release, or sensoring unit operation should be properly designed.

14.5.2. Nanoelectronics

The concept of nanoelectronics (or molecular electronics) is being developed to fabricate a variety of nanoscale components into working electronics, so we can ultimately overcome the inevitable physical limitation that we are going to face with the current silicon-based electronics.

Excellent achievements have been made over the past decades, which show a powerful possibility for nanoelectronics. The fabrication and performance of single-molecular transistors have been demonstrated (De Franceschi and Kouwenhoven, 2002), and it is shown that the carbon nanotube–based field-effect transistor (FET) outperforms silicon-based FETs (Wind et al., 2002). Also, logic-gate performance has been demonstrated with nanowire- and carbon nanotube–based systems (*Molecular Electronics*, 2001). Unit fabrications of jointing, crossing, deposition, and hybridization have been used for the nanofabrication of these systems. Unit operations of gating and switching and sensoring were demonstrated for their operation.

The details of the physics and principles of nanoelectronics performance (e.g., electron transport) and characterization of their engineering aspects are beyond the scope of this book. The biggest challenge at this stage is the massive parallel fabrication of those promising individual nanocomponents into actual working nanomachines, which means nanoelectronic devices that can communicate with the macroworld. Continuous development of new and effective unit fabrications and unit operations that most likely should be based on self-assembly will be critical (Heath and Ratner, 2003).

14.6. NANOMACHINES: INTEGRATED NANODEVICES TO WORK

As defined earlier in this chapter, nanomachines are integrated nanodevices that are designed to perform "work" and show "a communication capability" with the macroworld as a result. Since this is also an area that is rapidly advancing, it would be reasonable not to attempt to extract a conceptual scheme of nanomachines at this stage. No efforts will be made to classify them, either. However, some sense of their general features that are comparable to macromachines can be viewed briefly. While macromachines work at μm–m range of mechanical movement, nanomachines will work at nm–μm range. The main operational force for macromachines is mechanical, while the interplay between intermolecular and colloidal forces is critical for nanomachines. Classical mechanics can fully describe the operation of macromachines, but it will take both quantum and classical mechanics for the proper description of nanomachines. Components for the assembly of macromachines are of course macroscopic objects, but atoms, molecules, and nanoscale objects (nanocomponents) are used for nanomachines. Finally, even though the types of energy input and signal output will not have to be that much different, the mode of interaction with specific components and of generation principles usually will be quite different between macromachines and nanomachines.

Integration of nanodevices into fully working nanomachines will be the ultimate challenge. There is no credible report on this issue at the moment. Figure 13.9, showing the fabrication of nanointegrated systems, might be one possible picture of this. The rest of this section will describe issues that might be critical for operating nanomachines once they are fully integrated, whatever the actual scheme might be. This of course cannot be a full description; more issues will have to emerge as the field moves ahead.

14.6.1. Power Source

This is the issue of how to make nanoscale power sources, how to effectively integrate them, and how to deliver the power to the specific sites of the nanomachines.

Self-assembly principles again could have significant impact on this. It is envisioned that a nanoscale fuel cell can be used to deliver electrons to the specific sites of nanomachines though bio-mimetic self-assembly principles, for example, attachment of biomotors on inorganic-based nanostructures to be powered (Curtright et al., 2004). Artificial light-harvesting molecules have been developed by mimicking biological photosynthetic systems. They show a promising potential to convert sunlight into chemical energy as a part of nanomachines (Gust et al., 2001).

14.6.2. Synchronization

To be able to work as a machine, the different nanodevices within nanomachines should be able to work in a way that they are elegantly coupled. For example, a shafting unit operation of one nanodevice might have to be able to trigger another one such as a gating unit operation that is adjacent to it, and so forth.

14.6.3. Packing

This refers to how to pack the whole nanomachine, so it can be effectively protected and separated from harmful environments, but effectively reveal its work.

14.6.4. Communication with the Macroworld

This is how to exchange energy and signals, and so communicate with the macroworld. Nanodevices can work independently, but nanomachines should be able to communicate with the macroworld at all times. A couple of options they can have are: Work cooperatively (synchronization issue) between the nanodevices (homogeneously and/or heterogeneously) within nanomachines so the signal/energy can be amplified, or work sequentially so the flow of the signal/energy can be followed.

Many different types of nanomachines are envisioned for future applications (Freitas, 2005) in areas as widespread as medical, energy, environmental, and even societal issues. The functions, structural schemes, and application objects of the envisioned nanomachines show a wide diversity and variety, and they actually should be that way. Unit fabrication(s) based on principles of self-assembly is fundamental knowledge to be developed. This should be followed by the development of nanofabricated systems that can carry novel or improved properties, or nanodevices that can retain the capability to perform the desired unit operation(s). Applying proper unit fabrication or self-assembly processes is the next step in fabricating these nanofabricated systems or nanodevices to construct nanomachines. At this stage of manufacture, depending on the applications targeted, different degrees of complexity should be able to be controlled through the control of hierarchicality of the fabrication process, and, ultimately, this should be able to ensure mass-production capability.

14.7. SUMMARY AND FUTURE ISSUES

This chapter showed that a variety of seemingly different operations of nanodevices can be systematically described with a small number of unit operations. These unit operations are performed by the balancing of three element motions that are the result of the interplay of intermolecular and colloidal forces between the nanocomponents. Some views on the fabrication and the function of nanomachines also have been discussed.

The area of nanodevices has greatly advanced over the past decade. However, this also brings us a number of challenges to be addressed for the ultimate goal of progress toward nanomachines. Some critical issues that have inevitably originated from their operations at nanometer scale are as follows:

First is the *fidelity* issue, which is continuous reversible longer-time operation of nanodevices. Fundamentally, it is the issue of nanocomponents: to properly address the wearing problem of the nanocomponents, the problem of damage of nanocomponents possibly by wastes or byproducts (e.g., from catalytic reactions or oxidation/reduction-driven operations), and more. Irreversible self-assembly processes that can often occur by sudden strong imposition of intermolecular or colloidal forces (e.g., by force shock) or by just pure random chance also can be problematic for the operation of nanodevices. It is thus important to ensure how many cycles of unit operation(s) can be achieved before these types of disassembly of nanodevices occur.

Second is the *versatility* issue. Currently, most nanodevices are designed to be operated in solution. Obviously, they need to be operated in air and in vacuum, too, to expand their scope of applications.

Third is the *design–reality discrepancy* issue. There always can be unexpected nanoscale phenomena during the operation of nanodevices. Since many of the nanodevice designs are often inspired by their micro- or bigger-scale counterparts, this can sometimes work against the intended operation of nanodevices.

REFERENCES

Badjić, J. D., Balzani, V., Credi, A., Silvi, S., Stoddart, J. F. "A Molecular Elevator," *Science* **303**, 1845 (2004).

Balzani, V., Credi, A., Raymo, F. M., Stoddart, J. F. "Artificial Molecular Machines," *Angew. Chem. Int. Ed.* **39**, 3348 (2000).

Banerjee, I. A., Yu, L., Matsui, H. "Application of Host–Guest Chemistry in Nanotube-Based Device Fabrication: Photochemically Controlled Immobilization of Azobenzene Nanotubes on Patterned α-CD Monolayer/Au Substrates via Molecular Recognition," *J. Am. Chem. Soc.* **125**, 9542 (2003).

Browne, W. R., Feringa, B. L. "Making Molecular Machines Work," *Nature Nanotech.* **1**, 25 (2006).

Butt, H.-J. "Toward Powering Nanometer-Scale Devices with Molecular Motors: Single Molecule Engines," *Macromol. Chem. Phys.* **207**, 573 (2006).

Casasus, R., Marcos, M. D., Martínez-Máñez, R., Ros-Lis, J. V., Soto, J., Villaescusa, L. A., Amorós, P., Beltrán, D., Guillem, C., Latorre, J. "Toward the Development of Ionically Controlled Nanoscopic Molecular Gates," *J. Am. Chem. Soc.* **126**, 8612 (2004).

Chen, Y.-F., Banerjee, I. A., Yu, L., Djalali, R., Matsui, H. "Attachment of Ferrocene Nanotubes on β-Cyclodextrin Self-Assembled Monolayers with Molecular Recognitions," *Langmuir* **20**, 8409 (2004).

Credi, A. "Artificial Nanomachines Based on Interlocked Molecules," *J. Phys.: Condens. Matter* **18**, S1779 (2006).

Curtright, A. E., Bouwman, P. J., Wartena, R. C., Swider-Lyons, K. E. "Power Sources for Nanotechnology," *Int. J. of Nanotechnology* **1**, 226 (2004).

De Franceschi, S., Kouwenhoven, L. "Electronics and the Single Atom," *Nature* **417**, 701 (2002).

Fennimore, A. M., Yuzvinsky, T. D., Han, W.-Q., Fuhrer. M. S., Cumings, J., Zettl, A. "Rotational Actuators Based on Carbon Nanotubes," *Nature* **424**, 408 (2003).

Freitas, Jr. R. A. "Current Status of Nanomedicine and Medical Nanorobotics," *J. Comput. Theor. Nanosci.* **2**, 1 (2005).

Gust, D., Moore, T. A., Moore, A. L. "Mimicking Photosynthetic Solar Energy Transduction," *Acc. Chem. Res.* **34**, 40 (2001).

Heath, J. R., Ratner, M. A. "Molecular Electronics," *Phys. Today* **43** (May 2003).

Hernandez, R., Tseng, H.-R., Wong, J. W., Stoddart, J. F., Zink, J. I. "An Operational Supramolecular Nanodevice," *J. Am. Chem. Soc.* **126**, 3370 (2004).

Hess, H., Vogel, V. "Molecular Shuttles Based on Motor Proteins: Active Transport in Synthetic Environments," *Rev. Mol. Biotechnol.* **82**, 67 (2001).

Karlen, S. D., Garcia-Garibay, M. A. "Amphidynamic Crystals: Structural Blueprints for Molecular Machines," *Top. Curr. Chem.* **262**, 179 (2005).

Karlsson, R., Karlsson, A., Ewing, A., Dommersnes, P., Joanny, J.-F., Jesorka, A., Orwar, O. "Chemical Analysis in Nanoscale Surfactant Network," *Anal. Chem.* **78**, 5960 (2006).

Kay, E. R., Leigh, D. A. "Hydrogen Bond-Assembled Synthetic Molecular Motors and Machines," *Top. Curr. Chem.* **262**, 133 (2005).

Kim, P., Lieber, C. M. "Nanotube Nanotweezers," *Science* **286**, 2148 (1999).

Kinbara, K., Aida, T. "Toward Intelligent Molecular Machines: Directed Motions of Biological and Artificial Molecules and Assemblies," *Chem. Rev.* **105**, 1377 (2005).

Kottas, G. S., Clarke, L. I., Horinek, D., Michl, J. "Artificial Molecular Rotors," *Chem. Rev.* **105**, 1281 (2005).

Kwok, W. M., Ma, C., Phillips, D., Beeby, A., Marder, T. B., Thomas, R. L., Tschuschke, C., Baranovic, G., Matousek, P., Towrie, M., Parker, A. W. "Time-Resolved Resonance Raman Study of S_1 *cis*-Stilbene and Its Deuterated Isotopomers," *J. Raman Spectrosc.* **34**, 886 (2003).

Liu, Y., Mu, L., Liu, B., Kong, J. "Controlled Switchable Surface," *Chem. Eur. J.* **11**, 2622 (2005).

Lukin, O., Vögtle, F. "Knotting and Threading of Molecules: Chemistry and Chirality of Molecular Knots and Their Assemblies," *Angew. Chem. Int. Ed.* **44**, 1456 (2005).

Månsson, A., Sundberg, M., Bunk, R., Balaz, M., Nicholls, I. A., Omling, P., Tegenfeldt, J. O., Tågerud, S., Montelius, L. "Actin-Based Molecular Motors for Cargo Transportation in Nanotechnology: Potentials and Challenges," *IEEE Trans. Adv. Packaging* **28**, 547 (2005).

"Molecular Electronics: Nanocomputing Marches Forward," *C&EN* **7** (November 12, 2001).

Paxton, W. F., Sen, A., Mallouk, T. E. "Mobility of Catalytic Nanoparticles Through Self-Generated Forces," *Chem. Eur. J.* **11**, 6462 (2005).

Rapenne, G. "Synthesis of Technomimetic Molecules: Towards Rotation Control in Single-Molecular Machines and Motors," *Org. Biomol. Chem.* **3**, 1165 (2005).

Regan, B. C., Aloni, S., Ritchie, R. O., Dahmen, U., Zettl, A. "Carbon Nanotubes as Nanoscale Mass Conveyors," *Nature* **428**, 924 (2004).

Saibil, H. R., Ranson, N. A. "The Chaperonin Folding Machine," *Trends Biochem. Sci.* **27**, 627 (2002).

Seifert, U. "Configurations of Fluid Membranes and Vesicles," *Adv. Phys.* **46**, 13 (1997).

Shirai, Y., Morin, J.-F., Sasaki, T., Guerrero, J. M., Tour, J. M. "Recent Progress on Nanovehicles," *Chem. Soc. Rev.* **35**, 1043 (2006).

Shu, W., Liu, D., Watari, M., Riener, C. K., Strunz, T., Welland, M. E., Balasubramanian, S., McKendry, R. A. "DNA Molecular Motor Driven Micromechanical Cantilever Arrays," *J. Am. Chem. Soc.* **127**, 17054 (2005).

Simmel, F. C., Dittmer, W. U. "DNA Nanodevices," *small* **1**, 284 (2005).

Soong R., Montemagno, C. D. "Engineering Hybrid Nano-devices Powered by the F_1-ATPase Biomolecular Motor," *Int. J. Nanotechnology* **2**, 371 (2005).

Van Delden, R. A., ter Wiel, M. K. J., Pollard, M. M., Vicario, J., Koumura, N., Feringa, B. L. "Unidirectional Molecular Motor on a Gold Surface," *Nature* **437**, 1337 (2005).

Webster's New World™ College Dictionary, 4th ed. (IDG Books Worldwide: 2001).

Wind, S. J., Appenzeller, J., Martel, R., Derycke, V., Avouris, P. "Vertical Scaling of Carbon Nanotube Field-Effect Transistors Using Top Gate Electrodes," *Appl. Phys. Lett.* **80**, 3817 (2002).

INDEX

Self-Assembly and Nanotechnology: A Force Balance Approach, by Yoon S. Lee